MAPPING FROM AERIAL PHOTOGRAPHS

Aspects of Modern Land Surveying

Series editor: J. R. SMITH, A.R.I.C.S.

MODERN THEODOLITES AND LEVELS
by M. A. R. Cooper, B.Sc., A.R.I.C.S.

ELECTROMAGNETIC DISTANCE MEASUREMENT
by C. D. Burnside, M.B.E., B.Sc., F.R.I.C.S.

FUNDAMENTALS OF SURVEY MEASUREMENT AND ANALYSIS
by M. A. R. Cooper, B.Sc., A.R.I.C.S.

HYDROGRAPHY FOR THE SURVEYOR AND ENGINEER
by Lt. Commander A. E. Ingham, R.N.(Retd), F.R.I.C.S.

Mapping from Aerial Photographs

C. D. BURNSIDE, M.B.E., B.Sc., F.R.I.C.S.

Principal Lecturer in Land Survey
North-East London Polytechnic

A HALSTED PRESS BOOK

JOHN WILEY & SONS
New York

Published in Great Britain by Granada Publishing Limited
in Crosby Lockwood Staples 1979

Published in the U.S.A.
by Halsted Press, a Division of
John Wiley & Sons, Inc., New York

Copyright © 1979 by C. D. Burnside

Library of Congress Cataloging in Publication Data
Burnside, Clifford Donald
  Mapping from aerial photographs.

  1. Cartography. 2. Aerial photogrammetry.
I. Title.
GA109.B87   526   79–11497
ISBN 0–470–26690–2

Filmset in 'Monophoto' Times 10 on 11 pt by
Richard Clay (The Chaucer Press), Ltd, Bungay, Suffolk
and printed in Great Britain by
Fletcher & Son Ltd, Norwich

Contents

List of Figures

List of Black and White Plates

Preface

A more explicit title for this book would be 'An Introduction to the Main Theoretical Bases of Producing Topographical Maps from Aerial Photographs'. The text therefore is not concerned with providing technical instructions for the construction of maps but rather with the mathematical concepts on which their production is based. It has been written as an introduction to the main theoretical elements of photogrammetry. The restriction to the use of aerial photographs for the purposes of making topographical maps was introduced because it provided me with an obvious logical framework on which to develop the subject matter I had in mind. Of course, the mapping of other surfaces from other sources of photography is identical in theory and can often prove to be more simple in practice than the main application of photogrammetry discussed in this text.

The first two chapters are concerned with the nature of the basic data and examine the air survey camera and the procurement of suitable aerial photography. Having obtained the necessary photographs, the various ways in which map detail can be derived from them are examined systematically: from simple graphical constructions to sophisticated mathematical techniques. In this way, I have endeavoured to provide a text that will enable the reader to study the main topics of photogrammetric mapping in a manner that develops in a logical and systematic way.

A set of well-designed practical exercises are a highly desirable adjunct to any theoretical study of a subject of this nature and originally it was my intention to provide these in this volume. Unfortunately, it soon became apparent that their inclusion would expand the text to an unacceptable degree and also change the character of the book too much. However, it is realised that most students have a particular interest in the practical applications of photogrammetry. A series of exercises designed to illustrate various basic concepts can therefore do much to facilitate their appreciation and, at the same time, introduce the student to some of the practical aspects of the subject.

The text does not require any great familiarity with photogrammetric instruments and a minimum knowledge of these is assumed throughout. From this point of view, chapter eight, which is concerned with the design of analogue plotting instruments, was particularly difficult to write. A single chapter cannot hope to provide anything like a comprehensive treatment of the topic and it does not attempt to provide this. The intention has been to provide only a reasonably representative sample of instrument design features. Some detailed knowledge of photogrammetric equipment is most certainly necessary to any comprehensive study. Such information is however quite readily available through the medium of the excellent technical leaflets produced by many instrument manufacturers. A collection of these would make a valuable supplement to any text.

It may be noticed that many of the references given in the bibliography are to technical papers given at recent national and international conferences. By making use of these the reader should be able to make himself familiar with the latest developments in photogrammetry.

It will also be noticed that most of the diagrams in this book are simple two dimensional line diagrams. These I hope will prove to be clear and intelligible but they should have an additional advantage: when one is trying to understand a concept or proof of the type most frequently encountered in this subject one usually resorts to a diagram; if this is highly complicated then one cannot readily remember it or reproduce it and so one's understanding of the concept is often thwarted. Consequently, an attempt has been made here to devise and use diagrams of the simplest nature only. Where required, two simple diagrams have been used instead of one of more complexity. In this connection I would like to acknowledge the great help of Mr P. Sorrell who drew all the diagrams for me and and made many valuable suggestions as to their nature.

Rather than ask one unfortunate person to read the whole of my manuscript I asked a number of friends and colleagues to read various chapters. I am therefore most pleased to record my thanks to the following persons for their advice and criticisms; Dr A. L. Allen, Dr P. Dale, Mr G. B. Das, Mr B. Canzini, Mr J. R. Hollwey, Miss G. Sears and Mr S. Walker. Miss Sears and Mr Walker also gave me great assistance in the reading of the page proofs and so I am doubly grateful to them.

Finally I would also like to acknowledge the help given to me by Mr D. Wagstaffe who produced some of the photographs for me and also the following who so kindly and readily provided me with many of the photographs and diagrams included in the book:

Cartographic Engineering Ltd
The Editor, Photogrammetric Record
Survey & General Instruments Co. Ltd
Surveying & Scientific Instruments Ltd
Carl Zeiss (Jena) Ltd
Carl Zeiss (Oberkochen) Ltd
Wild Heerbrugg (U.K.) Ltd

September 1978   C. D. Burnside

1: Some geometric properties of cameras and photographs

1.1 Introduction

For the purpose of making maps from aerial photographs it is necessary to examine certain characteristics of the cameras used and the photographs produced by them. Since map making is perhaps a geometric process (if we give the literal meaning to the word) then clearly the geometric properties of the camera are of prime importance. In addition, we must also be interested in the quality of the images produced by the photographic process. Factors such as sharpness of detail, tonal range and contrast will not only affect our estimated location of an image point position but also our interpretation of what that image represents on the ground. Photogrammetric mapping is therefore both an art and a science; the art being a subjective process of photo interpretation coupled with the more scientific process of geometric analysis. In topographic mapping, the degree of interpretive skill required is often not very high for the requirement is to show the locations of a selection of familiar ground features. In this case, therefore, the emphasis is very much on the correct location of these features on the map sheet. On the other hand, in other more specialised forms of mapping (soil, geological surveys and so on) the emphasis is, more often than not, the other way round; a correct recognition of certain features being more important than great accuracy in plotted positions. To some extent this change of emphasis is reflected in the relative importance given to the camera geometry and to the qualities of the images produced by the photochemical process.

1.2 The geometry of a simple camera

In this chapter we shall be concerned mainly with the photogrammetric aspects of camera geometry. Certain other factors will be mentioned but for a more detailed treatment of these the reader will be referred to other texts. In the first instance, therefore, we consider the case of image formation by a thin lens when the object is some considerable distance from the lens. In such a case the sharpest image will be produced in the focal plane of the lens as illustrated in Fig 1.1.

Light arriving at the lens from distant objects such as P and T is parallel on arrival. The bundles of rays are therefore in the form of thin cylinders of maximum diameter, $A \cdot \cos \theta$ where A is the aperture of the lens and θ is the angle of incidence. The axial ray of each cylindrical bundle can conveniently be taken as representative of that bundle and in this text is termed the *chief ray*. We note that all such chief rays pass through a single point N situated at the centre of the lens. This point is called a nodal point. The chief ray (Pp) that lies normal to the lens plane and the focal plane is in fact the optical axis of the lens. The two chief rays illustrated in the diagram form the angle θ at the node N and then carry on undeviated to intersect the focal plane in the two points p and t. The positions of the images p and t of the distant object points P and T are therefore geometrically defined in this way. We note that all other rays of a bundle are deviated to some degree and in the diagram all have been shown to intersect the focal plane at the same point as their chief ray. This is an ideal situation that cannot be achieved with real lenses in practice. All lenses and combinations of lenses must exhibit to some slight degree certain aberrations

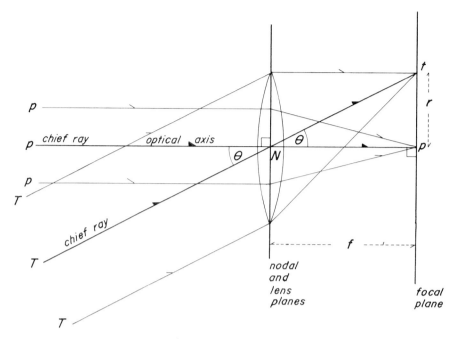

Figure 1.1. *The Geometric Optics of a Thin Lens*

(spherical aberration, coma, chromatic aberration, etc.) the effects of which are noted later in this chapter. At this point we note that their overall effect is to degrade the image from a sharp, well-defined point to a less well-defined area of confusion, the estimated centre of which may not be the point image defined by the geometry of our diagram. In the particular case of the air survey camera the object points are always at a considerable distance from the camera and so the plane of sharpest focus is the focal plane of the lens. The camera is therefore constructed so that the plane of the lens and the plane of sharp focus are parallel to one another and rigidly maintained at the required focal distance of separation. The point p is of special significance and is called the principal point of the photograph.

From an inspection of Fig 1.1 we see that the distant point subtends an angle θ at the nodal point N with the optical axis PNp where,

$$\tan \theta = \frac{r}{f} \tag{1.1}$$

Hence the value of this angle can be evaluated for the camera situation because r, the radial distance of t from p, is recorded on the photograph. However, in order to carry out this calculation we must know the values of the focal distance f and the position of p the principal point on the photograph. These data are therefore the essential parameters of the camera geometry that must be known to a high degree of accuracy for most mapping processes. Furthermore, it is also important that these parameters should remain constant over long periods of time under normal working conditions. That is to say, the structural geometry of the camera should be as stable as possible. The process of determining the parameters is called the camera calibration process.

From what has been said above we will now realise that the camera is in fact a recording goniometer. The data recorded by the photograph enable us to calculate the angles subtended at the station N by distant points such as T at the instant of exposure; the

2

reference direction being the optical axis pNP of the camera lens. Point N is taken to be the camera (or air) station.

1.3 The field method of camera calibration

There are a number of ways of carrying out the process of calibration and some of these are described later in this chapter. At this point we introduce the idea of field calibration because it requires no specialised equipment other than a theodolite and it reinforces the concept that calibration determines the relationship between angles in the object space and linear measurements taken off a plane photograph. In essence, we have to find the position of the node N with respect to the focal plane.

In the first instance we will take an ideal case in which all chief rays, regardless of their angle of incidence, are assumed to pass through a single nodal point situated somewhere on the optical axis of the lens. In practice this would suggest a thin lens devoid of any radial or tangential distortion and producing such image sharpness that the image point positions are not in any doubt. Later on, we will see what these terms imply in respect to the position of the node. This concept of a fixed nodal point regardless of the angle of incidence provides us with a model which we can profitably use as a standard to gauge other less perfect optical units.

The various stages in the calibration process are as follows:

(a) Set out a horizontal line of targets on the ground, the size and nature of which are such that they are capable of accurate bisection by the stadia of the theodolite selected, and that they also provide the most suitable photographic images for the accurate measurement of distances on the photograph.

(b) A theodolite of suitable precision is set up over a station mark some distance from the line of targets and approximately normal to them at the centre target. A series of horizontal directions are then observed and angles using the centre target as the reference object are then derived from the observations.

(c) The theodolite is now replaced by the camera, which is set up so that the nodal point is vertically over the station mark and with the optical axis horizontal so as to produce a line of image points across the centre of the photograph format. The location of the node will not be known with any great accuracy, say to a centimetre or so, hence the range should be such that this uncertainty introduces no appreciable error into the angular observations.

(d) A photograph is now taken of the row of targets. The camera is then rotated through $90°$ about the optical axis and another photograph taken. This will produce a row of images at right angles to those of the first photograph. When a detailed examination of the lens performance is required, more exposures will be made with rows of targets at other orientations.

(e) After careful processing, a series of radial distances are measured on each photograph using the centre target as a reference point. The results of this procedure are illustrated in Fig 1.2. To illustrate the method of calculation consider two targets T_i and T_i' symmetrically disposed about the centre target T_o. From the theodolite observations we have the angles θ_i and θ_i' and from the photograph we have the two associated radial distances r_i $(=t_o t_i)$ and r_i' $(=t_o t_i')$ as shown in Fig 1.3.

From Fig 1.3 we see that the quantities we require are Np, which is the focal distance f, and Δr, which is the distance of the principal point p from the centre target point t_o. The evaluation of these two quantities is best done by first calculating values for the angles α_i and α_i'.

3

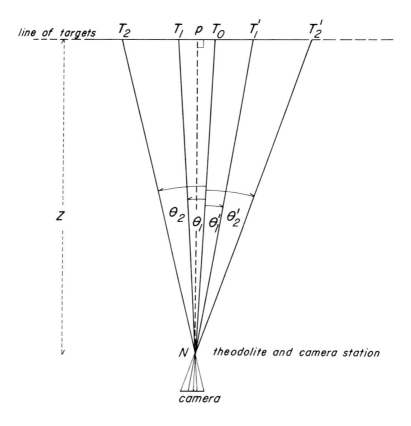

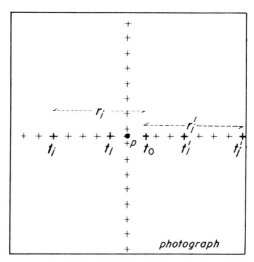

Figure 1.2. Field Method of Camera Calibration

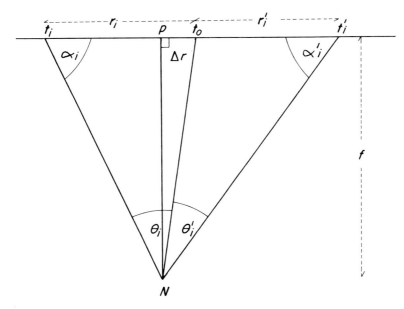

Figure 1.3. Calculation of Camera Parameters

We see immediately that

$$(\alpha + \alpha') = 180 - (\theta + \theta')$$
$$= k_1 \text{ (a constant)} \tag{1.2}$$

By the application of the sine rule to the two large triangles we have

$$t_oN = \frac{r \sin \alpha}{\sin \theta}$$

$$= \frac{r' \sin \alpha'}{\sin \theta'}$$

hence

$$\frac{\sin \alpha}{\sin \alpha'} = \frac{r'}{r} \cdot \frac{\sin \theta}{\sin \theta'}$$
$$= k_2 \text{ (a constant)} \tag{1.3}$$

Students of surveying will recognise this geometry as a special case of the three-point problem in resection where the three points lie in a straight line in this case.

There are a number of ways of solving the two equations (1.2) and (1.3) simultaneously. A direct substitution for α' in equation (1.3) is probably as good as any, hence

$$\sin \alpha = k_2 \sin (k_1 - \alpha)$$
$$= k_2(\sin k_1 \cdot \cos \alpha - \sin \alpha \cos k_1)$$

∴
$$\tan \alpha = k_2 \sin k_1 - k_2 \cos k_1 \cdot \tan \alpha$$

i.e.
$$\tan \alpha = \frac{k_2 \sin k_1}{1 + k_2 \cos k_1} \tag{1.4}$$

Having evaluated angle α and thence angle α' we are now able to calculate a value for the line Nt_o from either of the large triangles. Knowing this, Δr and f can then be calculated from the small triangle Nt_op.

Using the same pair of targets on the other photograph at right angles to the first we can repeat the calculation with two new values for r and r' and obtain a new set of values for Δr and f. We take the mean of the two f values as being the best representative value for that zone of the lens while the two values of Δr locate the position of p with respect to the two lines of targets.

In the same way, other pairs of targets will yield further values for f and the position of p. If the lens were distortion free then all such evaluations would fall within the limits of observational error. In fact, the position of the principal point should be constant in a well-constructed lens and the mean of all determinations would be taken as the best obtainable value. Calculated values of the focal length, however, will show some significant variation between lens zones and this is described as radial lens distortion.

1.4 Lens distortion
Radial distortion is one of the major components of the distortion present in a good quality lens. As indicated above, it implies that the effective focal length of the lens is changing very slightly from zone to zone and is symmetrical about the optical axis. Expressed in a slightly different way we can say that the position of the nodal point is not fixed but moves slightly along the optical axis for different angles of incidence. This is illustrated in Fig 1.4. The diagram, of course, greatly exaggerates the real situation for the variations found in the wide angle lens of a modern air survey camera are of the order of 10 μm only.

In presenting the results of calibration it is customary to assign a nominal value to the focal length together with a distortion curve showing the variations from this figure over the field of view. There are a number of criteria that can be used to arrive at a nominal value and these are often given special names such as 'calibrated focal length' or 'equivalent focal length'. When provided with a calibration curve it is therefore important to be aware of the criterion used for determining the nominal value. The following are four examples of the possibilities:

(a) The value could be an area weighted mean. This is calculated using a weighting for each zone proportional to the area of that annular zone on the photograph.

(b) The value could be chosen so that the maximum positive and negative values attained over the field of view are equal in magnitude.

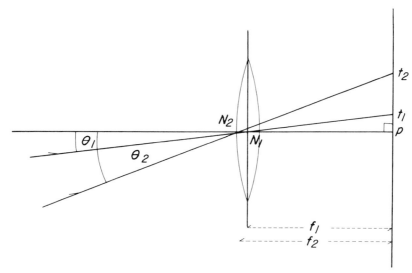

Figure 1.4. *Lens Distortion — Movement of Nodal Point*

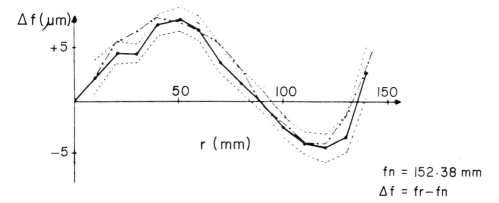

fn = 152·38 mm

Δf = fr − fn

Figure 1.5. Lens Distortion Curve for Modern Camera Lens

(c) The value might be selected so that the distortion is zero at some stated angle of incidence (say 45°) in addition to being zero on the axis.

(d) The value determined using rays close to the optical axis (i.e. paraxial rays only) may be selected.

Figure 1.5 illustrates a typical curve obtained for one of the most recent types of lens. In essence this curve depicts the departure of the real lens from a model of constant focal length 152.38 mm. The concept of radial lens distortion arises when we have need to assume that the lens is behaving in the model way when, in fact, it is not. Because the real image positions do not coincide with those of the model we say that the lens exhibits radial distortion. This is illustrated in Fig 1.6. In the diagram we see the relationship between the departure of the lens performance from that of the model (Δf) and the positional error (Δr) introduced into the image position by assuming the model performance. The point t is the

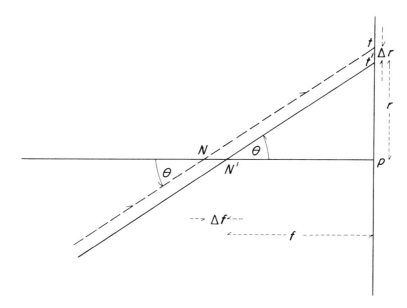

Figure 1.6. Radial Lens Distortion

7

real image point associated with the real node N, while t' is the predicted image position based on the model lens with a constant nodal point N'. The radial distortion present at the angle of incidence, θ, is therefore given by the expression:

$$\Delta r = \Delta f \cdot \tan \theta$$

It is a common practice in photogrammetric work to describe image displacements on the photograph as changes in scale over the format of the picture. This can be done in the case of lens distortion and we can in fact calibrate the camera by observing the changes in scale over the picture area if we set up our experiment with this in mind. Figure 1.7 shows the arrangement required, which is not as simple as previously. In this case the camera focal plane must be made parallel to the line of targets and the distance Z from the station to the targets accurately known. In such a case the variations in focal length are made apparent by the changes of scale that can be detected in various zones of the photograph. We therefore measure the distances between targets on the ground using the target T_o as a reference point; the camera station being located so that the image of this nadir point now defines the principal point p. From the diagram we see that the photo scale is given quite simply by:

$$S_1 = \frac{r_1}{R_1} = \frac{f_1}{Z}$$

for zone 1, and

$$S_2 = \frac{r_2}{R_2} = \frac{f_2}{Z}$$

for zone 2, and so on. Hence

$$S_1 - S_2 = \frac{f_1 - f_2}{Z}$$
$$= \frac{\Delta f}{Z}$$

i.e.
$$\Delta f = Z(S_1 - S_2)$$

In this case we have used f_2 as our nominal focal distance. See also section 1.5.3.

Finally on the topic of lens distortion we need to introduce the subject of tangential lens distortion. This defect can arise if the components of a compound lens are assembled so that the individual nodal points of the components are not exactly collinear. In the manufacture of air survey lenses great care is taken in their assembly and so the magnitude of this distortion is smaller than the radial component. In terms of our model lens, this form of distortion would occur if the chief ray Tt does not intersect the accepted optical axis Pp but lies slightly skew to it, the closest point being in the neighbourhood of N. Figure 1.8 shows the effect. The field method of calibration would detect this fault if the row of targets on the ground were accurately set out in a straight line for the points would not then be exactly collinear on the photograph. In general, one of the other methods of calibration mentioned in the next section would be more suitable. Some recent results of camera calibrations showing the magnitudes of the radial and tangential components present in modern air survey cameras are given in Ref 1.2.

1.5 Methods of camera calibration

In Section 1.3 the concept of calibration was introduced by describing a field method of calibration. The method is most suitable for terrestrial cameras. As far as aerial cameras are concerned the bulk and weight of the camera makes the procedure a little

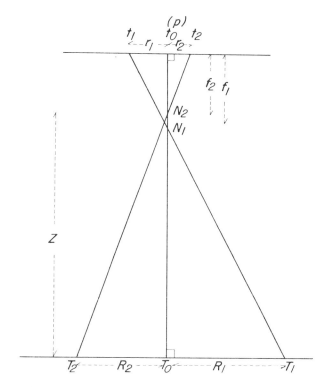

Figure 1.7. Lens Distortion and Changes in Photo Scale

cumbersome. Also, it should be noted, that the camera is held with its optical axis horizontal, while in use it would be supported in a vertical position. It is bad practice to calibrate any instrument in an unusual position. There are a number of laboratory methods of calibration; these either use fixed known angles of incidence in the object space (i.e. collimators) or fixed known distances in the focal plane (i.e. reseaux).[1.3]

1.5.1 CALIBRATION USING COLLIMATORS
In this case we have a bank of collimators rigidly set up so that they define an accurate set of directions converging at a single point. For example, the American NBS calibrator has

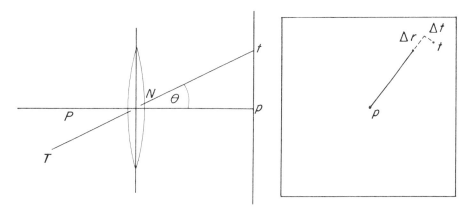

Figure 1.8. Tangential Lens Distortion

9

a central collimator and four banks of collimators at right angles in the form of a cross. Each bank contains six collimators fixed at 7.5° intervals precisely. In the focal plane of each collimator is a resolving power target. The camera under test should be set up so that its outer node coincides approximately with the point of convergence of the collimators and with its optical axis coinciding with the line of the central collimator. This can be done in the following manner. First, without the camera in position adjust a viewing telescope T so that it bisects the image of the centre collimator (see Fig 1.9). The camera is then introduced into position. When the position is such that the telescope again bisects the collimator image, but now passing through the camera lens, it means that we have detected the optical axis, that is, the chief ray that is neither deviated nor displaced on passing through the lens (see section 1.10). Photographic film now placed in the focal plane of the camera is exposed and the positions of 25 collimator images are recorded. The plate co-ordinates of the centre points of all collimator targets are measured with high precision. From these data the radial and tangential distortion characteristics are readily obtained. An examination of the resolving power patterns of the target provides a value for the resolving power on the axis and in six zones of the lens (see section 1.7).

1.5.2 CALIBRATION USING A RESEAU AND GONIOMETER
In this method a reseau plate is located in the focal plane of the camera. This is illuminated from behind and the directions generated in the object space are recorded using a goniometer that revolves about the outer nodal point. The location of this point is not critical provided the emergent rays are parallel or very nearly so, in which case a lateral movement of the goniometer will produce no apparent image displacement.

   In order to locate the optical axis in this case it is desirable for the goniometer telescope to be in the form of an autocollimating device. This simply means that by means of a thin, semisilvered glass surface (M) inserted into the eyepiece the cross hairs can be illuminated by the lamp (L) fixed at one side (see Fig 1.10(b)). With an optically flat mirror surface in the camera focal plane the image of the cross hairs reflected from this surface will be seen superimposed on the graticule (G) of the telescope eyepiece when the axis of the goniometer coincides with the undeviated and undisplaced chief ray that is by definition the optical axis.

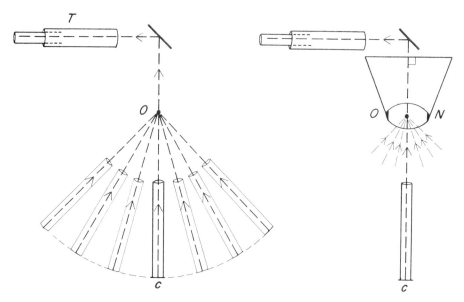

Figure 1.9. Camera Calibration Using Collimators

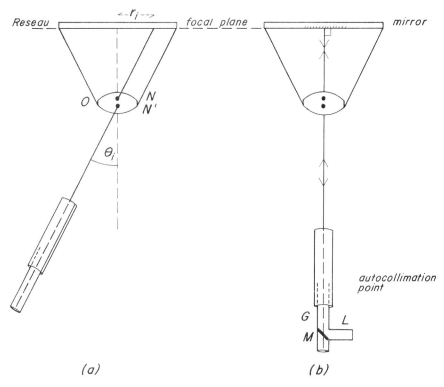

Reseau $\leftarrow r_i \rightarrow$ focal plane   mirror

O N N'

θ_i

autocollimation point

G L

M

(a)   (b)

Figure 1.10. *Camera Calibration Using a Goniometer*

1.5.3 CALIBRATION USING A TALL TOWER

This is a method developed by Hallert.[1.1] The camera was suspended from a tall tower 123 m high and a photograph taken of an array of ground targets on the ground below. Care is taken to level up the camera and position its nodal point vertically over the central target at a known height. The symmetrical array of targets must be in a horizontal plane and surveyed in to a high precision; 1 μm on the photograph being equal to 0.8 mm on the ground in the case of a wide angle lens. The method of calculation used was that involved with scale factors described in association with Fig 1.7. Of the methods so far described this one most nearly approaches the desirable situation in which the camera is calibrated under working conditions.

1.5.4 THE USE OF A TEST AREA FOR CALIBRATING PURPOSES

To test and calibrate a photogrammetric camera under actual working conditions requires the setting out of a large test area. The accurate surveying in of the control points is a task of some magnitude but their stability and maintenance are probably greater problems.

In this method the camera cannot be placed over a central mark or levelled up and so the computations involved are of a more complex nature. In the first instance, the camera position and tilt need to be calculated by one of the space resection methods (see section 10.3). Given the required distribution of control, using a least squares technique, values for the focal length and position of the principal point can be determined in addition to the camera station co-ordinates $(X_o Y_o Z_o)$ and the elements of camera orientation (κ, φ, ω). It should be noted that in this case changes in ground elevation are necessary, otherwise only a ratio f/Z_o can be determined but not f and Z_o (see Refs 1.2, 1.4, 1.5 for examples of this procedure).

11

Finally, for completeness, we note that the star pattern provides an excellent and dense set of direction vectors. From a knowledge of the time of plate exposure, the geographic positions and the astronomical co-ordinates of identified star images, the parameters of calibration can be determined. A description of this technique is to be found in Ref 0.1, vol 1.

1.6 Lens quality

The quality of a photographic image, that is its sharpness, depends on a number of optical and other factors. At this point we examine those optical properties of the lens that contribute to the quality of the image. We noted in paragraph 1.2 that a lens cannot bring into coincidence at a single point all the individual rays that make up a cylindrical or conical bundle of rays. The position of the plane of best focus is a compromise for all the rays that make up a bundle and for all bundles within the field of view. The main aberrations that contribute to the degrading of an image are well described in standard texts on geometric optics and need not be described in detail here.[0.2] We merely note them in passing as:

(a)   Chromatic aberration. Because the refractive index is a function of wavelength (i.e. colour) in a dispersive medium such as glass, the different wavelengths present in white light will be brought into focus at different focal distances.

(b)   Spherical aberration. This is a property of axial rays. Rays parallel to the optical axis but passing through different annular zones of the lens are brought into sharp focus at slightly different distances.

(c)   Coma. This is a property of oblique rays. Here again rays passing through different annular zones are brought into focus at slightly different points. In this case the focal distance is also a function of the plane in which the ray lies. The plane containing the chief ray of the bundle and the optical axis is called the meridional plane. The plane containing the chief ray but at right angles to the meridional plane is termed the sagittal plane. Coma is one of the most serious of aberrations for it is only symmetrical about one line.

(d)   Astigmatism. This is another property of oblique rays. An off-axis point image gives rise to two short line images situated at different distances. One of these, the sagittal line, is perpendicular to the optical axis and in the meridional plane. The other, the tangential line, is perpendicular to the chief ray of the bundle and at right angles to the meridional plane. The plane of best focus lies between the two lines at a point called the circle of least confusion.

By a suitable combination of elements of different shapes and glasses it is possible to reduce the effects of these aberrations considerably. Most of the effects increase rapidly with increased field of view and so the problems of designing satisfactory lenses with wide angles of view require the use of multielement units in their solution.

It was stated earlier in this chapter that the quality of an image can have two effects on the mapping process. Firstly, the image shape may be such that the observer selects the wrong position for what he estimates to be the correct image point. We now realise that this will most likely be recorded as lens distortion. Secondly, the quality of the image will affect the observer's judgement in the interpretation of the detail he is wanting to map. He may not be able to resolve certain detail with clarity and he may therefore have to omit it. Or, he may interpret it incorrectly. The ability of a lens to resolve fine detail is the second type of testing required in the assessment of lens performance.

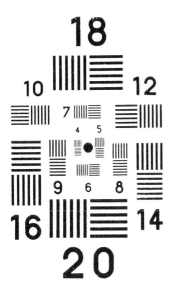
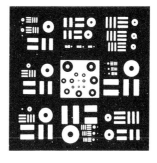

Figure 1.11. Resolution Test Cards

1.7 Resolution testing

At the present time it is customary to test the overall performance of a photographic unit (i.e. camera and film) by the use of test cards photographed in various parts of the field of view. Different testing laboratories have a preference for different types of test pattern printed on the card but all are utilised in the same way. The results of the tests are quoted as the resolution capabilities of the system in lines per millimetre over various zones of the picture format. Typical test cards used for air survey cameras are shown in Fig. 1.11.

Test card results indicate the characteristics of the entire system and therefore include factors relating to the nature of the test card and its illumination, the performance of the lens, the characteristics of the emulsion, the developing process and, finally, the acuity of vision of the observer assessing the results. If some statement is required concerning the performance of the lens alone then clearly the other factors will need to be discounted in some way. This in fact cannot be done with any precision and so, while lenses can be tested one against another over a limited span of time in a laboratory under a set of specified and controlled conditions, an absolute statement concerning the quality of a lens is not possible using this method.

1.8 The optical transfer functions

From the last section we see that resolution testing with the aid of test cards does not provide the universally reproducible measurement of lens performance that is desirable. The concept of the optical transfer function is a more fundamental approach to the solution of this problem and it seems likely that this technique will be developed and find an increasing application in the testing of photogrammetric camera lenses. These are two functions that find applications in optical work, the phase transfer function and the modulation transfer function. Changes in phase between input and output signals are related to lens distortion while changes in modulation transfer are associated with the resolving properties of the unit. Although both are being investigated for possible applications in photogrammetric optics, so far it is the latter that has been developed to the stage where practical test procedures have been devised and used. In this section we therefore confine our comments to an introduction to the nature of the modulation transfer function.

13

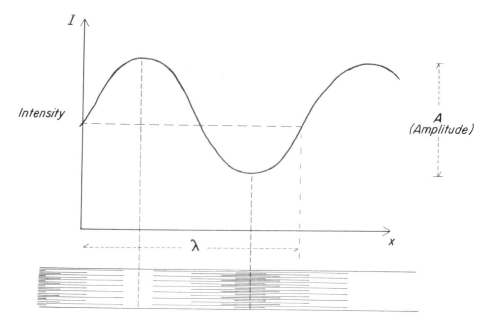

Figure 1.12. *Target with Sinusoidal Intensity*

Consider a target where the intensity of the illumination being emitted varies in a sinusoidal manner with distance along the axis. Figure 1.12 attempts to illustrate this but it should be appreciated that this is difficult to reproduce in a simple diagram of this sort. The wavelength is shown as λ and this is the distance over which one complete cycle of intensity variations are developed. Most often this expression is in terms of a spatial frequency of so many cycles per millimetre, i.e.

$$\text{Spatial frequency} = f$$
$$= \frac{1}{\lambda} \text{cycles/mm}$$

If an image of such a target is examined then an intensity pattern of the same sinusoidal nature will be produced. At unit magnification, the spatial frequency of the image will be that of the object but the amplitude of the intensity distribution will be different. The modulation transfer function is then given quite simply by the expression

$$\text{MTF} = \frac{\text{output amplitude}}{\text{input amplitude}}$$
$$= \frac{A_o}{A_i}$$

It will be appreciated that in the case of a lens the aberrations that must be present to some degree will influence the magnitude of this factor by an increasing amount as the spatial frequency increases. If we take the value of the function to be unity when the frequency is very low (say 0.1 c/mm) then by the time the frequency has increased to, say, 100 c/mm the amplitude might well be zero. A typical MTF curve for a wide angle lens is shown in Fig 1.13 and has been taken from Ref 1.6. On the axis the performance is best. Away from the axis the results fall off and the second curve shows this for 28° off-axis in the radial direction. The tangential direction for the axis of the targets would give a slightly different result.

14

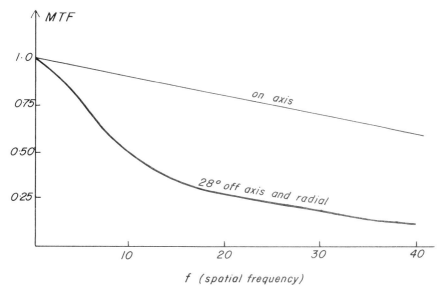

Figure 1.13. *Modulation Transfer Function for Wide Angle Lens*

The photographic images produced by a certain type of film and development process have their own MTF that can be determined separately by measuring the changes in film density produced by contact exposures with suitable targets. Any function derived from measurements carried out on a photograph is therefore a combination of the separate lens and film functions such that

$$(MTF)_{overall} = (MTF)_{lens} \times (MTF)_{film} \times (MTF)_{image\ movement}$$

In this way it is possible to isolate to some degree the various influences at work in the overall system.

The measuring instrument used in determinations of density changes is the densitometer. This measures the intensity of light passing through a small and very narrow slit aperture by photoelectric means. The instrument is employed to measure both the input and output density distributions. With regard to the input distribution, the production of targets that will give, either by reflectance or transmittance, the required sinusoidal variations in light intensity is difficult; a whole range of targets of varying spatial frequency being required in order to generate the required MTF curve. This is a particularly difficult task if we think in terms of producing targets suitable for photographing from an aircraft, for example. There are a number of other ways of producing the required distribution but one that is being used commonly is that involving the line spread function.

A single thin line of given amplitude can be represented by a Fourier series containing an infinite number of terms, all of the same amplitude and covering an infinite range of frequencies. A line (or a sharp edge of high contrast) can be regarded as an object composed of such a distribution of this nature and used as the object for analysis. Because of the varying nature of the MTF, different spatial frequencies will be imaged with different amplitudes, and so the line spread function is produced. Now, all the frequencies of the object were of the same amplitude: hence an analysis of the line spread function can provide the data necessary for the determination of the MTF. The mathematical analysis required is involved and various programmes have been devised for this. A practical graphical/digital method using this technique is given in Ref 1.7. Figure 1.14 illustrates the points made above. Ref 2.4 discusses the main parameters that govern image quality.

15

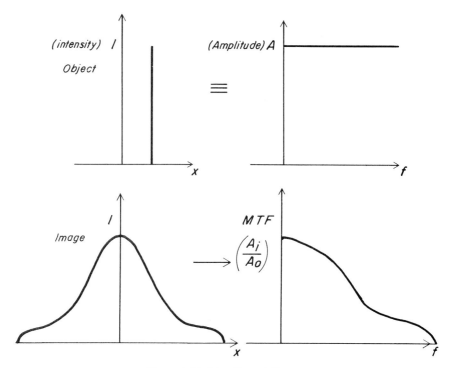

Figure 1.14. *Line Spread Function*

1.9 The air survey camera

In this section we examine the components and the characteristics of the modern air survey camera in so far as they have a direct bearing on our appreciation of the mapping process. From this point of view we shall comment on the following items:

(a)   The lens.
(b)   The lens cone and camera body.
(c)   Iris diaphragm.
(d)   Methods of film flattening and realisation of focal plane.
(e)   Film transport mechanism.

The requirements of a lens to be used for aerial mapping purposes are very demanding. In the first instance, the highest possible geometric accuracy coupled with excellent resolution is required over a wide angle of view. Secondly, the light gathering and transmission properties must also be excellent over the whole of the working format. The Earth as a photographic subject is rather dull, the tonal range is usually limited, and the illumination at the time of the photographic sortie may be poor. To meet the requirements a large and complex multi-element unit must be used. In Fig 1.15 is shown the optical components of two Wild Aviogon lenses. These are good examples of lenses used with the standard format size of 230×230 mm. They can be used with black and white, colour or infrared emulsions.

It should be noted that with the standard format there are at present three main focal lengths used in topographical mapping:

(i)   normal angle   maximum angle of receptance 57°   focal length 300 mm;
(ii)   wide angle   maximum angle of acceptance 94°   focal length 152 mm;
(iii) super wide   maximum angle of acceptance 123°   focal length 88 mm.

16

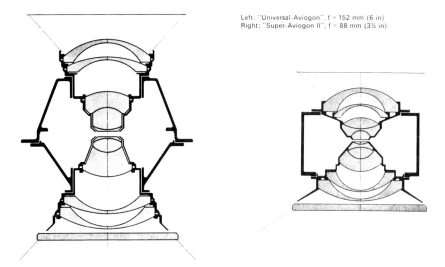

Left: "Universal-Aviogon", f = 152 mm (6 in)
Right: "Super-Aviogon II", f = 88 mm (3½ in)

Figure 1.15. *Components of Wild Aviogon and Super-Aviogon Lenses*

The listing also indicates their order of development. In the next chapter we will discuss in some detail the different applications of these types.

The lens elements are assembled in the lens cone which also has provision for taking the diaphragm and a between-the-lenses shutter. The cone is rigidly fixed to the camera body and in this way the lens is firmly held at the required distance from the focal plane. It is essential that this part of the camera geometry remains stable within a very small tolerance and that it suffers the minimum of changes due to either temperature fluctuations or changes in mechanical stress.

Because most photographs are taken from a fixed wing aircraft moving at high speed the shutter used must be of high efficiency and capable of exposure times down to 1/1000 s. For these reasons, a between-the-lenses shutter using continuously rotating blades, as developed by Zeiss, is now considered to be one of the best available, see Fig 1.16. Shutters using spring-activated blades are less efficient due to the influence of blade inertia on performance. Louvre shutters with low inertia blades situated just in front of the focal plane have been used but even when open some light is lost because the mechanism obstructs part of the light path. A focal plane shutter of the type used in some good quality

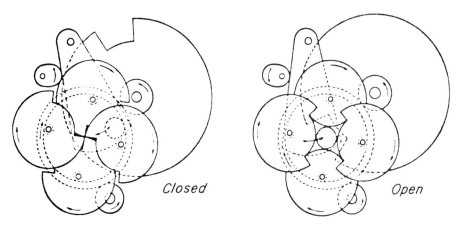

Closed   Open

Figure 1.16. *Zeiss Rotating Shutter Mechanism*

handheld cameras is not suitable for photogrammetric purposes if the camera is moving during the time of exposure — as it usually is. In this case the camera station has moved an appreciable amount during the interval of time required to expose the whole of the picture area.

The iris diaphragm used in air survey cameras is of the common interleaved type and is placed in a between-the-lenses position. The usual range of stop values (F-numbers) provided is from F/5.6 and smaller. The lens resolution is best when the aperture is smallest for then only the central part of the lens is in use. The image movement caused by the speed of the aircraft is least when the exposure time is least, that is, when the aperture is greatest. Usually it is this latter factor that needs to be taken into account when deciding upon exposure details. The F-value of a lens is defined as the focal length divided by the effective aperture:

$$\text{F/number} = \frac{\text{focal length}}{\text{effective aperture}}$$

As an example, if we consider a wide angle lens with a stop setting of F/8 then we have:

$$\text{The effective aperture } A = \frac{152}{8}$$
$$= 19 \text{ mm}$$

Now in the paragraph above we stated that the resolution would improve as the aperture was decreased but this is true only up to the point where diffraction effects can be disregarded. Beyond this point we note that the minimum angle of resolution, as given by Rayleigh's criterion, is as follows (see Ref 0.2):

$$\text{Minimum angular resolution } \Delta\theta = \frac{1.22\,\lambda}{A} \text{ rad}$$

Thus if we take λ, the wavelength of white light, to be about 0.56 μm we find that with aperture value of 19 mm quoted above we have

$$\Delta\theta = \frac{1.22 \times 0.56 \times 10^{-6}}{19 \times 10^{-3}} \text{ rad}$$

i.e. $$\Delta\theta = 3.6 \times 10^{-5} \text{ rad}$$

With a focal length as quoted this will give a linear resolution on the axis of Δr, where

$$\Delta r = f\Delta\theta$$
$$= 152 \times 3.6 \times 10^{-5} \text{ mm}$$

and

$$1/\Delta r = 183 \text{ lines/mm resolution}$$

With the film speeds in use at the present time, F/numbers are not likely to exceed F/8, and if we compare the above with the results of resolution testing we see that such diffraction effects are not of any great significance in the average photographic situation for photogrammetric purposes, where 80 lines/mm on the axis would be considered good.

An effective method of holding the film flat in the focal plane of the camera is obviously a matter of considerable importance. If it is not in this position at the moment of exposure then the overall sharpness of the photograph will be impaired and the parameters of calibration will not strictly apply. To achieve this correct positioning is one of the more difficult operations required of the camera mechanism. There are two methods of approach.

In the first method the focal plane is defined using a precisely ground steel plate. The surface of this plate is covered by a rectangular pattern of shallow grooves about 1.5 mm deep. Holes pierce the plate at a certain number of the groove intersections. If a vacuum is applied to the back of the plate then the film will be drawn on to it and into the correct position for taking a photograph. As an alternative to this, the lens cone and camera body can be made airtight and at the required moment the air pressure within the body increased by pumping in filtered air. In this case the pressure, higher than the external pressure, will force the film on to the focal plane. At the moment, the vacuum back system is the one most commonly employed in air survey cameras.

In the second method the focal plane is defined by the back surface of an optically flat glass plate. This plate is, in fact, the last element of the lens unit and must be designed as such. At the moment of exposure a pressure pad pushes the film into firm contact with the plate. The main difficulty in the operation would seem to be in squeezing all the air out of the interface in the short time available. A disadvantage of the method is that any blemishes that appear on the optical flat will also appear on every photograph. On the other hand, however, if the plate is etched with an accurate grid of crosses (a reseau) then this too will appear on every photograph and this can be used to advantage, particularly in analytical photogrammetry. In such a case accurate co-ordinate measurement need only be taken from the nearest reseau cross. In addition, any lack of film flatness can be detected and also changes in dimensions caused by expansion or contraction of the film at any time after the exposure.

Most air survey cameras at the moment use a plastic film as the emulsion support material. This polyester based material is very durable and has a dimension stability comparable to glass. It is for this reason that cameras using glass plates are no longer favoured. The main difficulties with the plate camera are the storage of plates in the camera before and after exposure, and moving the plates into position for exposure. Such cameras had limited magazine storage and a rather long minimum cycling time. Using film held in long rolls 70 m or more long, hundreds of exposures are possible without a change of magazine. A minimum cycling time of 2.5 s is often possible. The film transport mechanism is required to move film through the camera at about 6 m/min when working on a short cycle time without damaging it or subjecting it to a stress that might cause longitudinal deformation.

1.10 The geometric optics of thick or compound lenses

Where we have a thick compound lens such as the one illustrated in Fig 1.17(a) there would be a minor difficulty in applying the Gauss or Newtonian lens equations without some additional information. For example, from where would measurements be taken in the establishment of focal planes, planes of sharp focus and so on? In the case of the thin lenses that have been used to illustrate various optical properties so far, the measurements have in fact been taken from the centre of the lens. However, this simple expedient cannot work in the case of complex camera or projector lenses. Fortunately, with the aid of only a few extra concepts the elementary equations can still be applied.

Firstly, we note that any system of lenses, no matter how complex, can be defined by six parameters, namely the positions of the two focal points and the positions of the two nodal and principal points. These six points can be located by calculation or experiment because:

(i)   light rays parallel and close to the optical axis are brought into sharp focus at the two focal points F and F',

(ii)   the two principal points, P and P', locate the conjugate planes of unit positive lateral magnification;

(iii)   the two nodal points, N and N', locate planes of equal positive angular magnification

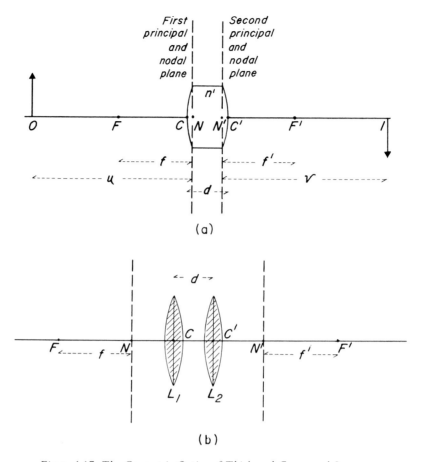

First
principal
and
nodal
plane

Second
principal
and
nodal
plane

n'

O F C N N' C' F' I

$\longleftarrow$ - - - - - f - - - $\longrightarrow$ $\longleftarrow$ - - - - - f' - - - $\longrightarrow$

$\longleftarrow$ - - - - - - - - u - - - - - - - - - - - $\longrightarrow$ $\longleftarrow$ - - - - - - - - - v - - - - - - - - - $\longrightarrow$

$\longleftarrow$ - d - $\longrightarrow$

(a)

$\longleftarrow$ - d - $\longrightarrow$

F - - - - - - f - - - N C C' N' - - - - - f' - - - $\longrightarrow$ F'

L_1 L_2

(b)

Figure 1.17. The Geometric Optics of Thick and Compound Lenses

    – that is to say, a ray of light entering the system through point N (or N′) emerges at the same angle of incidence to the optical axis from point N′ (or N).

If the refractive index of the medium in the image space is the same as that of the object space then, in fact, N coincides with P and N′ coincides with P′. Hence in the normal use of photogrammetric cameras there are just four parameters and not six.

    In the simple thin lens the two nodes (or principal points) were considered to be coinciding and at the centre of the material. However, in the case of thick lenses the nodal points do not coincide and, in fact, can lie anywhere along the optical axis from $+\infty$ to $-\infty$. Measurements in connection with the lens laws must, however, be taken from these two points in the case of thick or compound lenses; the unprimed dimensions of the object space being associated with the unprimed quantities F and N and so on. These simple rules also apply to a compound lens composed of a number of thin lens elements. Figure 1.17(b) illustrates this point.

    The positions of the nodal points can readily be found by experiment using a nodal slide and the property (iii) mentioned above. This indicates that the image formed by incident parallel light will not move laterally when the lens is rotated about the second nodal point (N′). This property is illustrated in Fig 1.18. Rotation about N′ with parallel light incident from the left will not alter the direction of the emergent ray N′F′. Similarly, N can be found with incident parallel light from the right. (This principle also finds an application in the construction of one type of panoramic camera.)

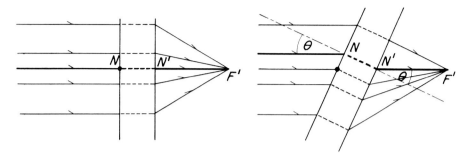

Figure 1.18. Location of Nodal Points using a Nodal Slide

To calculate the positions of the nodal planes for the single thick lens it can be shown that (see Ref 0.2)

$$CN = \frac{d \cdot \bar{f}}{f_2}$$

and

$$C'N' = -\frac{d \cdot \bar{f}}{f_1}$$

where $\bar{f}$ is the focal length of the combination, and f_1 and f_2 are the focal lengths associated with the first and second surfaces centred on C and C' respectively. The thickness of the glass CC' is denoted by d and, where the refractive index of the lens material is n', the focal length of the lens is given by:

$$\bar{f} = \frac{f_1 f_2}{f_2 + n' f_1 - d}$$

In the case of a compound lens composed of two thin lenses the above formulae still apply in a slightly modified form. In this case the centres C and C' are the two thin lens centres, d is their separation, and f_1 and f_2 are their two focal lengths. Hence we have

$$CN = \frac{d\bar{f}}{f_2}$$

$$C'N' = -\frac{d\bar{f}}{f_1}$$

where

$$\bar{f} = \frac{f_1 f_2}{f_1 + f_2 - d}$$

Note that the sign convention required here is that the positive direction is left to right and that $C'N' = -N'C'$, etc.

2: The photographic sortie

2.1 Introduction

In this chapter we shall be concerned with the main technical features of a set of aerial photographs, taken with a camera of a type discussed in chapter 1, for the purposes of producing a topographic map. Of course, to some extent these features will be influenced by the particular mapping technique to be employed, so this chapter will also serve as an introduction to some of the material to be discussed at later stages.

In order to produce a satisfactory map in the most economical way the first essential is to acquire a set of photographs appropriate to the task. To ensure this, then some form of technical specification must be drawn up for the photography and most mapping organisations will have one or more of such standard specifications for the types of aerial photography they usually require. An example of such a document is provided in appendix B, and an examination of this shows that the main technical requirements are as follows:

(a)   The type of camera to be used, its format size and focal length.

(b)   Camera calibration data indicating the optical performance of the lens unit.

(c)   The type of photographic emulsion to be used together with any special optical filters.

(d)   The nature of the emulsion support material.

(e)   A limitation on the amount of image blurring due to camera and aircraft motion.

(f)   The flying height above datum, or a mean flying height above ground level and hence (in combination with (d)) an average scale of photography.

(g)   The fore-and-aft overlapping of adjacent photographs in a strip and the overlapping of adjacent strips.

(h)   The maximum allowable tilt of the camera axis from the vertical direction.

(i)   The maximum allowable inclination of any air base.

(j)   The amount of crabbing allowable.

(k)   The direction(s) of the flight lines.

(l)   The time, date or season of the photographic sortie.

(m)   The amount of cloud or haze permissible at the time of exposure.

(n)   The provision of adequate air cover diagrams (ACDs) showing the location of each exposure.

(o)   The information required to be shown on each negative, on each strip of negative film and on each can of film.

(p)   The use of any auxiliary instruments to be employed in conjunction with the air survey camera.

In the following section we will now examine these various requirements in more detail.

2.2 Camera types

World wide there is still a great variety of camera types used for taking aerial photographs; see volume 1 of reference 0.1 for a comprehensive list of these. However, for normal mapping from aircraft there are three main types in use at the present time. All of these are roll-film cameras taking a standard picture size of 230 × 230 mm (9 × 9 in), the variation in type being confined to variations in the focal length of the camera lens. In the table below the angle of acceptance θ has been calculated in accordance with the formula:

$$\tan\theta = \frac{\text{semi diagonal length of format}}{\text{focal length of camera}} \qquad (2.1)$$

Type	Angle of acceptance (θ)	Focal length (f)
Normal Angle	57°	300 mm (12 in)
Wide Angle	94°	152 mm (6 in)
Super Wide Angle	122°	88 mm (3.5 in)

Table 2.1

The majority of these cameras do not use a glass reseau plate in the focal plane. The few cameras that do so are mainly of British manufacture. In analytical photogrammetry there are some advantages in using reseau photography, and this matter is discussed in later sections, see section 10.6 for example. Table 2.1 also indicates the chronological order of development, for it was not until about 1960 that lens technology had advanced sufficiently to produce a super-wide angle lens with acceptable resolution and distortion characteristics. From this point of view, one would always expect a lens with a narrower angle to have a better optical performance. However, there are other points to be taken into consideration when deciding on the best type of lens to use in a given set of circumstances. Figure 2.1 illustrates these and shows the three situations using the three types of lens to obtain the same scale of photography. The first thing we notice is that in order to obtain the same scale of photography in each case there is a reduction in flying height commensurate with the reduction in focal length value. (A more detailed statement concerning the relationship between flying height and focal length is given in section 3.2.) There are in practice a number of factors that might put a constraint on the flying height that can be chosen; some typical examples are listed below.

(i) The operational ceiling of the aircraft available for the photographic sortie will set a limit on the smallest scale of photography obtainable with a given camera.

(ii) In urban areas there is often a minimum permitted flying height that will limit the maximum scale of photography obtainable.

(iii) In certain locations, persistent haze or low cloud may set a limit on the maximum operating height.

(iv) Air turbulence will be greater at lower altitudes, and when flying over certain features it might be difficult to fly straight-and-level within the limits required by (h) and (i).

An examination of Fig 2.1 also shows the phenomena known as dead ground. It will be seen that this is greatest with the widest angle photography. This effect is most serious in urban areas over which there are many tall buildings. It is one reason why super-wide angle photography would not be considered suitable for such an area. Also, there is the additional disadvantage that with the wider angle of view the two aspects of the common

23

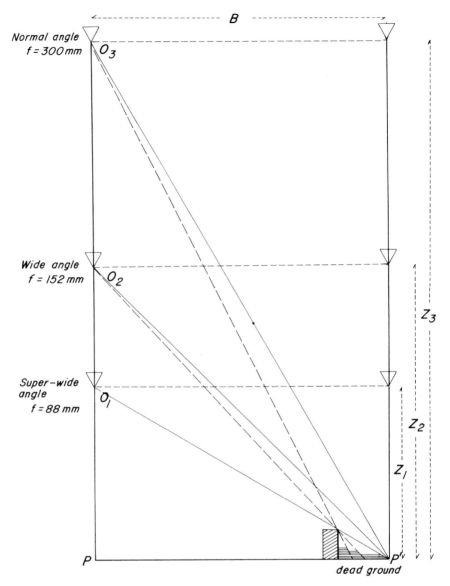

Figure 2.1. *Focal Length and Flying Height Relationships*

area can be so different that stereoscopic viewing of the two pictures becomes difficult. This can result in a lowering of the mapping accuracy. Finally, it should be appreciated that the geometric definition of any point, such as P′, depends on the parallactic angle α; the greater this angle is, the more precise will be the definition of the point. The super-wide angle photography clearly produces the most satisfactory angle of cut because the base/height ratio is biggest in this case. In the diagram, all three base lengths have the same value because all three photo scales are the same. Hence the base/height ratios are inversely proportional to the focal lengths. If this factor alone determined the precision of the mapping then clearly the super-wide angle photography would give the most satisfactory results. However, this is not the case in practice (see Fig 2.3 and section 2.7).

In any discussion concerning the selection of a suitable focal length, the fact that many analogue plotting instruments have a limited range of focal length settings should not be

24

forgotten. The cameras used will have to match the principal distance range of the plotting instruments available. At the moment, wide angle photography is by far the most common in use and all instruments can be expected to accommodate this. This is not the case with the other two types. In chapter 8 the main characteristics of analogue plotting instruments are discussed.

2.3 Camera calibration data

A comprehensive set of data concerning the optical qualities of the camera lens are provided with the camera by the manufacturers. These data will include a nominal focal length value together with distortion curves showing the variation of focal length (radial distortion) with radial distance from a defined principal point. Information will also be provided concerning the zonal values of the resolution or, more commonly perhaps in future, details of the modulation transfer function. From time to time it is necessary to check the more basic parameters of the camera, such as the nominal focal length, the radial distortion characteristics and the location of the principal point. Details of a recent check on these data are therefore requested in any specification.

It will be seen in later chapters that for some methods of planimetric mapping, based on graphical techniques or the use of optical rectifiers, the calibration information is not needed. However, for all other methods and for all height determination techniques the availability of such information is essential.

2.4 Emulsion types

For many years now, panchromatic black-and-white photographs have been used for mapping purposes. In recent years the use of colour emulsions has increased but as far as the mapping precision of identifiable points of detail is concerned the introduction of colour produces no discernible improvement. [2.1] However, the introduction of natural or false colours does greatly facilitate the interpretation of aerial photographs. The eye can distinguish between about 100 shades of grey but over 2000 colour shades. Hence the use of colour can provide more than twenty times the amount of information for the purposes of interpretation. For a familiar landscape (such as an urban area) there may be no advantage in using the more costly materials but for unfamiliar regions, where more interpretive skill is required in recognising important features, then colour emulsions can be used to advantage. For example, the location of shoal waters and deep-water channels as required for nautical charting can be greatly facilitated by the use of colour. Many examples of the uses of colour photography are given in Ref. 2.2.

When using colour film, which in fact is made up of three layers of emulsion, the three dyes selected are normally chosen with the aim of producing the most natural set of tones. However, the dye colours need not necessarily be chosen in this way. Sometimes, by the use of false colours, the interpretation of detail can be more accurately carried out.

Photographic emulsions can be made sensitive to wavelengths outside the range of the visible spectrum. At the moment, the longer wavelengths of the near infrared have a particular use in topographical mapping. The reflectance (and absorption) of a material depends on a number of physical and chemical factors including the wavelength of the illumination. A water surface, for example, has a high coefficient of reflectance (and low absorption) for wavelengths in the visible spectrum but a lower value for the reflectance of infrared wavelengths. Such a surface photographed with an emulsion sensitised to this latter region will therefore appear dark on the resulting positive print. This property finds an immediate application in the location and delineation of water areas as these will show up very clearly as very dark areas on this type of photography. A most striking example of this effect is the photography of a shore line of clear water in strong sunlight. On the infrared photography, the boundary between the dry land and the water surface is clearly

25

marked. On the other hand, with panchromatic black-and-white photography, the water line cannot be readily recognised and the presence of the water may not be detected until it is of some appreciable depth.

With regard to filters to be used at the time of photography, a minus blue filter (i.e. yellow) is almost invariably employed in aerial photography. Small particles suspended in the atmosphere will always cause a certain amount of light scatter, especially of the shorter wavelengths. Unless corrected for, this will result in a fogging of the picture. Occasionally, when photographing certain unusual landscapes, other types of filter can be used to advantage. When a lens is used in conjunction with a filter unit, this unit becomes part of the optical system and so the camera should be calibrated with the filter in position.

2.5 Emulsion support materials

From remarks made in chapter 1 it will be realised that it is important that the image points are formed in a plane at right angles to the optical axis of the camera lens. Clearly, in order to achieve this, the emulsion support material should be of uniform thickness over the whole of the format area. Furthermore it must be as dimensionally stable as possible if it is to be a faithful record of the image positions formed by the lens. To achieve these requirements the use of glass plates with optically flat surfaces is an obvious choice. However, in addition to their high cost there are other mechanical disadvantages (see section 1.9) and so for some years now plate cameras have not been available. Certain plastic films, notably those using a polyester-based material, do have a stability approaching that of glass and can be produced with the required uniform thickness. Tests have shown that film of this material can be manufactured with a uniform thickness of about 130 μm and with normal usage the change in size is limited to about 0.03%. With film cameras of course the flatness of the film at the moment of exposure depends on the efficient functioning of the flattening mechanism (see section 1.9).

2.6 Image movement

The exposure of the photographic emulsion takes place over a finite interval of time somewhere between $\frac{1}{100}$th and $\frac{1}{1000}$th of a second. In all probability it will be some few hundredths of a second. If the camera is being carried by a fixed-wing aircraft then the camera will move a certain distance in that exposure time. A ground point will therefore be elongated into a short line image on the photograph, the length and direction of which will depend on the ground velocity of the aircraft, photographic scale and exposure time. A maximum permissible value for this length is often stipulated, especially in low altitude large-scale photography where the effect is greatest. For example, an aircraft flying at 200 km/h is taking 1/3000 scale photography using an exposure time of 1/250 s. Hence,

$$\text{Ground distance travelled in exposure time} = \frac{200 \times 1000}{60 \times 60 \times 250}$$

$$= 0.22 \text{ m}$$

$$\text{On the photography, the image movement} = \frac{0.22 \times 10^6}{3000} \ \mu\text{m}$$

$$= 74 \ \mu\text{m}$$

In certain cameras used for specialised purposes the image plane of the camera can be caused to move during the time of exposure so that the image point remains stationary with respect to the photographic emulsion.

Image movement can also arise because of camera vibration. The camera must therefore be installed within the aircraft at a position of low vibration and also be mounted in a suitable antivibration mounting. This source of trouble can be more acute when a helicopter is being used as the camera platform. Special mountings have been developed for this purpose.

2.7 The selection of a suitable flying height

Having selected a suitable camera for the mapping project an appropriate photo scale must be selected. For economic reasons we must use the minimum number of photographs to cover the mapping area. That is to say we must choose the minimum photo scale and hence the maximum flying height at which the required accuracy of plotting can be confidently obtained. This of course depends on the techniques to be employed and to some extent on the nature of the terrain itself. It is also very much a matter of judgement based on extensive practical experience. However, it is important to have some idea of the capabilities of photogrammetric techniques from the very beginning, and so the following remarks are provided for this reason. It should be emphasised that they are not criteria to be adopted without further consideration.

(a)   When using mainly graphical techniques such as those discussed in chapter 4, a mapping scale close to that of the photo scale is most often employed.

(b)   In the production of mosaics and photomaps based on photographic techniques an enlargement of about four times photo scale is considered to be the maximum for photography of average quality. Enlargement beyond this point will produce a noticeably poor image quality.

(c)   By using first-order stereoscopic plotting instruments we achieve the most economical situation for the production of both planimetric and height information. The remaining paragraphs of this section are therefore devoted to this topic.

The overall enlargement from photo scale to machine plot scale can be quite large. In normal circumstances, an enlargement factor of up to six times photo scale could be satisfactorily employed. Occasionally, in certain circumstances, the factor might be as high as ten, depending on the requirement and the nature of the detail to be plotted.

It is important to realise that in mapping when both planimetric and height information are required, the specification for the height data is often more stringent than that for the planimetry. For example, in the case of a 1 : 2500 map sheet with 2 m contours an error of 2 m in a height value would be considered a serious mistake. On the other hand, a 2 m error in the planimetric position of a point (0.80 mm on the map sheet) might just be acceptable. The case of the 50 ft contour on a 1 : 50 000 sheet is even more striking. It is for reasons such as this that the specification for the height information to be provided will often dictate the minimum photoscale and maximum flying height that can be chosen for a project. The manufacturers of stereoscopic plotting instruments usually quote a value for the precision obtainable for spot height determinations under optimum conditions. This expresses the precision of the height determination expressed as parts per thousand of the flying height above the ground surface. For example, the precision of a first-order plotting instrument might well be quoted as follows:

$$e_z = \pm 0.1^o/_{oo} Z \qquad (2.2)$$

where Z is the flying height above ground level, and e_z is the error in the height value at natural scale.

If we now consider the height error at photo scale (that is to say, we consider a model of the ground formed at photo scale) we have:

$$e_z = \pm \frac{Z}{10\ 000} \times \frac{f}{Z}$$

i.e.
$$e_z = \pm 0.1^o/_{oo} f \qquad (2.3)$$

For wide angle photography and taking f as 152 mm, we have

$$e_z = \frac{152 \times 1000}{10\ 000}$$

$$= 15.2 \ \mu m$$

From this we deduce that for such a first order plotter the spot height precision at the photo scale is of the order of 15 μm. For wide angle photography this precision is in fact applicable to all three co-ordinates.

If a specification has been given for spot height values, then the above can be used directly to obtain the maximum satisfactory flying height if the capability of the plotting instrument has been quoted in the way described. Sometimes only the required contour interval is given, in which case a spot height precision of the order of one-fifth of the stated contour interval should be used when calculating the maximum flying height. It should be pointed out, however, that this method will give impracticable values for the flying height when the contour interval is greater than about 5 m. That is, when the photo scale is about 1:65 000 (taking $f = 152$ mm) or less. At such a photo scale, the recognition and correct interpretation of topographic detail can become difficult. There is therefore the danger that relatively important features might be missed. Now, such small-scale mapping will mainly be confined to undeveloped areas where small villages, footpaths and minor features form the important planimetric detail that must be recorded. Hence there is a limit to the smallness of the photo scale that can be used for such mapping; imposed by the overall resolution capabilities of the photographic systems in use at the present time. Theoretical considerations suggest that there will be no significant improvements in these for standard cameras in the future.[2.3, 2.4] Such considerations lead to the conclusion that there is no good practical reason for employing first order plotting instruments on small scale topographical plotting. (In mapping terms, 1:50 000 scale or smaller can be considered as small scale.) Second or third order instruments, of lower cost and precision are more than adequate for this purpose.

It should be noted that when using equation (2.3) above we substituted a focal length value appropriate to a wide angle camera. If we had substituted that of a super-wide angle camera then we would arrive at a value of 8.8 μm for the precision at photo scale. Because of adverse factors such as a lower lens performance and greater effects due to atmospheric refraction, this better value is not achieved in practice. Equation (2.2) should therefore be used to estimate a maximum flying height regardless of the focal length of the camera lens.

Finally, there is one further important point that needs to be made concerning the accuracy of photogrammetric plotting. The values quoted for the precision of an instrument are obtained under optimum conditions. That is to say, the targets used in the evaluation have maximum contrast and are of a near ideal shape and size. In this way, pointing or observational errors are reduced to a minimum. When using natural features, however, such points are very often far from ideal and so observational errors can be very significant. This factor can have a considerable influence on the planimetric and height accuracies that can be obtained. This matter is discussed in more detail in chapter 7. Results of tests on plotting accuracy are contained in the OEEPE report.[2.5]

2.8 Photographic overlap

For photogrammetric purposes the photography is taken in long strips, usually straight lines, such that adjacent photographs within a strip have a very large overlap and adjacent strips have rather less of an overlap. Figure 2.2 illustrates the situation. Stereoscopic photogrammetry requires the formation of a stereoscopic model from two adjacent photographs. The area of the model produced by the two photographs is limited to their common overlap. The whole of the mapping area needs to be covered by such stereoscopic models and it is important (particularly in aerial triangulation) that every model has a workable overlap with adjacent models. The photographic overlap must therefore be in excess of 50% and a working specification of 60% ± 5% is usually requested for adjacent photos in a strip. With a 60% overlapping of photographs of standard format 230 × 230 mm, the separations of principal points on the photographs will be 92 mm. Hence we can say that this will be the length of the air base (b) at the mean photo scale when this overlap specification is maintained.

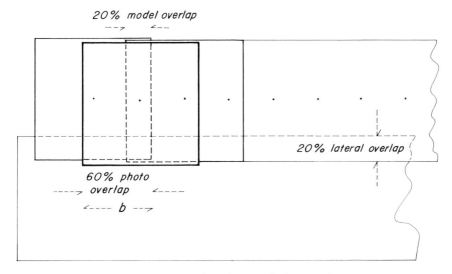

Figure 2.2. Overlaps for Aerial Photography

To ensure a workable lateral overlap of models of adjacent strips, an overlap of 20% ±5% between strips is usually requested.

We should note that if we consider a model at photo scale we can evaluate the B/Z ratio referred to in section 2.2 for any focal length value provided the fore-and-aft overlap and format are those specified. Figure 2.3 illustrates the situation.

In chapter 7 we shall be studying stereo models and their control requirements and at that time we will see that there are economic advantages in placing any control points required in the super-overlaps of the photographs — that is, in the overlapping areas of models. Furthermore these advantages are at a maximum if a point falls within the super-overlaps of two adjacent strips. To do this, the super-overlaps of the two strips must be coinciding and this can only be done directly if the positions of the exposure stations can be carefully controlled. This might be done if the photography required is of an area

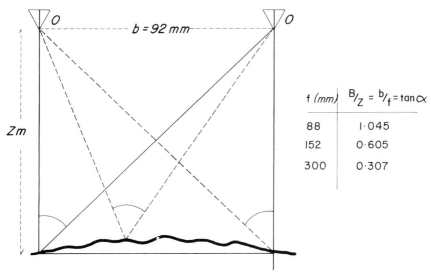

f (mm)	$B/Z = b/f = \tan\alpha$
88	1·045
152	0·605
300	0·307

Figure 2.3. The B/Z Ratio

29

already covered by map sheets. The required flight lines and approximate exposure stations (principal point positions) can then be marked up on these sheets. Any requirement to fly to such a rigid specification would of course be more costly. An alternative therefore is to specify a fore-and-aft overlap of $80\% \pm 5\%$. In this way, both the odd-numbered photographs and the even-numbered photographs will provide sets of photographs with the standard overlap, the only difference being the different ground positions of the areas of super-overlap. By a selection of the odd or even photographs of any strip a workable matching up of the areas can usually be made over most of the strip length. Figure 2.4 shows an ideal situation for control purposes; tie points and control points would be located in the centres of the super-overlaps with one point serving two strips and four models. The extra costs of this method are composed only of those of the extra negative film required and its processing.

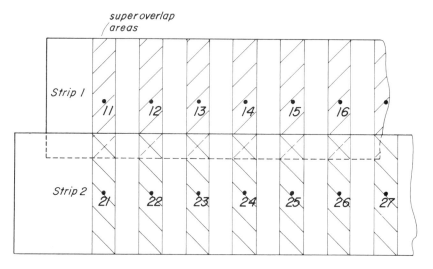

Figure 2.4. Photography with Ideal Overlapping of Photographs and Strips

In aerial triangulation (chapter 9) we will see that it is most desirable for the tie points, linking one model to those around it, to be similarly positioned in the super-overlapping areas.

The overlap requirements of aerial photography are important. If the average fore-and-aft overlap is too large (but less than 80%) then more photographs and more models are required to cover the area. This will mean more office work and possibly more groundwork in the form of photo control. Furthermore the resulting accuracies obtained from such models will be diminished since the photo base length will be less than the optimum 92 mm and hence the B/Z ratio will be smaller than necessary. If the side overlapping of strips is greater than specified then, again, the number of stereo models needed to cover the area will be greater. If any overlap falls below about 10 mm in width then the area in which a control point or tie point can be placed is very restricted, for points should not appear on the very edges of photographs where the resolution is worst and the distortion is greatest. Gaps between strips of photography can be expensive to fill in. Some form of rephotographing can be attempted to cover the missing areas but such a procedure generates much extra work in the office. Extra ground control might also be necessary. In some situations one must of course accept what can be obtained but when faced with using badly flown photography the possibility of rephotographing should always be considered. This might prove to be the more economic proposition and better photography will certainly improve the quality of mapping produced.

In order to achieve a satisfactory result on a photographic sortie, the use of certain navigational aids can be of great value in positioning the aircraft so that the various requirements are met. These instruments are the subject of section 2.17 at the end of this chapter.

2.9 The tilt of the camera axis
With near-vertical aerial photographs of the sort we are very largely concerned with in this text, there is a restriction on the amount by which the axis of the camera can be allowed to deviate from the vertical. The most common tolerance quoted is a maximum of 3° with sometimes a restriction of 5° on the maximum difference in direction between one exposure and the next. These tolerances are easy to comply with at the higher altitudes and experience has shown that at elevations of the order of 6000 m an average tilt of just over 1° can be achieved quite readily. The lower the flying altitude the more difficult it becomes to meet the specification due to the effects of air turbulation.

There are two main reasons for the tilt restriction. In the use of graphical methods for mapping (or more likely now the revision of mapping) the assumption is often made that the photograph is truly vertical and this will not give rise to any plottable error in the resulting planimetry if the tilt is less than 3°. This is part of the Arundel assumption for radial line plotting as described in section 4.3. More important, however, is the restriction imposed by certain types of photogrammetric plotting instruments in their ability to rectify photograph tilts of any great magnitude. As we shall see in chapter 8 when dealing with the various types of analogue instrument, one of the characteristics of many instruments using a mechanical method of reprojection is the limit of tilt that can be set on individual projectors. This limit is commonly 5–6°. If therefore photographs have tilts greater than this value the operator would be unable to complete the relative orientation process. That is to say, a stereoscopic model cannot be formed. Not all instruments, of course, have this particular lack of accommodation but at the present time the great majority of plotting instruments do use mechanical projection methods. Such a restriction does not apply to analytical type plotters.

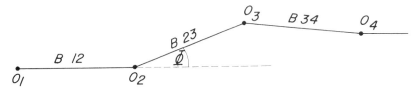

Figure 2.5. Inclination of the Air Base

2.10 The inclination of the air base
The air base (B) is the line joining two successive exposure stations in space. Ideally the aircraft should fly straight-and-level along the whole length of the strip, in a manner such as B_{12}. However, small changes in aircraft elevation must occur and so B_{23} is an air base with an inclination shown as Φ in Fig 2.5. Using an analogue plotting instrument, the individual photo tilts and the air base inclination must be faithfully reproduced in the instrument. The restriction arises therefore from the ability of the instrument to accommodate this situation. Many first- and second-order machines can accommodate a maximum of only 5° of inclination. Hence the value specified should be well within the capabilities of the instruments to be used for processing the photography.

2.11 The crabbing of photographs
This effect arises because the camera has not been maintained in correct alignment, in which case the appropriate sides of the photograph format have not been made parallel to

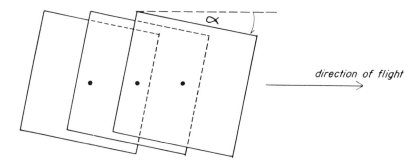

Figure 2.6. Angle of Crab

the ground velocity of the aircraft (see Fig 2.6). The problem occurs when the aircraft is subjected to variable cross winds. Due to the cross wind, the aircraft heading is inclined to the ground velocity direction. If the wind varies in speed the aircraft heading will also need to be varied to maintain the correct strip direction. At each variation, therefore, the camera must be rotated in its mounting to remove any crabbing; in practice it is rather easier to fly the sortie up and down the wind. In which case the constant attention to camera alignment is not necessary but care must be taken to ensure that the correct fore-and-aft overlapping is obtained.

The effect of any crabbing is to reduce the size of the stereo model and, perhaps more seriously, the overlapping areas of the stereo models. A limit of 5° is the usual restriction on the angle of crab permitted.

2.12 The direction of flight lines

Over limited areas, and with no other factors to consider, the flying should be such that the strip lengths are a maximum and the number of strips is a minimum. This is most likely to be the situation in large-scale surveys where the mapping requirement is confined to a comparatively small project or township area.

In the case of smaller scale mapping, some form of aerial triangulation will be employed to provide the control points required to set up each individual stereo model (see chapters 9 and 11). Thus the flight plan might well be determined by the shape and size of the blocks of aerial triangulation being used to cover the area. If the area is to be covered by a series of map sheets then the size of these may also have a bearing on the manner in which the photography is flown. In addition to the main photography there is sometimes a requirement for a number of cross strips as an aid to the aerial triangulation process.

As mentioned in section 2.11 above, the wind direction can have an influence on the chosen flight direction. In small-scale work where the aircraft is flying in the stratosphere, there is usually a strong and persistent east–west wind blowing. Hence much of this high altitude photography is flown in the east–west direction.

There is one other geographical factor that can influence the flight pattern and that is the general shape of the terrain being photographed. In section 2.10 we saw that there was a requirement to fly straight-and-level and this will produce a reasonably uniform photo scale if the terrain is flat. If the ground height varies by more than 12.5% from the mean value then the overlap will fall outside the $60\% \pm 5\%$ limit unless either the flying height is changed or the time interval between photographic exposures is altered. If changes in ground elevation are of an abrupt nature it will not be possible to compensate for these instantly, and in consequence the overlap specification will be lost. In practical terms, the constant adjustment of the time interval or change in flying elevation (within the air base inclination limit) is a situation to be avoided, if at all possible. This might be done by changing the direction of the flight lines to accord more or less with the linear features of

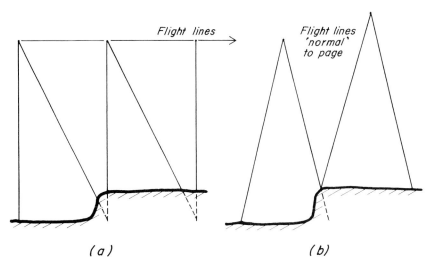

Figure 2.7. The Influence of Topography on Flight Line Direction

the topography being photographed. Figure 2.7 illustrates this situation. In this case it would be better to fly in a direction normal to the plane of the paper, taking care to ensure a satisfactory lateral overlapping configuration.

2.13 Time of the photographic sortie

The time of day should be such that long shadows are avoided since detail falling within areas of deep shadow cannot readily be interpreted. Early morning and evening photography should therefore be avoided. Sometimes photographing at a certain season may be advantageous. For example a season when the vegetation cover is a minimum is a good time for mapping rural topographic features. Frequently, however, it is the climate and the atmospheric conditions that will dictate the period when photography is possible. This is especially true in the case of high altitude work where the presence of cloud is more than likely to interfere with the photographing programme.

Sometimes the time of photography will be fixed by the occurrence of some particular feature that is to be depicted. For example, the high/low water line might be required to be shown on the sheet. In this case infrared photography of the area taken over a specified and limited time interval will be necessary. On the other hand, colour photography at low water can be used to detect sand banks, shoal waters, deeper channels, etc.

2.14 Cloud cover

The problem of cloud cover increases with the flying altitude and is particularly acute in temperate zones. For example, in Great Britain, it has been estimated that there are only a few days per annum when high level photography can be satisfactorily taken. It is therefore not unusual for such photography to be taken with small areas obscured by patches of cloud; a limit of 5% is therefore a common one specified for the amount of cloud cover that is acceptable. Loss of detail is more serious in some parts of a model than others. In the overlapping areas, what might be obscured on one model might well be clear on another. Loss of stereoscopic detail is most inconvenient over the areas of the six standard model points (see section 7.3 and Fig 7.5). In such cases relative orientation of individual models is made more difficult and the connection of one model to the next in aerial triangulation is hampered. The areas in the vicinity of the principal points are considered to be among the most important and so cloud cover over these areas is often not acceptable.

With a persistent low cloud base, the use of the shortest focal length camera available will provide the smallest scale photo cover possible.

Atmospheric haze can be a problem in certain dry areas. A minus blue filter will reduce the fogging but some loss of contrast is inevitable. Hence the resolution of detail will be less than normal. A rather lower flying altitude and larger photo scale could be used to compensate for this but this would raise the cost of the project by increasing the number of photographs required to cover the area.

2.15 Air cover diagrams

On the completion of the photographic sortie it is important that good air cover diagrams are provided showing as accurately as possible the position of each flight line and the principal point positions along each strip. These diagrams will be used over and over again in the course of planning and carrying out the photogrammetric mapping. It is therefore important to have good diagrams produced.

In the case of the remapping of an area, then the information can most usefully be drawn out on existing map sheets at perhaps half photo scale.

A most valuable air cover diagram can also be produced by putting together an uncontrolled mosaic made up of untrimmed photographs. The principal points are located and identified using pieces of white card, numbered and laid on the mosaic in the vicinity of all principal points. Some other features such as town names, approximate map sheet corners and so on can also be identified in a similar way. Then the whole assembly, together with a title and other basic border information, can be rephotographed at a reduced scale. Such diagrams will prove to be particularly valuable in the field when no other form of mapping is available. They are particularly useful as base maps when collecting field completion material.

2.16 Annotation of photographs and film rolls

In a large mapping project there will be a considerable number of photographs produced on a large number of rolls of film negative. It is essential that these and their containers be annotated so that any particular negative can be located with ease. Further, the information appearing along the edge of every print made should be sufficient for a complete identification of the print. Included with each film container should be a copy of the Sortie Report of the photographer giving the general details of flight times and conditions.

2.17 The use of auxiliary instruments

In addition to the camera itself and the associated photographic equipment, a number of other auxiliary instruments may be used on the photographic sortie. These may be carried in order to facilitate the carrying out of the photography in accordance with the specification, or they may be of such a type that they provide additional control information for use in subsequent photogrammetric processes such as rectification or aerial triangulation.

The first group of instruments are concerned with the navigation of the aircraft and in locating its position with an accuracy that facilitates a satisfactory coverage of the sortie area with the correct overlapping of strips of photographs. One of the earliest ways of doing this was to use a single radar beacon and maintain a series of flight paths at predetermined fixed distances from it. The resulting air-cover diagram was therefore composed of a series of circular arcs each overlapping by the correct amount.

With regard to the timing of the exposures, some form of intervalometer is required. One such instrument of this type projects a replica of the camera image onto a ground glass screen. A framing chain of variable speed moves across this screen and the operator adjusts this until it is synchronous with the moving ground image. The camera can then be set to give automatically exposures of anywhere between 20 and 90% overlap.

At the present time, the Decca navigator and similar systems can be used to provide an accurate control of the flight path, if the sortie area is covered by a series of control stations. In this case the proposed flight lines would be plotted on a chart provided with the Decca lattice system. The flight lines can then be translated into the appropriate Decca signals from which the aircraft can be navigated with a precision of about ± 25 m under favourable geometric conditions. For an explanation of this system of navigation see Ref 0.7. The Doppler navigation system is a method of dead reckoning using measured frequency shifts brought about by the aircraft velocity on radiated microwave signals. An accuracy of about 0.3% of the distance travelled can be achieved. [0.7, 2.6]

3: The geometry of the aerial photograph

3.1 Introduction

The geometric properties of the aerial photograph are of importance for a number of reasons. Firstly, an appreciation of this topic will allow us to produce usable maps from such photographs without the aid of sophisticated and expensive equipment. Secondly, such a study leads naturally to an understanding of the various forms of rectification whereby the effects of any tilt of the photograph are corrected for in analogue plotting instruments and optical rectifiers. In addition to the production of maps the properties to be discussed in this chapter are of particular importance in any appreciation of the techniques suitable for the revision of maps using aerial photographs.

In essence, the camera is a recording goniometer as far as geometrical considerations are concerned. With the aid of the data provided by camera calibration it is possible to recover the directions of the rays entering the camera at the moment of exposure from an examination of the recorded positions on the photograph. In this and the following chapter the directions we shall investigate first are those that form plane angles on the surface of the photograph itself. For this purpose we need only a knowledge of the position of the principal point. Later, in chapters 5 and 6 we examine the three-dimensional situation and at that point consider directions in space. It is interesting to note that it is the

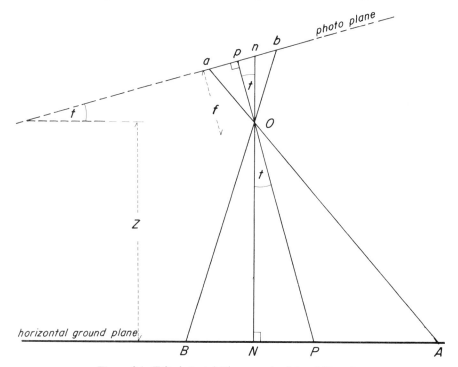

Figure 3.1. Tilted Aerial Photograph of Level Terrain

information provided by the camera calibration process that allows us to transform two dimensional photo co-ordinate data into three-dimensional vectors in space.

In the first instance we consider a tilted aerial photograph of a ground surface that is a horizontal plane. We then have the situation illustrated in Fig 3.1. The diagram is a vertical section through the nodal point O of the lens and contains the line of greatest slope on the photograph. In all the cases considered in this text the separation of the two nodes of the camera lens can be ignored. Also, because of the dimensions involved, all rays from ground object to camera can be regarded as chief rays. Within the camera we consider only the directions of the chief rays (see section 1.9 and Fig 1.1). Hence under such conditions the figure can be regarded as a diagram of a central perspective projection of an array of points in a ground plane onto a photo plane through the point O.

From an inspection of Fig. 3.1 we note that the distance from the principal point p to O is the principal distance and is equal to f, the nominal focal length of the camera. Those lines on the photograph at right angles to the plane of the diagram are in fact horizontal lines in space and are called photo, or plate, parallels. The vertical line through O defines the nadir point n on the photograph and N on the ground. The angle pOn is therefore the tilt of the photograph.

3.2 The vertical aerial photograph

A vertical aerial photograph of horizontal terrain is clearly the most simple of cases and the one that we consider first. In this case the nadir point n(N) coincides with the principal point p(P). In Fig 3.2 are shown both the negative and positive planes of projection. These are positive and negative in both a tonal, photographic sense and a mathematical sense. The points a and b are the images of any two ground points in the same horizontal plane H. We see that the scale is the same for all positions and directions on the photograph for all such points, and is given by

$$S = \frac{ab}{AB} = \frac{f}{Z - H}$$

where Z is the height of the camera above the datum surface. We note that the scale of

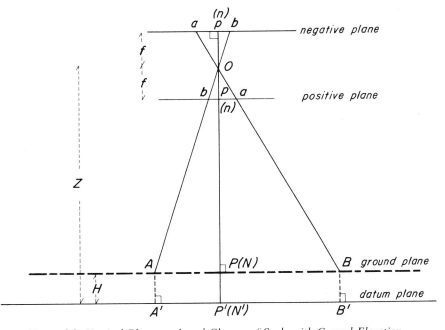

Figure 3.2. Vertical Photograph and Change of Scale with Ground Elevation

37

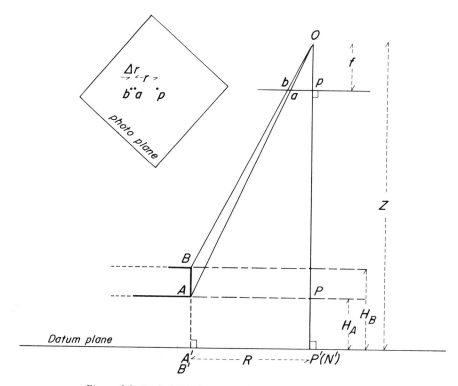

Figure 3.3. *Radial Displacement due to Changes in Elevation*

projection is inversely proportional to the height of the camera above the ground plane. This characteristic is one that is at variance with the orthogonal projection A′B′ of the points, which is the basis of most mapping projections. Figure 3.3 shows the consequence of this property. Because of their greater altitude all points such as B in the plane H_B are projected at a larger scale than points in the plane H_A. Hence the effect of a change in elevation $(H_B - H_A)$ has produced the image displacement ab in the photo plane.

Now

$$ap = \frac{f}{(Z - H_A)} \cdot AP$$

$$= \frac{f}{(Z - H_A)} \cdot A'P'$$

And

$$bp = \frac{f}{(Z - H_B)} \cdot A'P'$$

Therefore

$$ab = f \cdot A'P' \left[\frac{1}{Z - H_B} - \frac{1}{(Z - H_A)} \right] \tag{3.1}$$

∴

$$ab = f \cdot A'P' \frac{(H_B - H_A)}{(Z - H_B)(Z - H_A)}$$

∴

$$ab = \frac{ap \cdot \Delta H_{BA}}{(Z - H_B)} = \frac{bp \cdot \Delta H_{BA}}{(Z - H_A)}$$

i.e.

$$\Delta r = \frac{r \cdot \Delta H}{Z} \tag{3.2}$$

38

As an example, taking the maximum radial distance possible on a standard photograph to be rather less than $\sqrt{2} \times 115$ mm, with a flying height of 1000 m above ground level and an acceptable error in planimetric position of 0.5 mm maximum we have, using equation (3.2)

$$ab = 0.5 \text{ mm}$$

$$= \frac{\sqrt{2 \cdot 115 \cdot \Delta H}}{1000} \text{ mm}$$

i.e.
$$\Delta H = 3 \text{ m or } 0.3\text{°/}_{oo} \ H \text{ maximum}$$

An inspection of equation (3.2) indicates that the displacement Δr is a linear function of both the change in elevation and the radial distance from the nadir point, which in this case coincides with the principal point on the photograph. The image displacements introduced by changes in elevation are therefore radial from the nadir point on the photograph. From the mapping point of view, the consequences are therefore as follows. If all the ground points lie, within limits, in a horizontal ground plane then the geometry of the vertical photograph can be regarded as identical to that of a plane map. The scale of this mapping can be derived from a knowledge of the camera position or determined empirically from the known positions of any points falling within the area. If, however, the assumption cannot be made, there is still one invariant between ground and photo planes that can be used to derive accurate mapping, namely that angles measured about the nadir point in each situation are identical. In order to make use of this property, more than one photograph is necessary. This technique of radial line plotting is described in chapter 4.

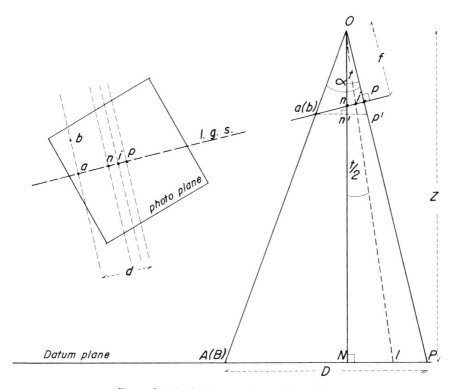

Figure 3.4. Scale Changes due to Photo Tilt

3.3 Scale changes on the tilted photograph

We now consider the situation where the ground plane is a horizontal one but the photograph has a tilt t as illustrated in Fig 3.4. Here again the diagram represents a vertical section through O orientated in the direction of the line of greatest slope on the photograph. The principal point and nadir point now no longer coincide. The lines shown in the photo plane at right angles to the principal line, that is the line of greatest slope (l.g.s.) passing through the principal point, are plate parallels. The horizon line, if it appears on the photograph is therefore a plate parallel. Because the parallels are horizontal lines in space they are in consequence lines of constant scale. The scale along any parallel, such as that passing through the points a and b, is given by

$$S_a = \frac{ab}{AB}$$

$$= \frac{Oa}{OA}$$

$$= \frac{On'}{ON}$$

since Oab and OAB are similar triangles in a plane at right angles to the plane of the diagram.

From Δs Oap and OAN we have

$$Oa = f \sec \alpha$$

$$OA = Z \sec (\alpha - t)$$

$$\therefore \qquad S_a = \frac{f \cos (\alpha - t)}{Z \cos \alpha}$$

$$= \frac{f}{Z} (\cos t + \tan \alpha \sin t)$$

i.e.
$$S_a = \left(\frac{f}{Z} \cdot \cos t \right) + d \left(\frac{\sin t}{Z} \right) \qquad (3.3)$$

where $d = f \cdot \tan \alpha$ is the separation ap of the parallel through a, from the principal point parallel. For any photograph the bracketed quantities are constant and so the scale along any parallel is a linear function of the distance d.

For the principal point parallel we have d equal to zero, hence

$$S_p = \frac{f}{Z} \cdot \cos t \qquad \text{i.e.} < \frac{f}{Z}$$

For the nadir point parallel we have

$$d = f \cdot \tan t$$

hence

$$S_n = \frac{f}{Z} \cdot \cos t + f \tan t \cdot \frac{\sin t}{Z}$$

i.e.
$$S_n = \frac{f \sec t}{Z} \qquad \text{i.e.} > \frac{f}{Z}$$

By inspection we can deduce that if

$$d = f \tan t/2$$

then
$$S_i = \frac{f}{Z} \cdot \cos t + \frac{f \tan t/2 \cdot \sin t}{Z}$$

i.e.
$$S_i = \frac{f}{Z}$$

This parallel through i is called the isometric parallel and is of particular importance. All photo planes can be regarded as planes tangential to a sphere of radius f. The isometric parallel is therefore the line of intersection between the horizontal plane of the vertical photograph and any inclined plane, as illustrated in Fig 3.5.

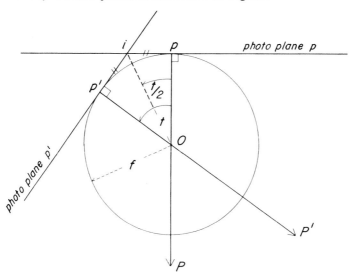

Figure 3.5. *The Isometric Parallel of a Tilted Photograph*

The horizon trace is a plate parallel of infinitely small scale. From Fig 3.6 we see that
$$Oh = f \operatorname{cosec} t \quad \text{and} \quad d = -f \cot t$$

hence
$$S_h = \frac{f}{Z} \cos t - f \frac{\cot t \cdot \sin t}{Z}$$
$$= 0$$

The horizon trace mentioned above is the image of a line in the datum plane at an infinite distance from P. In this text it will be termed the 'plane horizon' to distinguish it from the natural horizon that would appear on a high oblique photograph. This latter is at a finite distance OT as illustrated in Fig 3.6; the diagram grossly exaggerated for the sake of clarity. From the diagram we see that the angle of dip (θ) is given by
$$\cos \theta = \frac{R}{R+Z}$$

∴
$$1 - \frac{\theta^2}{2} \simeq \frac{R}{R+Z}$$
$$= \frac{1}{(1+Z/R)}$$

∴
$$\theta \simeq \sqrt{\left(\frac{2Z}{R}\right)} \, \text{rad}$$

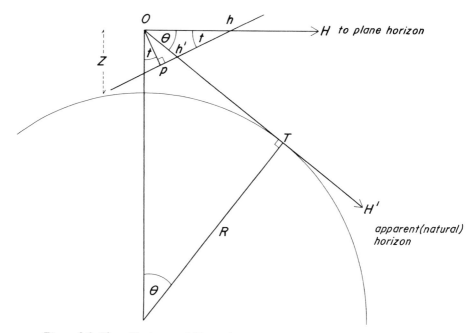

Figure 3.6. *Plane Horizon and Natural Horizon Trace on Oblique Photographs*

From the $\triangle$Ohh' we can calculate the distance hh' of the natural horizon, apparent on the photograph, from the plane horizon, for we have

$$\text{hh}' = \text{Oh} \cdot \frac{\sin \theta}{\sin (\theta + t)}$$

$\therefore \qquad\qquad\qquad \text{hh}' = f \cosec t \cdot \frac{\sin \theta}{\sin (\theta + t)} \qquad\qquad\qquad (3.4)$

When the natural horizon appears on a photograph it provides valuable information, for the principal line on the photograph can be established by dropping a perpendicular from the horizon trace through the principal point. Also, by measuring the distance h'p an approximate value for the tilt of the photograph can be calculated from

$$\tan t = \frac{f}{\text{h'p} + \text{h'h}} \simeq \frac{f}{\text{h'h}} \qquad \text{(for h'p is very small)}$$

Now, if we substitute this value for t in equation (3.4) we can then find a value for hh', which in turn can now be used to calculate a better value for t.

Concerning the scale in the direction of the line of greatest slope S_a' we see from Fig 3.4 that

$$S_a' = \frac{d}{D}$$

$$= \frac{\text{ap}}{\text{AP}}$$

$$= \frac{\text{ap}' \cos t}{\text{AP}}$$

i.e.

$$S_a' = \frac{On'}{ON} \cdot \cos t$$

$$= S_a \cos t \qquad (3.5)$$

3.4 Angular measurements in the tilted photo plane

All image displacements that take place over the format of a tilted photograph can be described in terms of changes in photo scale. In general these will lead to distortions and thus angular measurements at any point on the photograph will not agree with the equivalent angles measured on the ground plane. However, at just one point, the isocentre, the angles are identical as can be demonstrated with the aid of Fig 3.7. In this figure, the

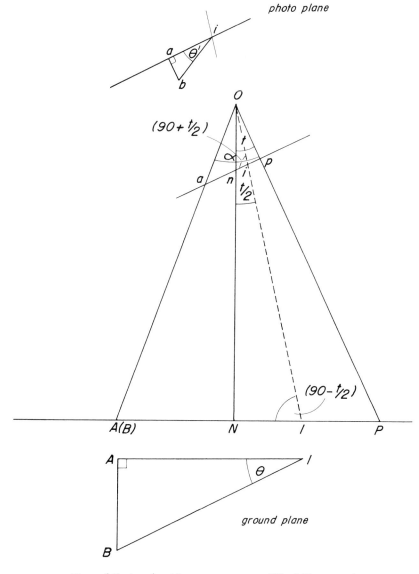

Figure 3.7. *Angular Measurements on a Tilted Photograph*

two points a and b are chosen to lie on the same photograph parallel. Hence the angles iab on the photograph and IAB in the ground plane are both right angles.

From Δiab in the photo plane and ΔIAB in the ground plane we have

$$\tan \theta' = \frac{ab}{ai} \quad \text{and} \quad \tan \theta = \frac{AB}{AI}$$

$$\therefore \quad \frac{\tan \theta}{\tan \theta'} = \frac{AB}{AI} \cdot \frac{ai}{ab}$$

$$= \frac{OA}{Oa} \cdot \frac{ai}{AI}$$

From Δs Oai and OAI we have

$$\frac{ai}{Oa} = \frac{\sin(\alpha - t/2)}{\sin(90 + t/2)} \quad \text{and} \quad \frac{OA}{AI} = \frac{\sin(90 - t/2)}{\sin(\alpha - t/2)}$$

hence

$$\frac{ai}{Oa} \cdot \frac{OA}{AI} = 1$$

Therefore

$$\theta = \theta'$$

If we now examine the angles formed about the principal point in exactly the same way, we have

$$\tan \theta' = \frac{ab}{ap} \quad \text{and} \quad \tan \theta = \frac{AB}{AP}$$

$$\therefore \quad \frac{\tan \theta}{\tan \theta'} = \frac{AB}{AP} \cdot \frac{ap}{ab}$$

$$= \frac{OA}{Oa} \cdot \frac{ap}{AP}$$

From Δs Oap and OAP we see that

$$\frac{ap}{Oa} = \sin \alpha \quad \text{and} \quad \frac{OA}{AP} = \frac{\sin(90 - t)}{\sin \alpha}$$

hence

$$\frac{\tan \theta}{\tan \theta'} = \sin \alpha \cdot \frac{\cos t}{\sin \alpha}$$

$$\therefore \quad \tan \theta = \tan \theta' \cos t \tag{3.6}$$

The above, of course, implies that the scale of projection along the parallel through a is greater by a factor sec t than the scale of the line ap, a result already realised in equation (3.5).

If we now consider angular values about the third cardinal point on the photograph, the nadir point n, it is easily demonstrated that such angles about this point are greater than the equivalent ground angles. The scale along the parallel is now less than the scale of the line an, by a factor of cos t, hence

$$\tan \theta = \tan \theta' \sec t \tag{3.7}$$

It should be noted that in the case of the principal and nadir points the angular relationships derived apply to acute angles measured from the principal line.

Sometimes we are required to estimate the error in an angle measured about some point on the photograph. Usually this will be the principal point, for generally this is the point most easily identifiable, and so we will take this as an example. We have found that

$$\tan \theta = \tan \theta' \cos t$$

hence

$$\tan \theta = \tan \theta' \left[1 - \frac{t^2}{L^2} + \frac{t^4}{L^4} - \cdots \right]$$

∴

$$\tan \theta' - \tan \theta \simeq t^2/2 \cdot \tan \theta'$$

$$\sin \theta' \cos \theta - \sin \theta \cos \theta' = \tfrac{1}{2} t^2 \cos \theta \cdot \cos \theta' \cdot \tan \theta'$$

∴

$$\sin (\theta' - \theta) = \tfrac{1}{2} t^2 \sin \theta' \cos \theta$$

But the difference in angular values will be small, hence we can approximate this expression to

$$\theta' - \theta = \Delta \theta$$

$$= \tfrac{1}{4} t^2 \cdot \sin 2\theta \tag{3.8}$$

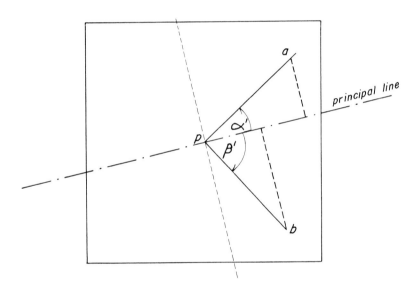

Figure 3.8. Angular Error at the Principal Point due to Tilt

The error in any angle measured about the principal point, such as that shown in Fig 3.8 is given by the expression set out below:

$$\text{Ground angle } APB = (\alpha + \beta)$$

$$\text{Photo angle } apb = (\alpha' + \beta')$$

where

$$(\tan \alpha + \tan \beta) = (\tan \alpha' + \tan \beta') \cos t$$

hence error

$$\Delta(\alpha + \beta) = \tfrac{1}{4} t^2 (\sin 2\alpha + \sin 2\beta) \tag{3.9}$$

3.4.1. THE EFFECT OF CHANGES IN GROUND ELEVATION

In all the above investigations the ground has been considered to be a horizontal plane. A further correction can be derived, for angles measured about the principal point, to

45

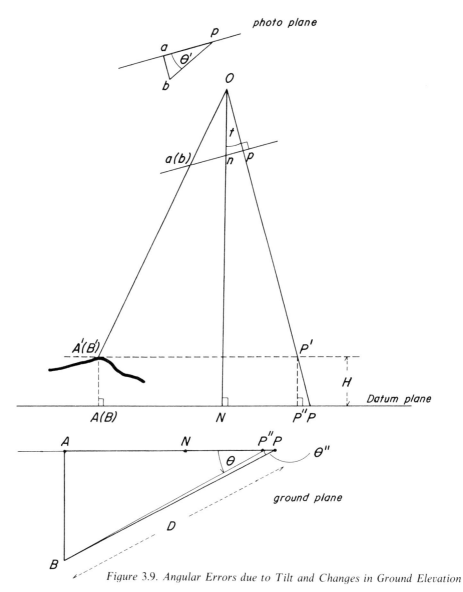

Figure 3.9. *Angular Errors due to Tilt and Changes in Ground Elevation*

account for the displacements introduced by changes in ground elevation. Figure 3.9 illustrates the situation. In the diagram the difference in height between point B′ and P in the datum plane is H. Firstly, therefore, we consider points in the plane A′B′P′. The horizontal ground angle measured about the point P′ is identical to that observed at the point P″, the orthogonal projection of P′ into the datum plane. Equation (3.6) therefore gives us the relationship between the angle AP″B and the photo angle apb. (A theodolite would of course produce the same ground angle if stationed at any point along a vertical line containing P″P′.) Hence we have

$$\tan \theta = \tan \theta' \cos t$$

But the principal point does not lie in the plane H but in the datum plane at P. The true ground angle is therefore APB (i.e. angle θ''). For the case shown where B′ is above P we have a further negative correction to apply to the photo angle given by

$$\Delta \theta'' = (\theta - \theta'')$$

46

Now, from $\triangle PP''P$ we see that

$$PP'' = H \tan t$$

and in the narrow $\triangle BP''P$ we have

$$\sin \Delta\theta'' = \frac{PP'' \sin(180 - \theta)}{D}$$

$\therefore$
$$\Delta\theta'' \simeq \frac{H \tan t \sin \theta}{D} \qquad (3.10)$$

Sometimes an angle of elevation E is introduced, such that

$$\tan E = H/D$$

Substituting this value in equation (3.10), we get

$$\Delta\theta'' \simeq \tan E \tan t \sin \theta$$

If the tilt is small we can approximate a little further to give,

$$\Delta\theta'' \simeq t \tan E \sin \theta \qquad (3.11)$$

Hence the total error between the photo angle at p and the ground angle at P is given by

$$\Delta\theta = \tfrac{1}{4}t^2 \sin 2\theta + t \tan E \sin \theta \qquad (3.12)$$

In an exactly analogous manner we can derive the expression for the error in the angle measured at the isocentre i due to a change in elevation H. In this case this factor is the total error in the angle and is given by

$$\Delta\theta = t \tan E \sin \theta$$

When we consider the effect of a change in elevation on angles measured about the nadir point n the effect is of course zero. The total error in this case is that due to the tilt as given in equation (3.7).

3.5 The perspective grid
The nature of the scale changes over the format of a tilted photograph of horizontal terrain gives rise to the formation of a perspective grid on the photograph, the nature of which is

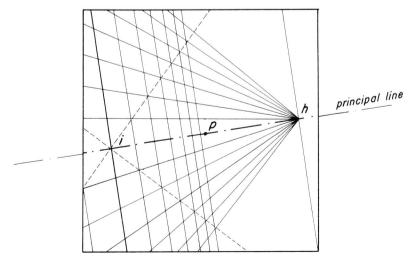

Figure 3.10. *A Perspective Grid with One Vanishing Point*

47

illustrated in Fig 3.10. The photograph is a high oblique on which the plane horizon is shown. Along the isometric parallel equal divisions have been marked off. Due to the linear change in scale, rectangular grid lines from these intercepts will converge to a vanishing point h. Because angles are projected correctly about i, the two diagonals set out at 45° from this point can be used to establish further parallels of the grid, which is rectangular on the ground. This particular grid is a special case as one axis of the rectangular grid on the ground is parallel to the line of greatest slope on the photograph. Hence we have generated one set of parallels and one set of converging lines on the photograph. In general this will not be the case. In general the tilt will resolve into two components in the directions of the two axes and therefore generate two sets of converging lines based on two vanishing points as illustrated in Fig 3.11. In this case the horizon trace is off the format of the photograph. The rectangular axes $(X_0\ Y_0)$ are now inclined at any angle θ to the principal line. These two axes cut the horizon trace in the two vanishing points hx and hy. The X grid lines converge on the former and the Y grid lines converge on the latter to generate the perspective grid.

One method of planimetric mapping is to transfer a rectangular map grid onto the photograph in this way using control of some form. When a perspective grid of sufficient density has been established, photo detail can then be transferred from one elementary area to another. In this way photo detail can be compiled on to the map sheet. The process is therefore one of piecemeal graphical rectification.

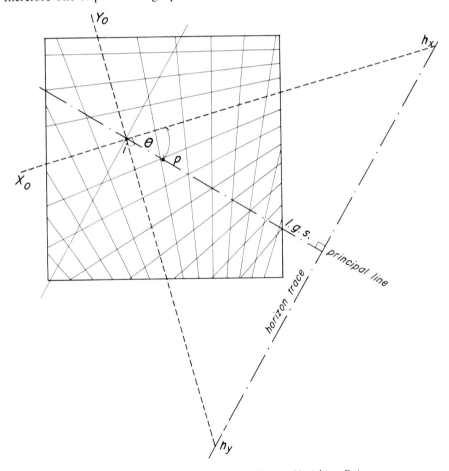

Figure 3.11. A Perspective Grid with Two Vanishing Points

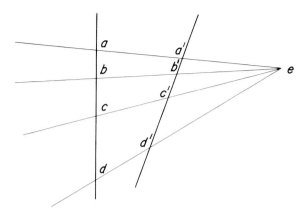

Figure 3.12. Cross Ratios for a Bundle of Rays

3.6 Cross ratios

In the process of projection or transformation many of the geometric characteristics can change (for example, constant scale factors and angular relationships) but some properties do not change and these are termed invariants. In the case of central perspective projection the invariants are the cross (or anharmonic) ratios. Referring to Fig 3.12, the four points a, b, c, d lie on a straight line in the plane that also contains the fifth point e. A cross ratio can be defined as,

$$(abcd) = \frac{ab}{bc} \cdot \frac{cd}{da}$$

$$= \text{constant}$$

Such a range of points with this property is produced when a pencil of rays from a vertex such as e is cut by a transverse abcd (or a' b' c' d'). The right-hand side is a constant in this case because the left-hand side can be expressed solely in terms of the angles at the vertex. By repeated application of the sine rule, we have

$$ab = be \cdot \frac{\sin aeb}{\sin eab} \qquad bc = be \cdot \frac{\sin bec}{\sin bce}$$

$$cd = de \cdot \frac{\sin ced}{\sin ecd} \qquad ad = de \cdot \frac{\sin aed}{\sin dae}$$

Hence, noting that sin bce = sin ecd, we obtain

$$(abcd) = \frac{\sin aeb}{\sin bec} \cdot \frac{\sin ced}{\sin aed}$$

$$= \gamma$$

Clearly this quantity is a constant for all transverses.

It is not difficult to find other combinations of transverse lengths that also lead to constants, not necessarily of the same value as the one quoted above. There are in fact factorial four (i.e. 24) possible combinations, giving rise to six different cross ratios related in the following way:

$$(abcd) = (badc) = (cdab) = (dcba) = \gamma$$

$$(abdc) = (bacd) = (cdba) = (dcab) = \frac{\gamma}{\gamma - 1}$$

$$(acbd) = (bdac) = (cadb) = (dbca) = 1 - \gamma$$

$$(acdb) = (bdca) = (cabd) = (dbac) = \frac{\gamma - 1}{\gamma}$$

$$(adbc) = (bcad) = (cbda) = (dacb) = \frac{1}{1 - \gamma}$$

$$(adcb) = (bcda) = (cbad) = (dabc) = 1/\gamma$$

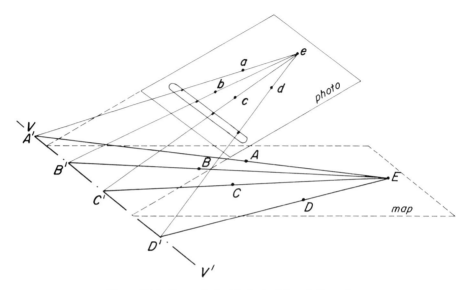

Figure 3.13. *Paper Strip Method of Detail Transfer*

This form of invariant arises in projective transformation because the axis of perspective (i.e. the line of intersection of the two planes VV' in Fig. 3.13) forms the common transverse between a pencil of rays in one plane and the homologous pencil of rays in the other. The significance of this property in mapping is as follows. Given four pairs of homologous points such as Aa, Bb, Cc, Ee in Fig 3.13, a fifth point d present in the photo plane but not in the map plane can be established in the latter. The process is most easily carried out graphically in the following manner. First, choose one pair of points (Ee) as the vertex for each pencil of rays. The rays are then drawn in from these vertices, i.e., ea, eb, ec and ed on the photograph, and EA, EB and EC on the map. A paper strip is now laid across the photo pencil at any convenient place to form a transverse. The cross ratio property is recorded in the form of four tick marks along the strip edge. The strip is then laid across the other pencil of rays in the map plane at any convenient point and arranged so that the three tick marks coincide with the three appropriate rays. The fourth mark indicates the direction of the ray ED in the map plane. This process must be carried out once more selecting another pair of points as a vertex so that the second ray direction makes a good intersection with the first. The point D has then been established.

To carry out a mapping operation using this technique a minimum of four control points not all falling in a straight line are required. Although accurate if the ground points are restricted to a horizontal plane, this method of point-to-point working is laborious. It is therefore most commonly used for supplying a limited number of extra control points that can facilitate some other method of plotting. This technique and others based on the various geometric properties discussed in this chapter find their applications mainly in the revision of maps and the mapping of very limited areas. Such techniques are the subject of the next chapter.

4: Graphical methods of producing planimetric maps

4.1 Introduction
Planimetric maps can be produced using techniques based on the theoretical considerations described in the previous chapter. They require little or no sophisticated equipment to apply but can be rather laborious in application. They are therefore most suitable for the production of maps of limited areas in situations where instruments, such as analogue stereo plotters, are not available. They also find an important application in the revision of maps where the use of large plotting instruments would be uneconomic and often inappropriate. The techniques are based on either the use of individual photographs or of pairs of overlapping photographs. The former can be described as rectification processes whereby the scale changes brought about by any tilt of the photograph can be eliminated. They work most satisfactorily when the ground surface approximates closely to a horizontal surface. The latter employ methods of radial triangulation and make use of the plane angles recorded on the photograph. These methods are most satisfactory when the photographs are vertical or very nearly so. In this ideal situation their accuracy is not affected by changes in ground elevation. To a great extent therefore the two basic techniques are complementary to one another. The rectification of photographs followed by some form of radial triangulation is capable of producing excellent results.

4.2 Methods of rectification
There are four basic methods by which the image displacements introduced by the tilt of the photograph can be corrected for, and these are as follows:

(a)   Graphical techniques.

(b)   Mathematical transformations.

(c)   Mechanical methods of reprojection.

(d)   Optical transformations.

4.2.1 GRAPHICAL METHODS OF RECTIFICATION
The most straightforward of these depends on the construction of a perspective grid over the format of the photograph. In this way the area is divided up into elements small enough to make possible a direct transfer of detail from the photograph to the map with little error. This may be done by eye or with the aid of a simple optical device by which changes of scale can readily be brought about.

The perspective grid can be constructed in a number of ways. For example, in the case of the revision of a map sheet, the rectangular grid of the map can be transferred to the photograph with the aid of a number of good common points of detail. If only a minimum number of four such points are available then a sufficient density of grid intersections can be located on the photograph using the cross ratio property. The four grid intersections of a rectangle would be the minimum requirement for the location of the two horizon points, Vx and Vy, see Fig 4.1. Further grid points can then be supplied by what is sometimes

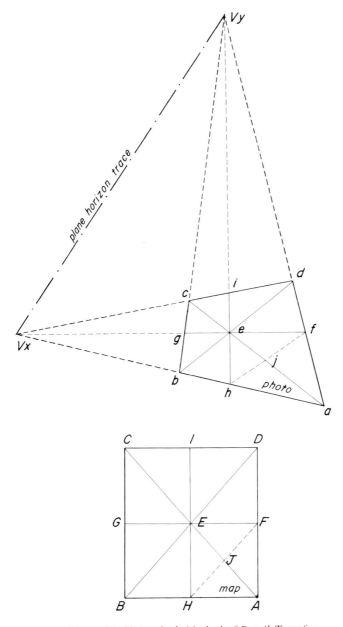

Figure 4.1. *Union Jack Method of Detail Transfer*

known as the Union Jack technique. Figure 4.1 illustrates how the point e, and thence f, g, h and i can be located using the two vanishing points Vx and Vy. If the element aefh is not small enough for an accurate transfer of detail from it to AEFH, then the process may be repeated by locating points such as j and so on. If the map does not have a grid, or the grid lines are not required, any four map points roughly in the form of a rectangle can be utilised. In this case, because the map lines no longer form a rectangle, both the map and the photo constructions will have pairs of intersecting points. On the photograph these are not now true vanishing points and do not lie on the horizon trace because they are no longer points at infinity on the ground. If the horizon trace appears on the photograph or

53

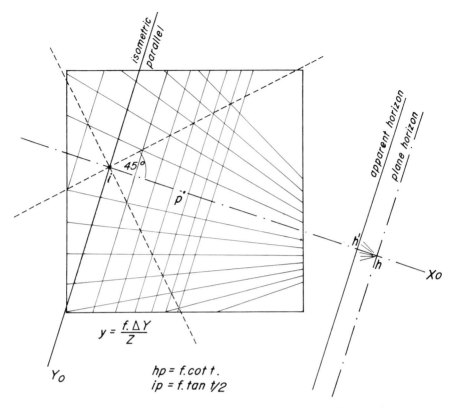

Figure 4.2. *Construction of a Perspective Grid on a Photograph*

can be established in some way, then the principal line can be drawn in and the isometric parallel located provided the focal length of the camera is known. A rectangular grid can now be constructed on the mapping sheet representing, at some suitable scale, a system based on the directions of the principal line and plate parallels. Figure 4.2 shows the method of construction. In order to find the position of the plane horizon the separation between this and the apparent horizon trace will need to be calculated as indicated in equation (3.4). Further details of such techniques can be found in texts such as Refs 0.1 (vol 2), 0.3 and 0.4.

4.2.2 MATHEMATICAL TRANSFORMATIONS

Although a technique of this sort cannot be described as graphical, a comment on the possibilities of the method is included here for the sake of completeness. With the introduction of quite powerful desktop computers such as the Hewlett–Packard 9800 and the Wang 2200 series, some of the graphical procedures of the past can now be usefully augmented or replaced by a more analytical approach. The derivation of the transformation formulae given here is based on the analysis carried out in section 3.

If a limited number of corrected photo positions are required a purely graphical method using the paper-strip technique and the cross ratios property could be employed. Equally well, a mathematical process of the form derived below could now be used, provided some method of measuring photo co-ordinates with sufficient precision is available. A well-made co-ordinatograph, of the type found in most well-equipped drawing offices, could for example, be used for this purpose. The basic formulae that describe a projective transformation process are as follows (see Refs 0.9 or 0.10 for example).

54

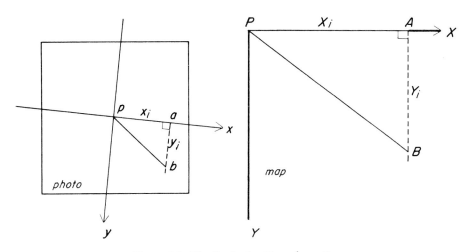

Figure 4.3. The Projective Transformation

$$X_i = \frac{a_{11}x_i + a_{12}y_i + a_{13}}{a_{31}x_i + a_{32}y_i + a_{33}}$$

$$Y_i = \frac{a_{21}x_i + a_{22}y_i + a_{23}}{a_{31}x_i + a_{32}y_i + a_{33}}$$

$$\left.\right\} \qquad (4.1)$$

Figure 4.3 serves to illustrate this process in a diagrammatic manner. From equation (3.3) we have

$$S_{ab} = \frac{y_i}{Y_i}$$

$$= \frac{f \cos t + x_i \sin t}{Z_i}$$

Hence

$$Y_i = \frac{Z_i y}{x_i \sin t + f \cos t} \qquad (4.2)$$

Also we have from equation (3.5)

$$\frac{x_i}{X_i} = S'_{ab} \cos t$$

$$= \frac{f \cos t + x_i \sin t}{Z_i \sec t}$$

Hence

$$X_i = \frac{Z_i x \sec t}{x \sin t + f \cos t} \qquad (4.3)$$

Now, in practice, although the principal point both on the photograph and on the ground might well serve as a local origin of co-ordinates, it is unlikely that the direction of the principal line will coincide with that of an axis. In fact, it will most likely be unknown. Hence, more generally, the photo co-ordinates will need to be expressed in some system of rectangular axes inclined at an angle κ to the principal line. Hence we have the transformation of photo co-ordinates associated with this rotation (see equation (A3) of Appendix A).

$$x = x' \cos \kappa + y' \sin \kappa \qquad y = -x' \sin \kappa + y' \cos \kappa$$

Substituting in equation (4.3) and (4.2), we get

$$X = \frac{(Z \sec t + \cos \kappa)x' + (Z \sec t + \sin \kappa)y'}{(\cos \kappa \sin t)x' + (\sin \kappa \sin t)y' + f \cos t}$$

(4.4)

and

$$Y = \frac{(-Z \sin \kappa)x' + (Z \cos \kappa)y'}{(\cos \kappa \sin t)x' + (\sin \kappa \sin t)y' + f \cos t}$$

Finally, if we also allow a shift in the origin of the ground system of co-ordinates $\Delta X, \Delta Y$ then we have

$$X' = X + \Delta X \qquad\qquad Y' = Y \Delta Y$$

(4.5)

If the shift is one representing only a change from the principal point position to the nadir point then of course

$$\Delta X = Z \tan t \text{ and } \Delta Y = 0$$

The air station O is frequently used as a local origin in photogrammetric analysis.

It will be seen that a combination of equations (4.4) and (4.5) readily leads to expressions of the same form as (4.1). By dividing both expressions by the coefficient a_{33} we find that we have ten unknown coefficients, and although these are not entirely independent (being functions of f, Z, t and κ) it is often convenient to treat them as such. In this case, five points with known co-ordinates in both systems are sufficient for their evaluation on the computer. Having evaluated the coefficients in this way, all required photo points can now be readily transformed into the map system.

4.2.3 METHODS USING MECHANICAL PROJECTION

The technique of using mechanical reprojection for rectification of single photographs has not been widely used in mapping. However, it should be realised that all stereo plotting instruments based on the concept of mechanical projection do in fact contain two such

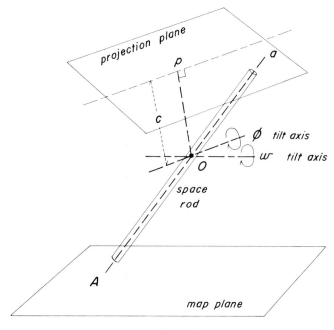

Figure 4.4. Mechanical Projection

rectifiers for the separate rectification of the stereo pair of photographs. At the present time, analogue plotting instruments based on such principles are predominant and so some discussion at this point will prepare the way for the more detailed examination of some of these instruments in chapter 8.

If we accept the fact that the purpose of the photogrammetric camera is to record directions from an air station, then the purpose of the projectors of most instruments is to recreate these directions as accurately as possible within the instrument. In mechanical reprojection, a mechanism that approximates very closely to the mathematical model of perspective projection is used. The perspective centre is represented by the centre of a universal joint while the direction of the light ray is replaced by a metal space rod (see Fig 4.4). From this diagram we see that if the projection distance c (i.e. the length of the perpendicular from the universal joint O to the plane of projection) is made equal to the focal length of the camera, and the plane of projection is tilted about O so that this plane takes on the same tilt as the camera, then a correct projection of all points such as a into A is obtained provided of course the space rod is straight. In theory this is a simple process but in practice there are some difficulties; the main one being that it is most difficult to locate the photograph in the projection plane and also define the point a, i.e. the point of intersection of the vector **AO** produced and the projection plane. In practice the

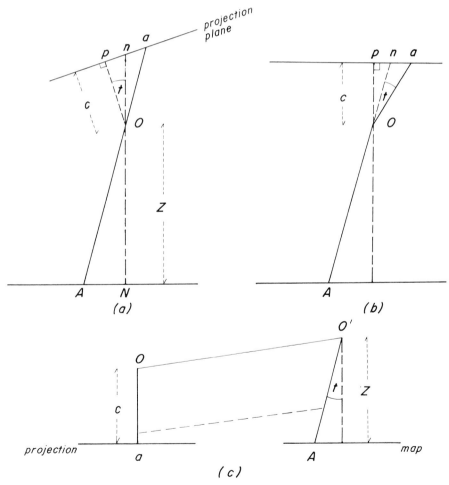

Figure 4.5. Mechanical Methods for Rectification

photograph and its viewing system are usually displaced away from the plane of projection but always remain parallel to it. Wild plotters are good examples of this point, and a rectified plot of a single photograph could be obtained on these or similar instruments in the rather unlikely event of this being required. Diagram (a) of Fig 4.5 illustrates the point being made. It can be arranged that the projection plane is held parallel to the map plane, and in this case the space rod can no longer remain straight but must be cranked at O through an angle equal to the photograph tilt (see Fig 4.5 (b)). This concept is a basic feature of the design of instruments such as the Kern PG2 and the Zeiss (Jena) Topocart. The cranking of space rods through precisely determined angles is not an easy thing to accomplish and both these instruments simplify the task to some extent by resolving the photo tilt into two components and carrying out separate rotations in the XZ and YZ planes for the two components φ and ω of tilt.

In the earlier models of the Santoni Stereomicrometer a three-dimensional cranking was carried out and this was facilitated by splitting the space rod into two parts: Oa and O'A of Fig 4.5 (c). The diagram shows in fact the φ component of tilt only. The ω component is obtained by rotating the space rod O'A about the longitudinal axis of the support beam OO'. This instrument is therefore one of the few that does in fact carry out a mechanical process of rectification because the planimetric plot is obtained from the right-hand photograph in this manner. More details of this instrument are given in section 8.5.4.

4.2.4 OPTICAL METHODS OF RECTIFICATION
Of the four methods listed above in section 4.2, the optical methods have been most commonly used in the past. The reasons for this are convenience and speed. So far the methods described have been of point-to-point working and in the past such methods have been regarded as unfavourable compared with the continuous solution, of which the photographic rectifier is an excellent example. With such an instrument all points are transformed instantly, when after the setting operations have been carried out, the photographic film on the copyboard is exposed. The theory and applications of this method are, however, extensive and these have therefore been described in some detail in the next chapter.

In this section we will describe only the more simple devices that attempt some form of

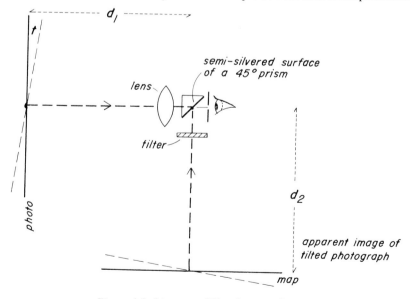

Figure 4.6. Diagram of Sketchmaster Device

optical rectification. When used over limited areas such instruments are capable of giving results of acceptable accuracy. The instruments fall into two main categories and are often of the Sketchmaster type using a semi-silvered mirror or some form of optical projector producing real images in an object plane (i.e. camera lucida or camera obscura devices).

Figure 4.6 shows in a diagrammatic way the main features of the Sketchmaster type of apparatus. The user views the map plane and the photo plane simultaneously using the small diameter peep sight and the 45° prism provided with a semisilvered surface. An overall scale change is effected by varying the ratio of the distances d_1 and d_2. In practice d_2 should be set at a comfortable viewing distance and changes in scale introduced by variations in d_1 only. In order to reduce possible errors due to parallax brought about by non-coincidence of the map and apparent photo planes, a meniscus lens can be selected so that the planes fall close together. The peep sight also helps to reduce this source of error by placing a restriction on the possible eye positions. The effects of small tilts can be compensated for by tilting the photo holder which is otherwise held vertical with a ball and socket type of mounting. The diagram shows the effect of a small tilting of the photo. The possibilities for parallax error are however increased. In use, it is important to ensure a good light balance between the map image and the apparent photo image and in order to aid this a range of grey filters is provided. When transferring detail from photo to map it is best to work over limited rectangular areas in which a reasonable fit can be obtained by trial and error. Such a pattern of control can be provided by some form of perspective grid or radial line plotting. In map revision existing detail points could form the basis of the control required.

The general arrangement of any instrument employing a projector lens is shown in Fig 4.7. Such equipment is used mainly for obtaining uniform changes in scale. The real image formed on the transparent drawing surface may be copied by drawing, or a photographic

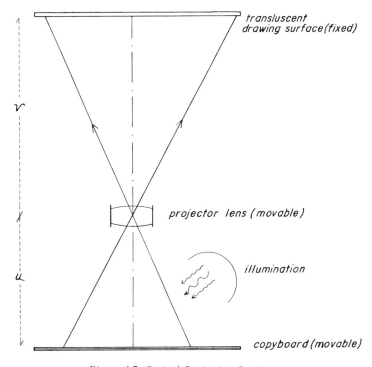

Figure 4.7. Optical Projection Device

copy may be produced. Scale settings and conditions for sharp focusing are done, by trial and error, in order to satisfy the Gauss equation for sharp focus

$$\frac{1}{u}+\frac{1}{v}=\frac{1}{f}$$

Some instruments have a copyboard that can be tilted through a small angle in order to introduce some degree of rectification. However, as the Scheimpflug condition is usually ignored (see section 5.4) the image will not then be in the sharpest possible focus.

4.3 The radial line assumption

From the results obtained in chapter 3 it will be noted that the effects of changes in ground elevation are radial displacements centred on the nadir point. In the absence of such changes, however, angular values remain correct when taken about the isometric point. In most cases, unfortunately, neither of these two points is readily identifiable. The only point that is readily available is the principal point. In general, therefore we have to make use of this latter point only. Provided the photo tilt is small ($<3°$) and the changes in ground elevation are not excessive ($<5\%Z$) then, as we shall see, we can assume that angles about the principal point are sufficiently correct for graphical radial triangulation. With wide angle photography and a tilt of $3°$ the isocentre is some 4 mm from the principal point. Equation (3.12) gives us the error in angles measured about the principal point

$$\Delta\theta=\tfrac{1}{4}t^{2}\ \sin\ 2\theta+t\ \tan\ E\ \sin\ \theta$$

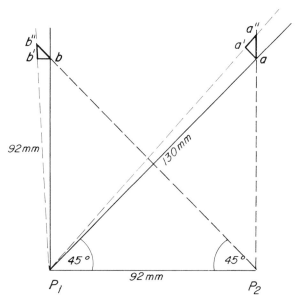

Figure 4.8. Radial Line Assumption Plotting Errors

Using this equation for directions of $45°$ and $90°$ we find the errors in these angles, in radians, as given in Table 4.1. These produce the radial displacements quoted for the situation shown in Fig 4.8 which would occur in a template produced at photo scale.

Azimuth angle	$45°$	$90°$
Angular error due to $3°$ tilt	0.0007	0.0
Angular error due to 5% elevation change	0.0022	0.0043
Total angular error	0.0029	0.0043
Length of radial	$p_1a = 130$ mm	$p_1b = 92$ mm
Radial displacement	$aa' = 0.38$	$bb' = 0.40$

Table 4.1

For the above, the angles of elevation were calculated as follows.

$$\text{Change in ground elevation} = 0.05Z \text{ m}$$

$$\text{Change in elevation at photo scale} = 0.05 \times 152$$
$$= 7.6 \text{ mm}$$

$$\therefore \quad \tan E \text{ for } 45° = \frac{7.6}{130}$$

$$= 0.0584$$

$$\therefore \quad \tan E \text{ for } 90° = \frac{7.6}{92}$$

$$= 0.0826$$

Figure 4.8 shows the situation where angular errors occur only on the first photograph, i.e. about the principal point p_1. The directions about p_2 of the second photograph are taken as correct. In such a case the points of radial intersection would fall at the points a'' and b'' where

$$aa'' \simeq \sqrt{2} \cdot aa' = 0.54 \text{ mm and } bb'' \simeq \sqrt{2} \cdot bb' = 0.57 \text{ mm}$$

From the above it will be seen that the limits set by the Arundel assumption are realistic for graphical work.

An inspection of the two sources of error shows clearly that the elevation factor is the major source of error. Hence, if it were possible, radial triangulation from the nadir point would give a more accurate result; the only source of angular error being one due to photo tilt.

4.4 Radial line plotting

By this technique points of detail falling in the overlapping area of a pair of photographs can be fixed by the intersection of radial lines drawn from the two principal points. On each photograph, the base line (i.e. the line joining the two principal point positions on each photograph) is taken as the reference direction. In Fig 4.9 they are therefore shown as collinear lines. The spacing of the principal points determines the scale of the resulting plot. In the situation shown in the figure, two control points A and C plotted at the required scale are controlling the scale of the plot. In order to establish the base line on each photograph it is necessary to first identify p_1 and p_2 on each by means of the fiducial marks. Those two points must then be transferred to the other photograph to provide the points $(p_2)_1$ on photo 1 and $(p_1)_2$ on photo 2. This can be done by carefully examining the common photo detail in these areas. On photo 1 the position of p_1 in relation to the surrounding detail is carefully noted. The position of p_1 on photo 2 is then estimated from

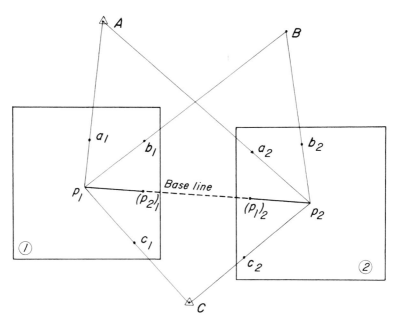

Figure 4.9. *Radial Line Plotting*

an inspection of that same pattern of detail as it appears on photo 2. A more accurate method of carrying out this transfer can be done by the stereoscopic viewing of the two areas and pricking through the position of the point on the unmarked photograph. If this is done accurately then the two principal point marks will fuse to form a single point of detail that apparently rests on the ground surface (see section 6.5). Using special equipment, such as the Zeiss Snapmarker or the Wild PUG devices, an accurate stereoscopic transfer of points can be carried out. The minimum control information required to fix the scale of the plot is the known length of an identified line. In practice this information is more likely to take the form of the co-ordinates of the terminal points, as shown in the figure. In addition to scale, two such points provide the correct azimuth of the plot and locate it in the co-ordinate frame of reference.

To carry out a plot with no equipment of any kind, the angular information from each photograph is best traced on to two pieces of tracing paper. With their two base lines collinear, one trace is moved over the other until the intersections at the control points give the best mean fit. The pairs of angles for all other detail points such as B, are transferred to the plot and the intersecting radials drawn in. The accuracy of the result depends mainly on the photo tilts present. With two vertical photographs there are no approximations being made because in this case the nadir points are coinciding with the principal points and so radial displacements take place about the latter.

In practice, with photographs containing a wealth of detail it would be time-consuming to plot each individual detail point. The technique employed therefore is to plot points of good firm detail that define small rectangular areas. The detail within such areas can then be filled in more rapidly by inspection. If the plot is not at a scale close to photo scale a simple aid to facilitate changes of scale is of great value. The maximum size of the areas that can be selected and filled in in this way depends on the photo tilts present and the changes in ground elevation. It will be realised that radial plotting cannot be carried out in a satisfactory manner for that band of detail falling within about 30 mm each side of the base line; the intersections within this region being very weak. However, this is also an area of small displacements and so this area close to the principal points need only be the subject of an overall change in scale. (One of the optical devices mentioned in section 4.2.4

is most suited to this purpose.) To control the plotting, a series of intersected points should be provided along the edges of the area.

The radial line plotting instrument is one that greatly facilitates radial line plotting. In this instrument the two photographs are base lined and fixed onto two flat tables. Each photograph is fixed so that its principal point coincides with the centre of the table about which a cursor revolves. The photographs and cursors are then viewed stereoscopically, and the plotting arm is moved so that both cursors pass through the point of detail to be plotted. With a little practice, the arm can be moved so that the point of intersection of the two cursors always coincides with the detail to be plotted. The scale of the plot can be varied from $\times\frac{1}{2}$ to $\times 2$ by a simple adjustment of the linkage mechanism. Plate 5 is a photograph of the Hilger-Watts radial line plotting instrument.

4.4.1 RADIAL LINE TRIANGULATION

Because of the 60% overlapping of adjacent photographs in a strip of photography there is an overlap between adjacent models formed. This means that if the scale of the first model is fixed then it should be possible to transfer the scale to the next model using points in the strip of common detail. In theory it should therefore be possible to carry this scale forward for all models of the strip. It is due to the fact that adjacent models contain a common photograph and also have a slight overlap of area that the process of aerial triangulation is possible. In this section we are only concerned with a planimetric aspect of this process which can be carried out using the technique of radial line plotting.

In all triangulations the object of the process is to provide a network of control points from which the plotting of detail can be carried out. In photogrammetric work the convention is to use six standard control points on each model as illustrated in Fig 4.10. Ideally, all points should fall in the centres of areas of overlap; they should also fall within the common overlap of three adjacent photographs, i.e. the super-overlapping area. In addition, the wing points should fall close to the centres of the areas of lateral overlap between strips. Points 3, 1, 5 of the model shown are therefore common to the previous model and points 4, 2, 6 are common to the next model of the strip. Points 3, 4 and 5, 6 are common to models in strips above and below respectively.

There are a number of mechanical methods by which the radial line concept can be utilised. All are in essence the same; we describe here one of the most commonly used techniques, involving the use of slotted templates made from thin semirigid plastic sheeting. Accepting the premise that directions from the principal point are correct, these directions are transferred to the templates and slots radial from the principal point are then punched in them. The following briefly outlines the various stages of their production. Figure 4.11 shows a prepared template and the method of assembly.

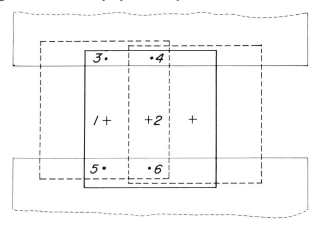

Figure 4.10. The Six Standard Photo Points (Von Gruber Points)

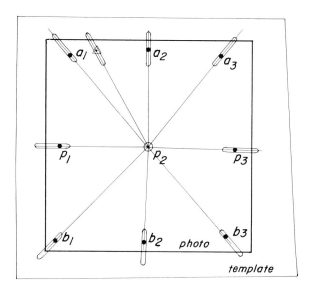

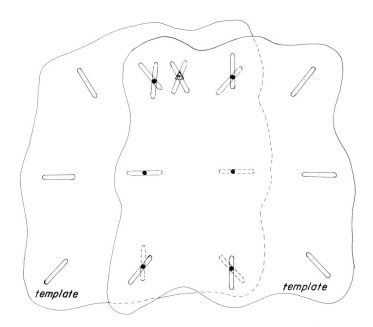

Figure 4.11. *Radial Line Triangulation with Slotted Templates*

(a) The principal point is located on each photograph with the aid of the fiducial marks. The points should be marked using a needle with a very fine point.

(b) By examination, preferably stereoscopically, all principal point positions are transferred to adjacent photographs.

(c) The other four tie points now have to be selected so that, as far as possible, they occupy the best geometric positions. The points should be fine points of photo detail that appear as such on all the appropriate photographs. The selection of these points is time consuming and requires both skill and patience. Much of the accuracy depends on the selection and accurate transfer of tie points.

(d) The template sheet, preferably a little larger than the 230×230 mm format of the photograph, is then laid over the photograph and the nine points accurately pricked through onto the template.

(e) The central principal point is then punched out with a special punch to give a circular hole of a given diameter centred on the point position. Using this hole as a radial centre, slots are now cut using a special radial line cutter so that these accurately represent the radial directions from the principal point. For work at approximately photo scale, the slots will be cut so that the tie point position falls at about the middle of the length of slot. For other scales, the slots will be displaced inwards or outwards as appropriate. To facilitate the positioning of the slots the radial directions are scribed on the template using a sharp pointed needle.

(f) In addition to the slotting of tie points there will also be slots cut for all ground control points identified and marked up on the photographs and templates. When all templates have been cut they are usually trimmed of surplus template material in order to reduce friction between template surfaces.

(g) The positions of all the control points are now plotted as accurately as possible on the prepared laydown board at the required scale of the laydown. Pins are fixed into the board at these points and the special studs then placed over them. The templates are now assembled, one with the other, using studs this time to mark the intersections of slots of adjacent models. In the process of assembly it is important to ensure that friction between overlapping templates and the board itself is reduced to a minimum. Any areas of high friction will adversly affect the nature of the adjustment that is being carried out mechanically.

(h) When satisfied that the templates have taken up their best position, each free stud is secured to the board by means of a pin inserted in its hollow stem. In this way the adjusted tie point positions are recorded on the laydown board. The nine points so fixed can now be used as control for the plotting of detail from individual photo-graphs or pairs of photographs.

4.4.2 CONTROL REQUIREMENTS FOR TEMPLATE ASSEMBLIES

If we assemble a block of templates as a free network with no control whatsoever, then the result will provide a network of control points at some constant but unknown scale and azimuth. The first requirement is therefore to introduce some form of control that will determine these two parameters. Clearly two co-ordinated points situated as far apart as possible will provide this. Such a minimum distribution provides no form of check and so one point in each corner of the block must be regarded as a minimum. In many ways such a block can be regarded as a basic unit. The question then arises: what is the maximum size of such a unit if a certain accuracy is required? Unfortunately there can be no explicit answer to such a question. The matter is, however, discussed in Refs 0.3, 0.4 and 0.8.

Using the empirical formula quoted by Tracey for an approximately square block of eight photographs in four strips, we have

$$e = 0.16 \sqrt{\frac{t}{c}}$$

i.e.

$$e = 0.16 \sqrt{\frac{32}{4}}$$

$$= 0.45 \text{ mm}$$

where e is the mean arithmetic error in mm, t is the total number of templates, c is the number of control points.

For many purposes a mean error of this order is quite acceptable for graphical results and this suggests that a block of this size might be regarded as a realistic one. However, in more recent years the quality of flying has been improved, especially at the higher altitudes, with more stable aircraft and modern instrument flying techniques. In such cases investigations have shown that the average photo tilt that can now be expected is of the order of 1.25°. Using such photography acceptable results have been obtained with less control than that suggested above. In a practical situation some idea of the effectiveness of any control distribution within a block can be gauged from the rigidity or otherwise of the template laydown at points farthest from the control. In recent years, it has been shown that planimetric control around the perimeter of a block is sufficient. This point is discussed in section 11.7.2.

More often than not, any control requirements are met by the introduction of co-ordinated photo points (see chapter 13). It should be realised, however, that ground control can be introduced in other ways and sometimes these can prove to be a more economic proposition. For example, the scale of a laydown can always be controlled at any point by introducing a line of known length. This can be done by simply punching two holes at the scaled length in a strip of template material and assembling the normal templates over these two extra studs. Sometimes, in areas of very sparse control, a long line with terminal points well away from the base line can be photo identified and measured (especially with EDM equipment) in a very short time. On the other hand, to provide them with co-ordinates might take very much longer.

In a similar way the azimuth of a line can be introduced into a laydown. In this case the terminal points of a long line are identified on the photograph and an azimuth determination made on the ground. A square template is then slotted with the direction of the slot at the correct azimuth, and this is fixed to the laydown board in the correct orientation using a simple X–Y slide arrangement. This will allow the template to move but not rotate. Again two studs are inserted and the normal templates assembled over these two 'floating' studs. In this way the correct azimuth can be maintained in a laydown.

The use of slotted templates provides a simple graphical method of carrying out a non-conformal transformation of each set of tie points: that is to say, the shape of the set of points is allowed to change in addition to any change of scale and shift in position of their centroid. Unfortunately each transformation is to some extent an approximation and so under adverse conditions, beyond the limits stated by the Arundel assumption, there will be some discrepancy between the results of one transformation and another. This will show up in the laydown procedure as a tendency of the templates to buckle on laying down. However, before assuming any difficulties from this source, the selection and transfer of the tie point positions should be carefully checked. A check should also be made on the plotted positions of control points.

If the radials could be cut from the nadir and not the principal point the situation would be much improved for here the only error is due to the small angular error given by equation (3.8). The maximum error with a 3° photo tilt is less than 2.5 min of arc. This on a radial line of 130 mm length would produce a positional error of about 0.1 mm. At the moment, however, there is no simple way of locating the positions of the nadir points to the accuracies required.

5: Photographic methods of map production

5.1 Introduction

The transfer of information from photograph to map sheet, point by point, takes time. There is therefore some attraction in devising methods by which the photographic images can be used directly as map detail. In this chapter we will therefore examine the various methods that have been used for the production of planimetric photomaps.

The chief characteristics of such maps might be set out as follows:

(a)   The speed of production should be quite fast, for no time is spent on the transfer of detail or in fair drawing. We would therefore expect a reduction in costs and production time.

(b)   All of the photographic detail appears on the map sheet. This could confuse and mislead the user unfamiliar with such maps, for the clarity of interpretation associated with the use of the standard conventional signs is missing. On the other hand, a specialist user might well find the sheet more useful, for much of the information he requires might not be considered suitable material for inclusion on the typical line map.

(c)   The planimetric accuracy of the map will depend on the tilt of the photograph, the changes in ground elevation present and on the nature of any rectification process, if such is attempted.

(d)   The clarity of the map will depend on the nature of the terrain, the photographic conditions at the time of exposure and on the quality of the photographic processes employed in the production of the map. The finest quality of detail will be obtained if a continuous tone process is used but the materials required to do this are expensive. Using a half-tone process, much cheaper map copies can be produced by printing on litho paper but this will introduce a reduction in image quality.

(e)   To make the map sheets more useful, names, grid lines and other information of this nature will be added to the sheet.

(f)   To improve the clarity of the map some line work may be introduced, various colours may be used and a number of overlay techniques employed. Taken to extremes, the advantages of (a) might well be eroded to a considerable extent.

5.2 Photo mosaics

Using photography with the standard overlapping, only about 30% of each photograph, in the vicinity of the principal point, need be used in the production of a photo mosaic. This is an area where the effects of tilt and changes in ground elevation are at a minimum. If both effects are of moderate value then the resulting errors in detail position might be acceptable for certain purposes without recourse to any rectification (correcting) process. Flying with a lateral overlap, closer to the fore-and-aft overlap of 60% would of course further improve the situation but it is doubtful whether the increased cost of doubling the photo cover required could be justified in many circumstances.

In making up the mosaic every effort is made to disguise the edges of individual photographs. To achieve this the photographs are cut along the edges of linear features where these are present and in smooth rounded curves across areas where they are not. By making oblique cuts in the photograph paper and carefully tearing the photographs along these cuts a thin feathered edge can be produced that enables one photograph to be stuck over another in a most effective way. In the production of the photographic prints, some type of automatic dodging is desirable in order to ensure a uniform range of tonal values from one print to the next.

To fix one photograph over another using only common points of detail as a guide will not produce a satisfactory map sheet. Such a process of optimistic extrapolation will rapidly introduce unacceptably large errors in position and orientation. Some method for the containment of errors is therefore essential and a slotted template assembly (as described in chapter 3) is one simple way of providing this requirement. In its simplest form this would entail producing a laydown at an estimated mean photo scale with no ground control whatsoever. This would produce a homogeneous array of control points over which the mosaic can be assembled. Each photograph would be fixed down to give the best overall fit to these points and points of common detail. On completion, the mosaic could then be rephotographed to transform it to some required scale value, provided a minimum of data were available for this purpose. Before photographing, place names and other information of a similar nature could be introduced using print on white card. Also the mosaic(s) could be assembled within the neat lines of a map sheet system and with the provision for the usual border information. If only a limited number of copies of the sheets were required then these could be produced photographically, though the cost per sheet would be very high. On the other hand, the use of a half-tone screen would enable many copies to be reproduced using a lithographic printing machine.

This method of mapping can be used for the depiction of areas where a portrayal of a wealth of information is more important than high positional accuracy and for exploratory and reconnaissance surveys over regions of limited ground relief. The method has also been used for the systematic mapping of large areas of flat terrain of low economic value.

5.3 Controlled mosaics using rectified photographs

The positional errors arising from any tilts of the photographs can be eliminated by methods of optical rectification. Displacements due to changes in ground relief will of course remain. In order to carry out this transformation some control data are necessary. These can take the form of co-ordinated photo control points or information from auxiliary instruments that provide tilt and flying height details of each exposure (see chapter 12). To begin with, if we consider the first situation then here again a slotted template assembly can be used to provide the necessary control points. In this case a number of ground control points should be incorporated in the laydown to ensure an accurately scaled laydown. This would result in the provision of nine fixed points on each photograph. As in fact only three are theoretically necessary for rectification, clearly there is here an adequate provision of control for the purpose. Most practical techniques of optical rectification employ four points in an approximately square disposition. If the ground is not a horizontal plane then an exact fit to control is not possible and in this case residual errors are reduced to a minimum by trial and error.

In some cases an adequate distribution of control points will be readily available. For example, in a map revision programme this would most often be the situation. However, regardless of the origin of the control there are basically two methods of procedure that can be adopted. Most commonly the rectifier will be set up directly by using some empirical process so that the projected photo images finally coincide with their correctly plotted map positions to within some satisfactory degree of precision. This technique is discussed in more detail in section 5.7. Alternatively, the flying height and camera tilt can be determined analytically or by using some analogue device; one such method being described in section 5.8. The values so determined are then set on the optical rectifier.

For the situation where the rectification parameters of a selected number of photographs are required for some specialist purpose then there are a number of techniques available such as those developed by Anderson,[5.1] Church [5.2] and others.[0.1] However, with the availability of a modest computer the problem is probably best solved now by a direct analytical approach using the space resection formulae developed in section 10.3. Using an equation such as (10.4) the required parameters can be calculated provided the ground co-ordinates of at least three well-distributed points are available together with the camera calibration data.

For the situation where readings from auxiliary instruments are available, the rectifier could also be set up correctly using this source of information alone. Instruments such as the horizon camera, solar periscope and airborne profile recorder are described in chapter 12, and in that section the methods by which the aircraft tilt and height above the ground surface can be determined are described in some detail. In section 5.6 the use of such information is described, but in order to appreciate the nature of the problem in setting up a rectifier using any form of data we must first examine the rectification process.

5.4 Optical rectification

The optical transformation of a tilted photograph can be carried out by projecting the original negative through an optical unit onto a tilted copyboard and thereby producing a rectified positive print. In order to produce an image in sharp focus over the whole of the format and with the correct perspective (i.e. that equivalent to a vertical photograph taken at the same camera station) certain optical and projective conditions must prevail. These are as follows.

1. The condition for the sharp focusing by the lens of points situated on the optical axis must be satisfied. This condition can be expressed in the form of either the Gaussian or Newtonian formula. The object, image and lens planes are considered to be parallel to one another and normal to the optical axis of the lens in the simple application of these equations.

2. The Scheimpflug condition must be satisfied and this is concerned with the situation where object, image and lens planes are not parallel to one another or normal to the optical axis. In this case sharp focusing is only achieved over the whole of the image plane when all three planes intersect in a straight line.

3. The projective condition states that the perspective obtained by projecting from one plane to another through a given perspective centre can be preserved when using a different perspective centre and principal distance only if the vanishing line condition is obeyed.

The projective condition (3) is a requirement to be met by all methods of rectification – graphical, mechanical and optical. It is therefore a fundamental geometrical requirement and so we will consider this first of all. The situation is illustrated in Fig 5.1 which shows a vertical section through O in the direction of the line of greatest slope. In the diagram the datum plane is a plane parallel to the ground plane and contains the ground points plotted at the required mapping scale, the value of this being determined by the length of ON. In the case of the air survey camera, p is the principal point of the photograph lying on the optical axis of the camera pOP, with distance Op being the focal length of the camera. It should be noted that the point h, the vanishing point, is the image of a point on the horizon

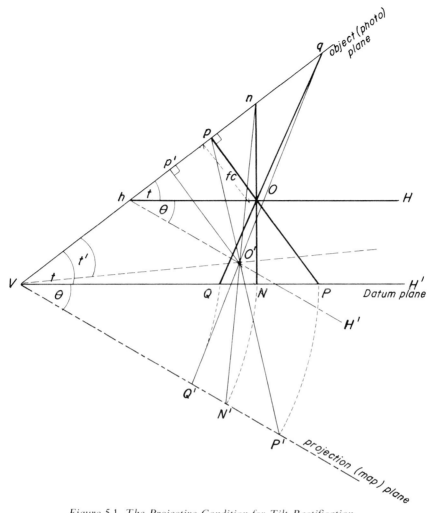

Figure 5.1. *The Projective Condition for Tilt Rectification*

and so the line hOH must be parallel to the datum plane meeting it in H. The line qOQ is the line joining any point Q in the datum plane with its image point q.

It is required that the projection of all points in the object (photo) plane through a new perspective centre O′ onto a projection (map) plane VQ′N′P′H′ shall produce a perspective identical to the original one associated with centre O and the datum plane VQNPH. If we define the principal point as the foot of the perpendicular from the perspective centre on to the object plane then the new principal distance is O′p′.

The desired result will be obtained if the scale of the perspective in the projection plane is everywhere identical with that of the datum plane. It can be demonstrated that this will be the case under the following circumstances:

(a)   The point O′ lies on the arc of a circle centred on h with radius hO, and

(b)   the angle OhO′ is identical to the inclination θ of the projection plane to the datum plane.

From the diagram we see that this construction provides the necessary conditions for we

71

now have two pairs of similar triangles, namely, qhO, qVQ and qhO', qVQ'. Hence we have

(i)   The scales of projection in the plane of the paper are identical for

$$\frac{hO}{hO'} = 1$$

$$= \frac{VQ}{VQ'}$$

$$= \frac{VN}{VN'}$$

$$= \frac{VP}{VP'}$$

etc.

(ii)   The scales of projection (λ) in planes at right angles to the paper are also identical for,

$$\lambda_Q = \frac{Oq}{OQ} \quad \text{and} \quad \lambda_{Q'} = \frac{O'q}{O'Q'}$$

but

$$\frac{Oq}{OQ} = \frac{hq}{Vh}$$

$$= \frac{O'q}{O'Q'}$$

hence

$$\lambda_Q = \lambda_{Q'}$$

From this result we see that the required perspective can be reproduced using a principal distance different from the original value, provided we change the relative tilts of the two planes by an amount θ and at the same time displace the principal point p of the original perspective so that it now lies at a distance p'p along the line of greatest slope from the new principal point position p'.

   The following relationships can be derived from an examination of the diagram,

$$\sin t = \frac{Op}{Oh} = \frac{f_c}{Oh} \tag{5.1}$$

where f_c is the original principal distance, i.e. the focal length of the camera.

$$\sin (\theta + t) = \frac{O'p'}{O'h}$$

$$= \frac{c}{Oh}$$

$$= \frac{c \sin t}{f_c} \tag{5.2}$$

where c is the new principal distance.

   Also we note that, from Fig 5.2, we have

$$p'p = (c - f_c) \tan (t + \theta/2) \tag{5.3}$$

Hence in the situation where we are given the parameters c, f_c and t the positions of O' and p' can be determined.

72

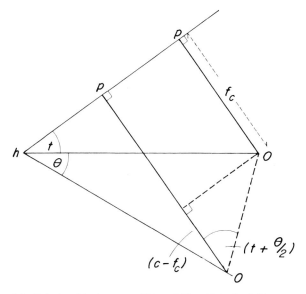

Figure 5.2. *Relationship between Old and New Principal Distances*

It will be noted that the above evaluations depend primarily on a knowledge of the angle of tilt t and the direction of the line of greatest slope. Ground control data does not provide us with this information directly, although it can always be derived. Auxiliary instruments, however, can provide such information when they are used in conjunction with the camera. In general, however, it is true to say that the information is not usually available in the form required for the above treatment.

It should be noted that the above would apply to any form of mechanical rectifier. Although there are few such instruments in existence, the Santoni stereomicrometer is one example of an instrument that does carry out a process of mechanical rectification on the right-hand photograph. Also, in fact, analogue plotting instruments using mechanical projection must carry out the rectification process on both photographs in some way or other (see chapter 10). In such instruments, if the need did arise for using them with a principal distance different from the focal length of the camera then the above analysis would apply.

Mechanical rectification as such is unusual because point by point working is time-consuming whereas optical rectification combined with the photographic process is very swift. We therefore now need to consider in some detail optical rectification and the optical requirements listed at the beginning of this section. Figure 5.3 illustrates these conditions for a lens of focal length f_p. For simplicity a thin lens with coinciding nodes has been used. To fulfil the Scheimpflug condition the three planes are shown to intersect in the line V at right angles to the paper. Using the Newtonian form of the equation for sharp focus, we have for the two axial points (l, L)

$$x_l \cdot y_l = f_p^2$$

From the diagram we see that

$$\frac{x_l}{x_q} = \frac{h'l}{h'q} \quad \text{and} \quad \frac{y_l}{y_q} = \frac{SL}{SQ}$$

But Δs O'QS and qO'h' are similar, hence

$$\frac{SQ}{SO'} = \frac{O'h'}{h'q}$$

73

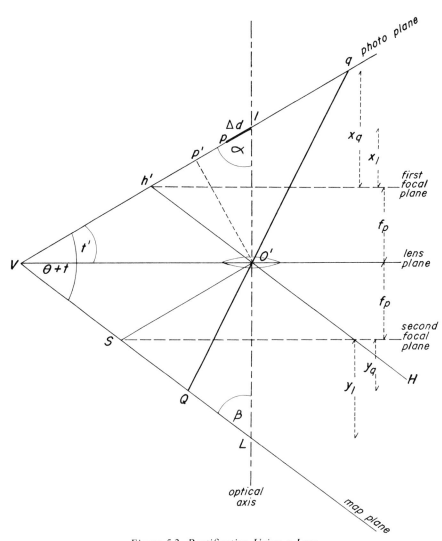

Figure 5.3 Rectification Using a Lens

Because Δs SLO′ and h′O′l are also similar, we have

$$\frac{\text{SL}}{\text{SO}'} = \frac{\text{h}'\text{O}'}{\text{h}'\text{l}}$$

Hence

$$\frac{\text{h}'\text{l}}{\text{h}'\text{q}} = \frac{\text{SQ}}{\text{SL}}$$

i.e.

$$\frac{x_l}{x_q} = \frac{y_q}{y_l}$$

i.e.

$$x_q y_q = x_l y_l$$

$$= f_p^2$$

This result indicates that if the axial points are in sharp focus then all other pairs of points will be in sharp focus provided the three planes intersect as shown in the diagram. If therefore we set up an optical projector in such a manner, with the photo negative in the object plane then a transformed set of sharp images will be produced in the projection plane. The nature of this transformation will depend on the positioning of the negative (i.e. the original perspective) in its plane and the inclinations of the three planes. In practice, with a fixed optical axis this latter need involve only a consideration of one tilt only, probably that of the projection plane. To obtain the correct perspective transformation the vanishing point condition must also be fulfilled for this alone determines the correct position of the negative in its plane.

From an examination of Figs 5.1 and 5.3 we note that the point h on the photograph must therefore be made to coincide with point h′ of the rectifier. Hence

$$O'h' = Oh$$

$$= f_c \operatorname{cosec} t \qquad (5.4)$$

This requirement determines the value of the angle α in Fig 5.3 and hence the inclinations of all three planes, and so we have

$$\frac{\sin \alpha}{O'h'} = \frac{\sin (\theta + t)}{O'l}$$

$$= \frac{\sin (\theta + t)}{f_p + x_l}$$

i.e. $$\sin \alpha = \frac{f_c \operatorname{cosec} t \cdot \sin (\theta - t)}{f_p + x_l} \qquad (5.5)$$

On the right-hand side of this expression, the focal length (f_c) and the tilt (t) are fixed parameters of the air station, while f_p is a fixed value for the rectifier. The value of θ is determined by the vanishing point condition while x_l is determined by the scale (λ_m) required in the projection plane. For this latter we note that

$$\lambda_m = \left(\frac{f_p + y_l}{f_p + x_l}\right) \cdot \lambda_l$$

$$= \frac{y_l}{f_p} \cdot \lambda_l$$

$$= \frac{f_p}{x_l} \cdot \lambda_l \qquad (5.6)$$

where λ_l is the photo scale along the parallel through l. If the photograph is a near vertical then

$$\lambda_l = \frac{f}{Z} (\text{very nearly})$$

If a more accurate value is required, then this can readily be found from considerations set out in chapter 3 and a knowledge of Δd as defined below.

In order to correctly locate the principal point p on the photo in the object (photo) plane we need to calculate the distance pl and to do this we can proceed as follows. From Fig. 5.3 .

$$h'l = \sin \beta \frac{O'h'}{\sin \alpha}$$

$$= \frac{\sin \beta}{\sin \alpha} \cdot f_c \operatorname{cosec} t$$

From Fig 5.1

$$hp = f_c \cot t$$

Now
$$\Delta d = pl$$
$$= h'l - hp$$
$$= f_c \cdot \frac{\sin \beta}{\sin \alpha} \operatorname{cosec} t - f_c \cot t$$

i.e.
$$\Delta d = f_c \cot t \left(\frac{\sin \beta}{\sin \alpha} \cdot \sec t - 1 \right) \tag{5.7}$$

If the tilting mechanism of the rectifier utilises two components such as Φ and Ω then we would need to resolve the displacement Δd into two components:

$$\Delta x = f_c \cot \Phi \left(\frac{\sin \beta_x}{\sin \alpha_x} \sec \Phi - 1 \right) \tag{5.8}$$

and
$$\Delta y = f_c \cot \Omega \left(\frac{\sin \beta_y}{\sin \alpha_y} \sec \Omega - 1 \right) \tag{5.9}$$

5.5 Mechanisms for automatic focusing and vanishing point condition

When attempting to set up an optical rectifier for the purposes of carrying out a required transformation there can be as many as eight orientation elements to be adjusted, depending on the construction of the instrument. However, we now realise that these movements are not all entirely independent of one another and so a number of mechanisms have been developed to reduce the number of adjustments that need to be made by the operator. For example, most rectifiers have automatic devices for ensuring that sharp focusing will exist at all times. These ensure that the Newtonian lens law and the Scheimpflug conditions are always fulfilled, and in this way the number of operator adjustments can be reduced.

The Newtonian law can be satisfied in a number of ways and the simplest of these is probably the Pythagorean right-angle inversor shown in Fig 5.4. Referring to this figure we see that the arm gjh has a right angle at the pivot point j, a distance f from the optical axis. Rotation of the arm causes the sliding sleeves at g and h to move along the optical axis in agreement with the equation, for we have

$$x^2 + f^2 = (gj)^2$$
$$y^2 + f^2 = (hj)^2$$
$$x^2 + y^2 + 2f^2 = (x + y)^2$$

i.e.
$$f^2 = xy$$

Another form of inversor commonly found in rectifiers is the one due to Peaucellier illustrated in Figs 5.5(a) and (b). In Fig 5.5(a) the mechanism is shown in the position of equal conjugate focii, in part (b) of the figure a more general position for the scissor mechanism is depicted. The lens law will be obeyed provided the dimensions are correctly set out for (a). Hence we have, from (a)

$$f^2 = a^2 - b^2$$

and from (b)

$$z^2 = a^2 - c^2$$

76

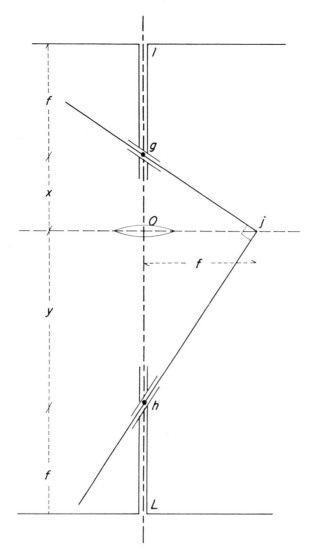

Figure 5.4. Pythagorean Inversor

and

$$(z-x)^2 = b^2 - c^2$$

$\therefore$

$$2xz - x^2 = a^2 - b^2$$

$$= f^2$$

but

$$2z - x = y$$

$$xy = f^2$$

The Scheimpflug condition can be automatically introduced using a Carpentier inversor of the type shown in Fig 5.6. From this figure we see from the construction that

$$\frac{c}{d_1} = \frac{OV''}{u} \quad \text{and} \quad \frac{c}{d_2} = \frac{OV'}{v}$$

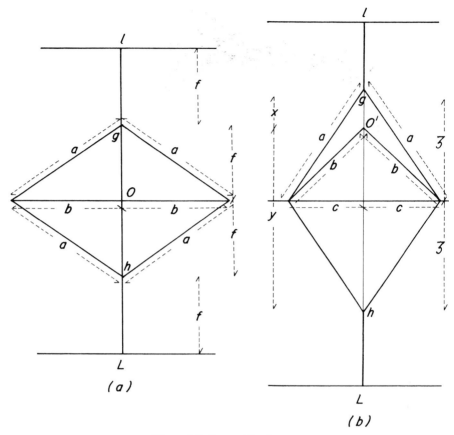

Figure 5.5. *Peaucellier Inversor*

But we have also two similar triangles formed by the guide rail l'O'L' such that

$$\frac{d_1}{d_2} = \frac{u}{v}$$

Hence the points V' and V" must coincide (at V) so that OV' and OV" are of equal length. The three planes lV, OV and LV therefore intersect in a line through the point V.

A combined Carpentier–Peaucellier inversor is used in both the Wild E2 and the Bausch and Lomb rectifiers. The Zeiss (Jena) instrument uses a Carpentier inversor in conjunction with a band inversor for the production of sharp imagery. This latter instrument also has an automatic control mechanism for the introduction of the correct displacements of the photograph in the negative plane. In response to the tilting of the easel plane, the components of tilt Φ and Ω are used in an electromechanical solution of equations (5.8) and (5.9). In this particular instrument the operator adjustments have therefore been reduced to just three; the two components of the tilt of the easel plane (Φ and Ω) and an adjustment of the lens–easel distance required for scale changes. More commonly, the operator has five adjustments to make, the vanishing point condition being carried out manually. The selection of the movements depends to some extent on the design of the instrument.

A refinement found in some of the more elaborate instruments is an automatic dodging unit. This monitors and controls the exposure required for each elementary area of the

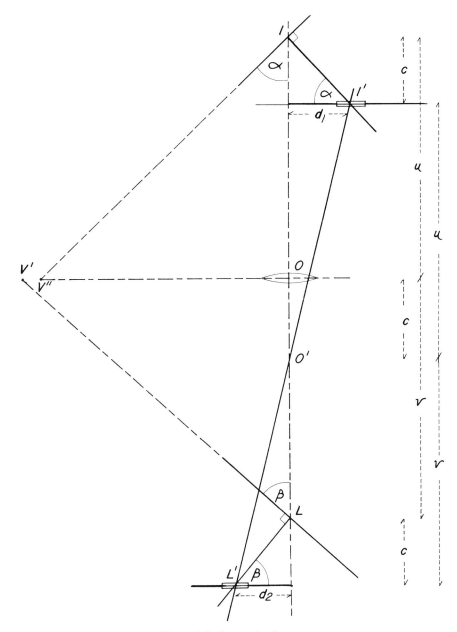

Figure 5.6. Carpentier Inversor

negative, thus automatically ensuring a satisfactory tonal range over the photographic print so produced.

5.6 The use of auxiliary data for optical rectification

In the first instance, let us consider the case where direct settings of the rectifier controls are possible. This would be the situation if data from auxiliary instruments were available for these could provide us with a value for the flying height of the aircraft above the ground nadir point (i.e. by using APR) and the tilt of the camera at the time of exposure (i.e. by using the horizon camera or solar periscope). This information together with known values for the focal length of both camera and rectifier would enable us to calculate the required rectifier settings.

If we consider a typical rectifier having a fixed lens plane and vertical axis and where all tilt adjustments are made using the easel tilt setting; then, to calculate the easel tilt β, we see from Fig 5.3 that similar Δs lh'O' and lVL give

$$\frac{VL}{O'h'} = \frac{lL}{lO'}$$

And from $\Delta VO'L$ we have

$$VL = O'L \sec \beta$$

From Fig 5.1 we see

$$Oh = f_c \operatorname{cosec} t$$

$$= O'h' \text{ (of Fig 5.3)}$$

$\therefore$
$$\frac{O'L \sec \beta}{f_c \operatorname{cosec} t} = \frac{lL}{lO'}$$

$\therefore$
$$\cos \beta = \frac{lO' \cdot O'L}{lL} \cdot \frac{\sin t}{f_c}$$

$\therefore$
$$\cos \beta = \left(\frac{u \cdot v}{u+v}\right)\frac{\sin t}{f_c}$$

But
$$\frac{1}{f_p} = \frac{1}{u} + \frac{1}{v}$$

$\therefore$
$$\cos \beta = \frac{f_p}{f_c} \cdot \sin t \qquad (5.10)$$

In some instruments this calculation and setting will be carried out using the two components of tilt of the easel plane and the two components of camera tilt β_x and β_y.

We note that the corresponding tilt of the negative plane, automatically introduced in this case, is given by

$$\frac{\tan \alpha}{\tan \beta} = \frac{O'V}{O'l} \cdot \frac{O'L}{O'V}$$

$$= \frac{O'L}{O'l}$$

The correct positioning of the photograph in the negative plane requires that the line of greatest slope of the photograph coincides with the line of greatest slope of the negative plane and that the principal point of the photograph is located the correct distance (Δd) down the line of greatest slope, from the point l on the optical axis of the rectifier lens. Knowing the direction of the camera tilt, the first part of this requirement can be taken care of when placing the negative in the rectifier if it is of a type where the negative rotates in its plane to take up a correct orientation. The second requirement is accomplished by calculating the required displacement (Δd) from equation (5.7), i.e.

$$\Delta d = f_c \cot t \left(\frac{\sin \beta}{\sin \alpha} \sec t - 1\right)$$

In an instrument where the tilt is resolved into two components then the negative is placed in a fixed orientation in the instrument and the two components of displacement (ΔY and ΔX) can be calculated using equations (5.8) and (5.9).

Finally, in order to achieve the required scale factor (λ_m), we note that for points on the optical axis we can determine the easel distance setting from equation (5.6), hence

$$y_l = f_p \cdot \frac{\lambda_m}{\lambda_l}$$

Hence the easel setting y_l can be determined provided λ_l is known, i.e. the photo scale along the parallel through l. If the tilt is small, as it would be in the case of near vertical photography, then l can be regarded as sufficiently close to the parallel through the nadir point, in which case we have

$$y_i \approx f_p \cdot \frac{\lambda_m Z}{f_c}$$

where the value of Z is provided by the auxiliary instrument.

If the tilt is large and the above approximation is not valid then the scale along the photo parallel through l must be determined from the known height value. In this case the value of Δd is required. For example, given the flying height Z, the scale along the principal point parallel through p can be calculated. The scale λ_l can therefore be found by substituting the known value of Δd in equation (3.3).

$$\lambda_l = \left(\frac{f_c}{Z} \cdot \cos t \right) + \Delta d \left(\frac{\sin t}{Z} \right)$$

5.7 Ground control for optical rectification

Perhaps more usually, a rectifier will be set up to carry out the required transformation by an empirical method using a number of control points plotted in their correct map positions. Although in theory only three planimetric points are required for this purpose (provided the focal length of the camera is known) such a solution can be unstable and so four points are more commonly used in practice.

The exact procedure that will be adopted will depend to some extent on the design of the instrument used and the number and nature of the adjustments required to be set by the operator. However, in order to appreciate any of these methods the effects of each of the possible adjustments must be understood. Figure 5.7 therefore illustrates these various effects and the comments below explain them.

(a) SWING ΔK

In this case the rotation of the photograph in the negative plane causes the projected positions to move along elliptical arcs roughly proportional in length to their distances from the lens. In the diagram the trapezium ABCD has been distorted so that the lines D'A' and C'B' now have a new position for their vanishing point on the horizon line. The lines B'A' and C'D' now have a vanishing point at a finite distance on the vanishing line.

(b) DISPLACEMENT Δd (along the line of greatest slope)

The lines in the new figure A'B'C'D' retain their original vanishing points but the separation of the parallel lines has decreased if Δd is a movement up the line of greatest slope. A movement down the slope will move the lines into areas of larger scale and so the separation of the lines will increase.

(c) MAGNIFICATION λ

The figure ABCD is a square centred on in the negative plane. Image displacements

81

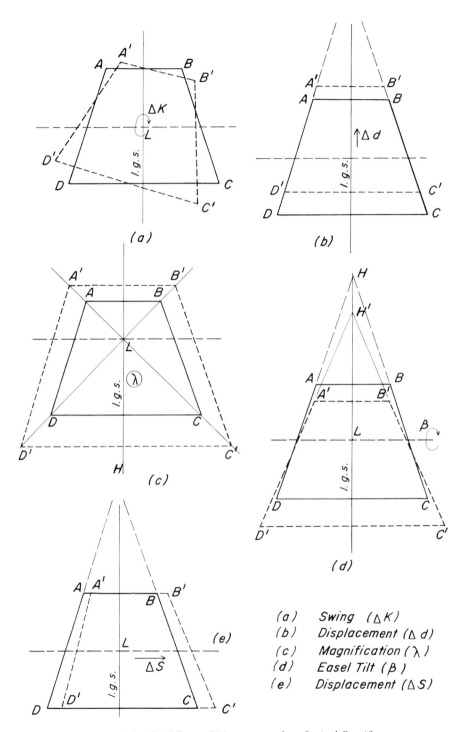

Figure 5.7. The Effects of Movements of an Optical Rectifier

(a)   Swing (ΔK)
(b)   Displacement (Δd)
(c)   Magnification (λ)
(d)   Easel Tilt (β)
(e)   Displacement (ΔS)

are therefore radial from the axial point L. If the easel tilt is small then the vanishing point of the lines CB and DA is a great distance from L in the direction of the line of greatest slope. Within an acceptable degree of approximation, the lines C'B' and D'A' can be regarded as parallel to the original lines CB and DA. The lines AB, DC, A'B' and D'C' are all photo parallels and will remain so. Hence the figure can be regarded as unchanged in shape provided the tilt is small.

(d)   EASEL TILT β
Here the increased easel tilt has brought nearer the vanishing point H' of the lines C'B' and D'A', hence their angle of convergence has increased.

(e)   DISPLACEMENT Δs (along the axis of tilt)
This displacement of the photograph in the negative plane is at right angles to the line of greatest slope. It does not therefore introduce any change in the vanishing point positions. Points AA'BB' are collinear as are DD'CC'. There is therefore an increase in the convergence of the lines C'B' and D'A' as the photograph is moved in either direction away from the central position hence the movement produces a shear effect. This movement will not be required when the control points all lie in a horizontal plane. It is only needed when attempting to fit to a control located on sloping terrain.

When carrying out a procedure for fitting the rectified image points to control it is generally advantageous, after each adjustment, to bring one pair of points into coincidence by rotating the plot and changing the magnification. An analysis of the remaining displacements will then suggest the nature of any further adjustments required. With practice the whole process usually takes 15 to 20 min in a straightforward case.

5.8 The use of control points in map revision
The source of the control points for rectification will often be points taken from existing maps, especially when the process is being used for the purposes of map revision. If these

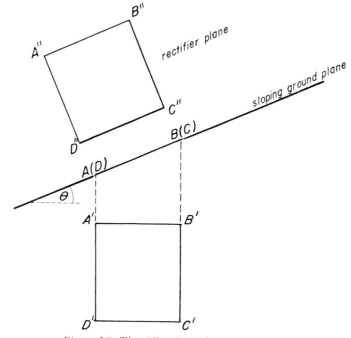

Figure 5.8. The Affine Transformation Problem

83

points show any significant changes in their ground elevations then there are two possible methods of approach.

The first method would be to select four points over a limited area such that they do lie, more or less, within a plane, horizontal or inclined. Each area is therefore the subject of a slightly different transformation. It should be noted however that if the ground has the form of a tilted plane then a single stage rectification cannot produce the correct perspective, although an acceptable approximation to the correct solution might be found if the slope is not too great. This difficulty arises because the transformation required is an affine one. Figure 5.8 illustrates the problem for a square figure ABCD on inclined terrain; the current map positions of these points being depicted by the rectangle

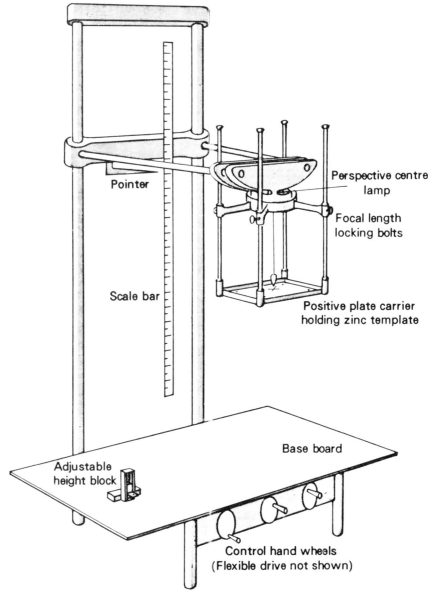

Figure 5.9. The Bloggoscope for Tilt Determination

A'B'C'D'. When adjusted, the rectifier is able to carry out a projective transformation on the photo points A"B"C"D" to produce the square figure but it cannot produce the map rectangle which requires a uniform scale change in the AB direction smaller than that of the BC direction by a factor of cos θ. To achieve the required result a multistage technique must be used to carry out a shear and variable magnification transformation (see Ref 0.1).

A second method of allowing for variations in the ground height of control points is to take into account the image displacements introduced using either an analytical or analogue technique and in so doing determine the correct camera tilt as opposed to finding a projector tilt that gives the best overall fit in the circumstances. In this way a rectified print is produced that has the correct scale for all points in the selected datum plane but exhibits the usual height displacements for all other points. An analogue device used by the Ordnance Survey of Great Britain for applying this technique is illustrated in Fig 5.9. Using this device, the correct settings for SEG V rectifiers are provided with the minimum amount of time spent on each print.

In Fig 5.9 the adjustable height blocks are set to the appropriate scaled heights above the selected datum level and placed over the control points plotted on a transparent plastic sheet. The photo positions of the controls are transferred to a thin metal template the same size as the photograph and appear as small holes in the material. The simple projector is adjusted to have a principal distance equal that of the camera and so the correct directions of the rays are reprojected. The projector tilt and elevation are adjusted until all the rays strike their height blocks in the correct positions. The elevation of the projector is then read off and the tilt of the projector is recorded by locating the position of the nadir point on the template with the aid of a plumb bob suspended from the perspective centre. For further details see Refs 5.3 and 5.4.

5.9 Differential rectification

For the methods described so far in this chapter minimum changes in ground elevation are essential. However, in the last decade or so a new category of instrument has been developed in which the changes in photo scale introduced by changes in ground elevation are also eliminated. The process is described as one of differential rectification and the product produced by the orthophotoscope is termed orthophotography. In these instruments small areas are individually corrected for the effects of both relief and camera tilt. The tilt correction process may be the continuous one described above while the differential scale change required is introduced for each element in response to any variation in height from the datum value. Information regarding the height of each element is derived from a stereoscopic model produced by an analogue plotting instrument set up in absolute orientation. Although the treatment of plotting instruments is not considered until chapter 8 it is convenient to introduce a description of the orthophoto instruments at this stage. It should be appreciated, however, that some knowledge of analogue plotting instruments is required in an understanding of the working principles of this category of instrument.

5.10 The working principles of orthophotoscopes

Although an instrument of this type was first patented in 1927 by R. Ferber, it was not until some 40 years later that the principle was taken up and developed on a commercial scale. The Ferber instrument was based on the Ferber–Gallus plotter, which employed optical projection by two projectors onto a plotting surface. In more recent times, the earlier types of orthophotoscope also used optical projectors based on the development work carried out by Bean[5.5] (see Fig 5.10). The Kelsh K-320 orthoscan is one of the instruments of this type, and its working principles are shown diagrammatically in Fig 5.11. The instrument uses three standard Kelsh projectors (see chapter 8); the centre one being used to produce

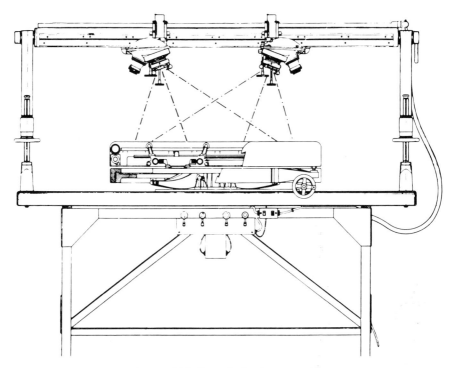

Figure 5.10. Bean Orthophotoscope

the orthophotograph. It can therefore serve to illustrate the working principles of those instruments that produce an orthophotograph by direct optical projection within the model space, see Fig 5.11. The left-hand and central projectors are first brought into absolute orientation (see section 7.6) and the scaled and levelled stereo model so produced can be delineated by a small viewing platen that is moved in the X, Y and Z directions in the model space. When correctly set to the height of the model surface at any point the images projected onto the platen by the projectors are rectified images at model scale. The correct tilt of a projector brings about the changes in scale required to compensate for the effects of camera tilt. Placing the platen at the correct Z level on the model surface at the point ensures that the magnification of that point on each photograph is such that the projected images are at model scale. Hence both images are identical. In the surface of the platen is fixed the face of a coherent flexible glass fibre ribbon of rectangular cross section 2×27 mm. The elementary area utilised in the production of the orthophotograph is therefore in the form of a narrow rectangular aperture whose maximum dimensions are slightly less than the above. The dimensions of the working aperture used can be varied to suit the nature of the terrain. By means of the flexible glass fibre optic the image can be directed, through an exposure head, on to the photographic film, housed in a light-proof container just below the instrument datum table.

In operation the platen and the exposure head are moved automatically and systematically over the area of the model and the photographic film respectively. To do this the exposure head and platen are made to move in the Y-direction and then step over in the X direction at the end of each line. During the line scan the instrument operator has to maintain the platen at the correct surface level at all times. The instrument is of the anaglyph type and so the left-hand projector is fitted with a red filter and the central projector is fitted with a blue filter. By viewing the platen using spectacles with similar filters the operator sees a three-dimensional model on the surface of the platen. Because orthochromatic film is used, only the blue image is recorded on the film. The exposure

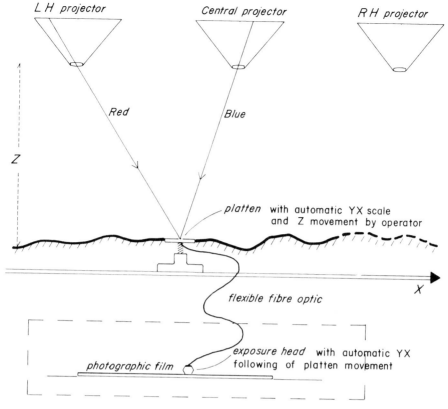

LH projector   Central projector   RH projector

Red

Blue

Z

platten with automatic YX scale
and Z movement by operator

X

flexible fibre optic

photographic film

exposure head with automatic YX
following of platten movement

Figure 5.11. Kern K-320 Orthoscan Instrument

head moving systematically across the film, in unison with the movement of the platen, produces one half of the orthophotograph of the centre photograph. Figure 5.12 illustrates this scanning process and also indicates that the speed of travel (v) is a function of the aperture width (w). At all times the correct exposure must be maintained, hence we have

Correct exposure time $T = w/v$

In order to disguise the edges of the scan lines the aperture can be made in the form of a narrow trapezium with end angles of $45°$. Because of this the total exposure time for the overlapping area is still T. The right-hand projector and the central projector can also be set up in absolute orientation and, in a similar manner, the second half of the transformation of the central photograph carried out. There are a number of instruments of this type that employ optical projectors and the anaglyph technique. The Kelsh, however, is rare in the fact that it employs a fibre optic. Most instruments project images directly on to the film through some form of moving blind with a rectangular aperture in it. The SFOM Orthophotograph 693 is a device of this type and can be fitted to a Kelsh plotter or similar instrument.

In other instruments, the third projector is housed in a light-proof container quite separate from those used to form the analogue model. In these instruments the third projector and diapositive form a duplicate unit of one of those used for model formation. After absolute orientation of the model, the tilt values of that projector are transferred by the operator to the third projector. When used in the orthophoto mode the floating mark of the plotting instrument is made to follow the YX scan pattern at some selected speed; the floating mark representing the centre of the exposure aperture. At the same time the

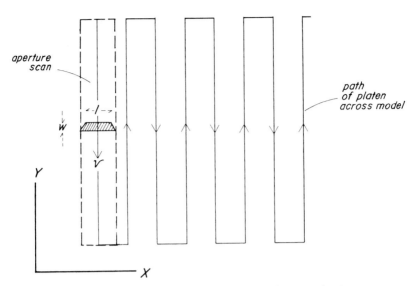

Figure 5.12. Scanning Process for Orthophoto Production

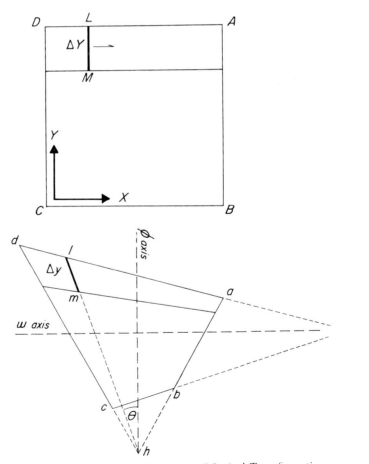

Figure 5.13. Rectification by means of Optical Transformations

exposure head of the orthophotoscope follows a similar path across the surface of the photographic film. The floating mark is maintained by the operator on the model surface by continuously adjusting the Z movement and these movements are automatically transferred into Z movements of the third projector. By this means the diapositive is rectified for the changes in scale due to camera tilt and the scale changes brought about by changes in ground elevation. Among the instruments of this type are the Gigas–Zeiss GZ-1, which uses a projector similar to those of the Zeiss stereoplanigraph and the Zeiss (Ober) DP Ortho-3 which is an anaglyph instrument with the third projector connected to a common Z column. In this case all three projectors move up and down in unison. The Gigas–Zeiss instrument employs an auxiliary lens unit to ensure sharp focus over a very wide Z-range.[5.6] On the other hand, the Zeiss DP3 instrument, like all instruments of this kind, has to rely on using a very small lens aperture to produce a working depth of focus. Such arrangements are alternatives to the fulfilment of the Scheimpflug condition, see section 5.4.

It should be noted that the projectors of the plotting instrument need not be of an optical type. For example, the Kern PG2 instrument has mechanical-type projectors and has an orthophoto attachment which employs an optical projector with a 152 mm focal length. However, some instruments using mechanical projectors use a quite different technique for carrying out the differential rectification process. In these instruments the rotation and scale change needed for the transformation are brought about by optical units other than projecting lenses. Figure 5.13 illustrates the requirement. In part (a) of the diagram we have a rectangular grid in the model (or ground) space. In part (b) of the figure is shown the perspective grid pattern produced on a tilted photograph with tilt components φ and ω in the X and Y directions respectively. Any line element on the photo such as lm therefore requires a rotation (θ), a scale change (λ) and displacements $(\delta x, \delta y)$ to bring that element to the same size, orientation and position of the equivalent element LM. The element is brought into the correct position by the rectification process carried out by the plotting instrument but additional optical units in the form of a Dove prism and a pancratic optical unit are required to bring about the required rotation and change of scale. The photo image to be transformed in this way is obtained by placing a semisilvered surface at some convenient point in the optical train of one side of the binocular viewing system of the instrument. This image is therefore identical to that seen by the operator which, in this type of instrument, is unrectified, for the viewing microscopes are held normal to the plane of the photograph as in a simple stereoscope. Figure 5.14 shows the arrangement as used in the Wild PPO-8 orthophoto attachment for their A8 instrument.[5.7] The image from the right-hand microscope R1 is split into two portions at BS. Half of the light continues through the Dove prism D1 into the eyepiece optics, R2. The other half of the light is directed into the orthophotoscope attachment and passes through the grey wedge GS, which controls the intensity of the light to give the correct exposure. The Dove prism D2 is caused to rotate through an angle $\theta/2$ to produce a rotation of the image. The Galilean telescope optics GT produce the required change in the scale of the image. The final zoom lens component produces a further change in scale in response to the variations in height of the model surface. The photographic film is held on the rotating drum.

From the geometric considerations discussed in chapter 3 it can be shown that, to a sufficient degree of accuracy,

$$\theta = \omega_R \frac{X}{Z}\left(1 + \frac{\Phi + \varphi_R}{\rho} \cdot \frac{X}{Z}\right)$$

and

$$\lambda = \frac{Z}{f}\left(1 - \frac{\Phi + \varphi_R}{\rho} \cdot \frac{X}{Z} + \frac{2\omega_R}{\rho} \cdot \frac{Y}{Z}\right)$$

Where ρ is the conversion factor from radian to circular measure.

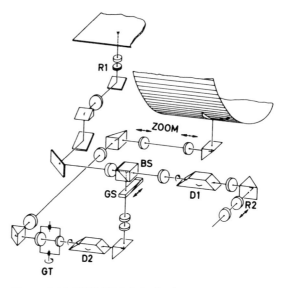

Figure 5.14. Wild PPO-8 Orthophoto Arrangement

To set up the instrument the operator therefore first carries out an absolute orientation on the plotting instrument. The elements of orientation of the right-hand projector (φ_R, ω_R) and Φ together with the focal length value are then entered into a small computer unit. As the floating mark automatically tracks through the model space the X, Y co-ordinates are continuously fed into the computer. The Z values provided by the operator keeping the floating mark on the model surface are also fed into the computer. The output from this activates the optical units that provide the rotations and changes of scale required to satisfy the equations given above.

The Zeiss (Jena) orthophoto unit for the Topocart instrument works in a similar way, although it should be noted that no provision for image rotation is included. Provided the camera tilt is less than a few degrees its effect can be disregarded.

In the later Wild OR1 orthophotoscope a similar transformation to that of the PPO-8 is carried out but the transformation formulae employed are much simpler. In the process of scanning the model both the XY model co-ordinates and the homologous xy plate co-ordinates are recorded continuously along the profile. This is made possible on the Aviomap instrument AM-U because of the facility of being able to record left-hand plate co-ordinates by means of encoders fitted to the left-hand projector. The instrument stores these profile co-ordinates and at the end of a double run it is able to compute the parameters of transformation for equivalent pairs of points, as follows:

	Line 1	Line 2
In the model:	X_i, Y_i	$(X_i+\Delta X)$, Y_i
On the photo:	x_i, y_i	$(x_i+\delta x)$, $(y_i+\delta y)$

Hence:
$$\tan \theta = \frac{\delta y}{\delta x} \quad \text{and} \quad \lambda = \sqrt{\left(\frac{\Delta X^2}{\delta x^2 + \delta y^2}\right)}$$

In addition to the many orthophotoscopes based on analogue equipment there are a few available for use with analytical plotters. One such instrument is the Orthophotoprinter of the OM1 AP/C analytical plotter. Another instrument employing an analytical plotter and an image correlation module is the Gestalt photo mapper of the

Hobrough Company. In this instrument the operator assists in the carrying out of the relative orientation processes and in the general setting-up of the instrument. However, after that, the instrument automatically produces the orthophotograph. In this instrument the elementary areas take the form of small hexagonals of about 50 mm square. A paper by Blachut[5.8] gives a comprehensive review of the orthophoto instruments available in 1972.

5.11 Properties of the orthophotograph

The process of differential rectification can produce an end product geometrically similar to a map if the process is carried out with care. The dimensions of the aperture need to be selected to suit the nature of the terrain. If there are slopes in the direction of the scanning lines (generally the Y-direction) then the speed of scan should be slower and in consequence a narrower width of aperture is necessary. Slopes in the transverse direction can cause a loss of quality if the length of the aperture is too great, see Fig 5.15. However, the shorter this length the greater will be the number of lines required to cover the area. The time to produce an orthophotograph can therefore vary considerably depending on the nature of the terrain. For flat country a scanning time of 15 minutes or so may be possible but for mountainous areas the time taken will be considerably longer.[5.9]

The orthophotoscope cannot produce the correct orthogonal representation if there are any vertical surfaces and abrupt changes in elevation present. In such cases there will not be a unique Z value for each XY position, hence the scale change required is not unique. This situation arises for example in the case of large structures and buildings. Either the roof of the building can be at model scale or the ground below it. Usually it is the ground level that is chosen and so the roof feature is produced at larger than model scale. Away from the nadir point this change of scale will become apparent as an image displacement. This effect can be reduced by using only the central portions of photographs and it has been suggested that flying with a 80% fore-and-aft overlap is one way of further reducing this effect. Of course using only the central areas of photographs will also reduce errors in position due to any error in the tilt setting of the projector.

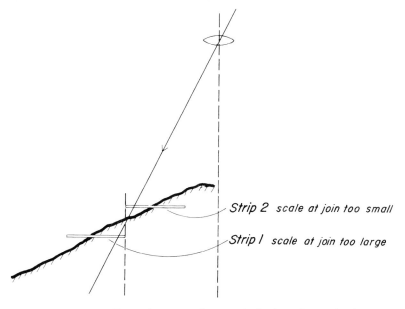

Figure 5.15. Effects of Terrain Slope on Orthophoto Image Quality

Orthophotographs are assembled into map sheets in the usual way but there are some advantages in having sheet areas coinciding with photo areas if the flying can be controlled to this degree. Orthophoto maps are an obvious choice for small-scale mapping of areas of low economic value or for situations where coverage is required in a hurry. The maps have also proved to be popular with map producers for the mapping of urban areas. One of the main attractions in this case is the considerable amount of fair drawing that can be avoided in the depiction of buildings and dense urban detail in general. However, at the very largest scales (say 1/1000–1/2000) the effects of height displacements and dead ground are very noticeable and can detract from the appearance and usefulness of the map.

There is usually a requirement for the addition of contour information on all types of map, and in the case of orthophotomaps this can be provided in a number of ways. One method used is to simply scribe the contours using the stereo-plotter and a scribe coat material on the plotting table after the exposure of the orthophoto film. Alternatively, the profile information obtained for each scan line can be stored digitally and then processed by a large computer to produce contours using an automatic plotting table.[5.10]

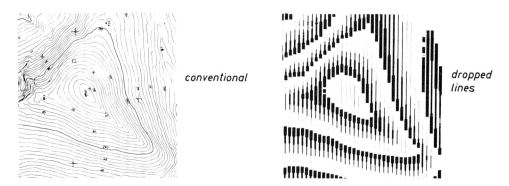

Figure 5.16. Conventional and Drop-Line Contours

Finally, there is an analogue technique that produces 'drop-line' contours. For these there is a separate projector that projects a spot of variable diameter onto a sheet of photographic material. The spot moves in unison with the plotter and exposure head. The spot diameter and therefore the line width produced is made to vary directly with changes in model elevation. Points falling within a given elevation range produce a line of given thickness. As only a limited number of line widths can usefully be employed, the contour interval must be a coarse one. Examples of conventional and drop line contours are shown in Fig 5.16.

The paper by Visser and others[5.11] describes various characteristics and applications for orthophotomaps.

6: Heights from parallax measurements

6.1 Introduction
So far we have been concerned entirely with the methods by which planimetric information can be derived from aerial photographs. However, perhaps the most important property of the photographs is their ability to indicate changes in ground elevation when examined as overlapping pairs. The reason why this height information can be obtained is simply due to the fact that changes in ground elevation produce changes in photo scale and these in turn produce measurable relative image displacements on the two photographs (see sections 3.2 and 3.3). Unfortunately, other factors such as camera tilt, inclination of the air base, lens distortion and so on will also produce scale changes and image displacements, and so before accurate height values can be obtained the effects of these other factors must be allowed for in some way. When two overlapping photographs are viewed under a simple stereoscope the relative image displacements present will produce the effect of a three-dimensional model. This property is a most useful one for it enables us to locate identical points on each of the photographs. Stereoscopic viewing is therefore utilised to a considerable extent in practical photogrammetry, although, in terms of the theoretical structure of photogrammetry, it is not of any great importance. In the first instance we will examine the ideal situation where all disturbing factors are missing and the image displacements present are due entirely to changes in ground elevations. This is the situation we would have with two truly vertical photographs taken at the same height above datum level as illustrated in Fig 6.1. This diagram is one of fundamental importance, for it illustrates some of the basic concepts of stereo-photogrammetry. It therefore warrants a particularly detailed discussion.

6.2 Co-ordinate transformation
The photographing process results in a co-ordinate transformation process. The ground point A, with co-ordinates (E_A, N_A, H_A) in some geographical system of axes, will produce photo point a′, whose position can be expressed in terms of a co-ordinate system associated with the first photograph. Similarly, a second transformation is carried out when the second photograph is taken from the second air station. Now, photogrammetric analysis is concerned with the nature and results of such transformations and, to examine these, the three independent systems of co-ordinates present must be combined so that a more homogeneous situation prevails. In order to do this the following conventions are introduced.

(a)   The left-hand perspective centre (air station) O, is taken as the local origin of the ground and model system of axes.

(b)   The origin of the plane photo co-ordinates is the principal point p on the photograph in each case.

(c)   The direction of flight, that is the vector $\mathbf{O_1O_2}$, is taken as the positive X-axis and x-axes in all cases. The co-ordinates of the right-hand perspective centre O_2 are therefore (B, O, O) in the local ground system, if the air base length is B.

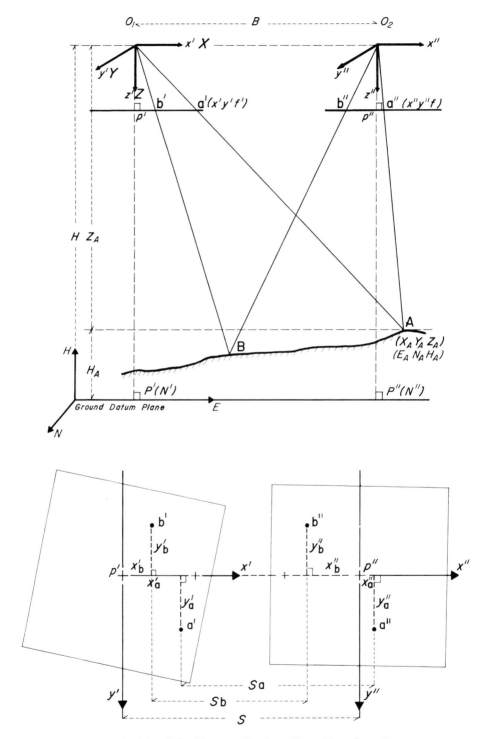

Figure 6.1 (a) and (b). Photographic Co-ordinate Transformations

(d) All systems of co-ordinates are right handed and rectangular with the positive Z-axis and z-axes downwards. This ensures that Z values are always positive. (Differentiate clearly between Z ranges which are usually measured downwards from the air station and height values which are usually measured vertically upwards from some geographical datum surfaces.)

If we now look at each transformation in turn we can easily arrive at the following results, remembering that the photographs are verticals. Taking any two ground points such as A and B of Fig 6.1 we have the following transformation formulae for the left-hand photograph:

$$
\begin{aligned}
x'_a &= \left(\frac{f}{Z_A}\right) X_A \quad : \quad x'_b = \left(\frac{f}{Z_B}\right) X_B \\
y'_a &= \left(\frac{f}{Z_A}\right) Y_A \quad : \quad y'_b = \left(\frac{f}{Z_B}\right) Y_B \\
\\
z'_a &= \left(\frac{f}{Z_A}\right) Z_A(=f) \quad : \quad z'_b = \left(\frac{f}{Z_B}\right) Z_B(=f)
\end{aligned}
\qquad (6.1)
$$

For the second photograph we have a similar set of equations for the two points

$$
\begin{aligned}
x''_a &= \left(\frac{f}{Z_A}\right)(X_A - B) \quad : \quad x''_b = \left(\frac{f}{Z_B}\right)(X_B - B) \\
y''_a &= \left(\frac{f}{Z_A}\right) Y_A \qquad : \quad y''_b = \left(\frac{f}{Z_B}\right) Y_B \\
\\
z''_a &= \left(\frac{f}{Z_A}\right) Z_A(=f) \quad : \quad z''_b = \left(\frac{f}{Z_B}\right) Z_B(=f)
\end{aligned}
\qquad (6.2)
$$

An inspection of the two sets of equations (6.1) and (6.2) shows immediately that the only changes in photo co-ordinates present are in the x values. This must of course be the case because we have selected a situation such that the only difference between the first and second transformations is a shift of photo origin and this has been defined as the positive direction of the x', x'' and X axes.

6.3 Parallax and the parallax equation
If we now define a quantity known as the parallax of a point as the differences between its two co-ordinates, we then have:

P_y is the Y-parallax of any point, hence

$$
\left.
\begin{aligned}
(P_y)_a &= y'_a - y''_a \\
&= 0 \\
\\
(P_y)_b &= y'_b - y''_b \\
&= 0
\end{aligned}
\right\}
\qquad (6.3)
$$

In the case we are considering all points will have a zero value, hence

$$(\Delta P_y)_{ba} = 0$$

P_x is the X-parallax of any point, hence

$$
\begin{aligned}
(P_x)_a &= x'_a - x''_a \\
&= \left(\frac{f}{Z_A}\right)[(X_A - (X_A - B)] \\
&= \frac{fB}{Z_A} \\
(P_x)_b &= x'_b - x''_b \\
&= \left(\frac{f}{Z_A}\right)[X_B - (X_B - B)] \\
&= \frac{fB}{Z_B}
\end{aligned} \tag{6.4}
$$

In this case, we see that the difference in ground range (Z) has produced a change in x-parallax given by,

$$
\begin{aligned}
(\Delta P_x)_{ba} &= (P_x)_b - (P_x)_a \\
&= fB\left[\frac{1}{Z_B} - \frac{1}{Z_A}\right] \\
&= \frac{fB\,\Delta Z_{AB}}{Z_A Z_B}
\end{aligned} \tag{6.5}
$$

We note that all points in the same XY plane have the same x-parallax value and so this difference represents the change in parallax associated with the separation ΔZ_{AB} of the two planes Z_A and Z_B.

It should be realised that if the overlapping of the photograph is 60% for a set of points lying in a horizontal ground plane then the parallax of these points must be 92 mm (i.e. 40% of 230 mm). Parallax values are therefore always of this order of magnitude for photographs taken under the standard conditions set out in chapter 2, the exact value depending on the overlapping achieved in practice. In aerial photogrammetry the changes in ground elevation are usually small in comparison with the mean flying height, and so we can say with sufficient accuracy in this case

$$
(\Delta P_x)_{ba} = \frac{fB\,\Delta Z_{AB}}{Z_A^2} \tag{6.6}
$$

This is the basic parallax equation from which changes in ground elevation can be derived from changes in the x co-ordinates measured on two photographs. Figure 6.1(b) shows the two photographs fixed down so that their base lines are collinear and defining the x-axis as required. The separation (S) of the two photographs has been arbitrarily selected but in practice this would be chosen so as to give a comfortable viewing position when using some form of stereoscope. We should note, however, that

$$
\begin{aligned}
(\Delta S)_{ba} &= S_b - S_a \\
&= -(\Delta P_x)_{ba}
\end{aligned} \tag{6.7}
$$

and so in practical work it is this quantity that is more easily measured and therefore used in practice. In the absence of other equipment a simple linear scale very carefully used could provide acceptable values of the quantity ΔS in certain circumstances.

96

Equations (6.6) and (6.7) are not in a form best suited to practical calculations, and a more satisfactory equation can be derived by noting that

$$\Delta H_{AB} = -\Delta Z_{AB} \quad \text{and} \quad Z_A = H - H_A$$

where H is the flying height of the aircraft above ground datum level. Hence we have

$$\Delta S_{ba} = -\frac{fB}{(H - H_A)^2} \cdot \Delta H_{BA}$$

Making use of equation (6.4), we can write

$$\Delta S_{ba} = -\frac{(P_x)_a}{(H - H_A)} \cdot \Delta H_{BA}$$

Taking due regard to signs, we have finally

$$\Delta H_{BA} = \frac{(H - H_A)}{(P_x)_a} \cdot \Delta S_{ab} \tag{6.8}$$

To use this formula the x-parallax of a reference point A is required and this can be most accurately measured if the point lies on or near the base line.

If the height of a datum point is not available then less accurate estimations of changes in ground elevation can be obtained by finding a mean value for the parallax and using an estimated mean flying height above the ground surface.

In order to appreciate more fully the implications of equation (6.8) it will be necessary to carry out an analysis based on the theory of errors. We can, however, immediately see one important characteristic of the equation if we write it in the slightly abridged form,

$$\frac{\Delta H}{H} = \frac{\Delta S}{P}$$

If we can measure the quantity ΔS with a precision of, let us say, $\pm P/10\,000$ then ΔH can be calculated with that same precision of $\pm H/10\,000$ (i.e. $\pm 0.1^o/_{oo} H$). As stated above, the value of P will be of the order of 90 mm. Hence ΔS will need to be determined to $\pm 9\ \mu m$ and, as will be seen later in this chapter, such an order of precision represents something near the limit in x-parallax measurements. This comment should not, however, be confused with the accuracy of the height determinations. These depend on a number of additional factors. In the case under discussion here the determinations will have a very low accuracy if assumptions made in arriving at equation (6.8) are not entirely correct. To calculate the total error introduced into the determination of ΔH by errors in the various quantities of the equation, we note that

$$e_{\Delta H}^2 = \left[\frac{\partial(\Delta H)}{\partial H} \cdot e_H \right]^2 + \left[\frac{\partial(\Delta H)}{\partial H_A} \cdot e_{HA} \right]^2 + \left[\frac{\partial(\Delta H)}{\partial P_a} \cdot e_P \right]^2 + \left[\frac{\partial(\Delta H)}{\partial(\Delta S)} \cdot e_{\Delta S} \right]^2$$

where $e_{\Delta H}$ is the error arising in ΔH due to errors $e_H, e_{HA}, e_P, e_{\Delta S}$ in the other quantities of the equation. All those various errors should of course be expressed in the same form throughout and are assumed to be entirely independent.

When realistic values are calculated for the coefficients of the above equation it will be found that the first three are small compared with the fourth term; indicating that a much higher precision is required for the value of ΔS than the other parameters. Below we consider a case where the flying height (H) is 1000 m and the maximum change in ground elevation is 10% of this value. The parallax value (P) is taken as 92 mm.

$$\frac{\partial(\Delta H)}{\partial H} = \frac{\Delta S}{P} \simeq \frac{\Delta H}{H}$$

$$= 0.1$$

$$\frac{\partial(\Delta H)}{\partial H_A} = -\frac{\Delta S}{P} \simeq -0.1$$

$$\frac{\partial(\Delta H)}{\partial P} \simeq -\frac{H \cdot \Delta S}{P^2} = -\frac{H}{P} \cdot \frac{\Delta S}{P} \simeq \frac{(1000)^2}{92} \times 0.1 \simeq 1087$$

$$-\frac{\partial(\Delta H)}{\partial(\Delta S)} = \frac{\partial(\Delta H)}{\partial(\Delta P)} \simeq \frac{H}{P} \simeq \frac{(1000)^2}{92} \simeq 10\,870$$

6.4 Stereoscopic viewing

It was mentioned earlier that differences in parallax could be determined using a scale but such a method is not likely to give precisions of the order of 10 μm. However, with the aid of a stereoscope and a parallax bar a parallax acuity of about this order of magnitude can be obtainable by a trained observer under favourable conditions. The measurement principle involved here is one of prime importance in photogrammetric mapping procedures, for although at this point we are considering only the use of simple equipment, the same principle is used in all forms of stereo-photogrammetric instruments. Stereoscopic viewing and the stereoscopic measurement of parallax therefore, merit our attention at this point.

Up to a distance of about 500 m an average observer can estimate range by appreciating the value of the parallactic angle subtended at his eye base by the object. It seems that a good observer with an eye base of about 65 mm can appreciate differences in parallactic angles as small as 4″ arc or thereabouts. However, wide variations from this value frequently occur in practice.[6.1,6.2,6.3] In Fig 6.2, therefore, the observer should be able to detect differences in the range of two points such as A and B if the parallactic angles λ_A and λ_B differ by more than 4″.

When we examine two aerial photographs stereoscopically each pair of image points will generate a parallactic angle, the value of which will depend upon the X-parallax value

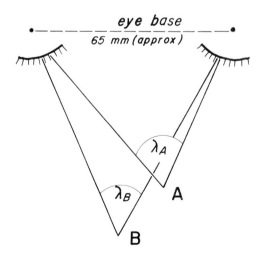

Figure 6.2. Estimation of Range by Changes in the Parallactic Angle

of the point. The value of the angle will of course also depend on the eye to photo viewing distance, and in the absence of any lenses a minimum value for this will be determined by the viewer's least distance of distinct vision. (For the average observer this is about 250 mm.) In the case of vertical photographs, if the viewing distance is equal to the camera focal length, then the relief model seen will be more or less undistorted; the horizontal and vertical scales being the same. If enlarged prints of the photographs are used then we should note that the photo air base (b') and the effective focal length (f') of the photographs have also been enlarged. If the viewing distance remains unchanged then the observer will see a model with exaggerated relief for the vertical scale is now greater than the horizontal (see section 7.15.3). With regard to the x-parallax measurements themselves there should be some increase in the precision of any ΔH determinations for in equation (6.8) the value of P_a has also been enlarged (it equals b' approximately). However, in practice it has been found that only a limited amount of enlargement is of practical use. Beyond $\times 2$ or $\times 3$ magnification, the deterioration in image quality and the reduced field of view are such that no increase in precision can be expected.

Many stereoscopes employ lenses to provide a certain amount of magnification, and the remarks made above in connection with any increased precision obtainable also apply to this case. Figure 6.3 shows a thin convex lens as it might be employed in a stereoscope eyepiece to provide a limited degree of magnification. Usually, the image is formed at infinity in order to provide the most restful form of viewing and to achieve this the eye is held close to the lens with the photograph held in the focal plane of the magnifier. The angle now subtended at the eye by the object ap is given by $\tan \theta = \mathrm{ap}/f'$. Hence the angular magnification is given by

$$M = \frac{\text{angle subtended with lens}}{\text{angle subtended without lens}}$$

$$= \frac{\mathrm{ap}}{f'} \times \frac{\text{least distance of distinct vision}}{\mathrm{ap}}$$

i.e.

$$M = \frac{\text{least distance of distinct vision}}{f'}$$

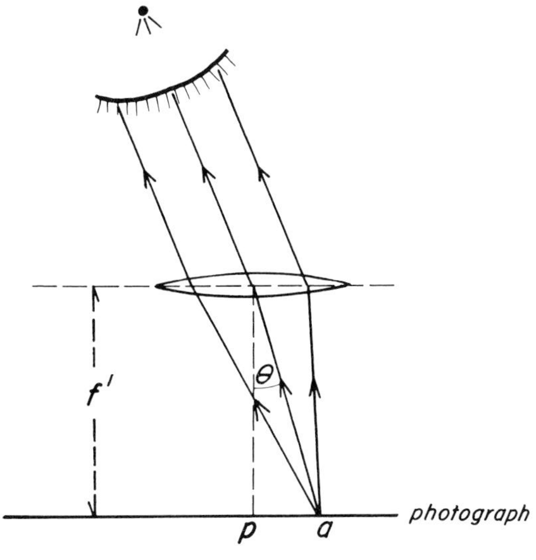

Figure 6.3. Angular Magnification of Simple Viewing Lens

99

If for example the photography had a focal length of 150 mm and the magnifying lens also had a focal length of the same value then the model seen would be undistorted with a magnification of 1.67 (i.e. 250/150).

In practice, viewers sometimes find some initial difficulty in seeing a stereoscopic model. The process is aided if the photographs are accurately base lined and the viewer maintains his eye base parallel to the base line. The photographs should be set with the most comfortable separation so that both images are being viewed normally. It also helps if photographs are arranged so that the shadows fall towards the observer and the lighting is from the front.

6.5 The stereoscopic measurement of parallax

The basic principle used in the stereoscopic measurement of parallax is the introduction of an artificial mark into the stereoscopic relief model seen by the observer. This mark formed from images introduced into the left and right-hand photo viewing systems has, in the case of the parallax bar, a variable x-parallax value. By varying the x-parallax of the mark it appears to move in a vertical direction within the model space. In some form or other this 'floating mark' is used in all stereoscopic photogrammetric instruments. The x-parallax measurements made with aid of the parallax bar clearly illustrate the mechanism of this important concept. Figure 6.4 illustrates the main features of the instrument.

6.6 The parallax bar

The two index marks are usually small black circles of the order of 50 μm in diameter, although small crosses or annuli of similar dimensions may also be employed. These marks are etched onto the bottom surfaces of the glass plates. The separation of these two plates is controlled and measured by a micrometer screw-gauge device attached to the right-hand plate holder. The least count on the gauge is usually 10 μm. While a model is being viewed with the aid of a stereoscope, the parallax bar is introduced so that the right-hand glass plate covers that part of the photograph under inspection. The separation of the two plates is then adjusted (with the aid of the clamp of the left-hand glass plate) so that the left-hand plate covers the identical area of detail on the left-hand photograph. In this way an artificial point is introduced into the photo imagery of both photographs.

The situation is illustrated in the form of the series of diagrams of Fig 6.5. For simplification the only point of photo detail shown is a simple pattern such as might be formed by a crossroads junction. The two dots of the parallax bar are shown in a variety of

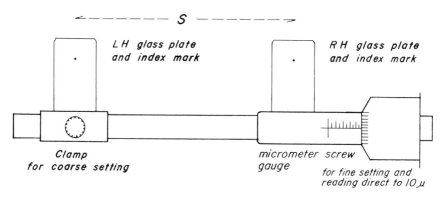

Figure 6.4. Parallax Bar

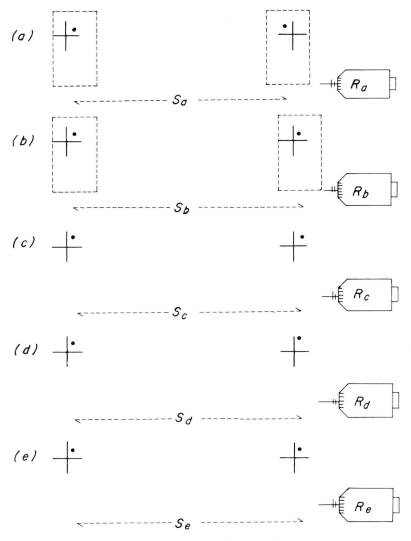

Figure 6.5. *Floating Mark Technique and Parallax*

positions with respect to this detail. When these simple stereograms are examined under a stereoscope the following results should be observed.

In Fig 6.5(a) the separation of the detail and dots differ by about 5 mm, i.e. their difference in x-parallax is about 5000 μm. This is excessive for most people and their accommodation is such that they cannot simultaneously fuse both the detail and dot images. The detail images are the stronger and so, unless one tries very hard to do otherwise, the detail will fuse to form a stereoscopic image but two dots will be seen separated in the x-direction. This indicates that the separation S_a does not approximate to the correct setting.

In Fig 6.5(b) the vernier gauge has been adjusted to R_b so that the separation S_b is much closer to that of the photo detail. However, it is in fact still smaller by an amount equal to about one dot diameter. Now this difference is such that most people have the required accommodation in their vision to be able to fuse both the detail and dot images simultaneously. The observer should therefore see one image of the detail and just one dot

101

image floating above the top right-hand limb of the road junction. The dot apparently floats above the surface because S_b is slightly too small. The x-parallax of the dots is therefore a little large so the parallactic angle subtended at the eye base is also a little large. The brain interprets this latter fact as a slight difference in range, with the dot closer to the observer than the point of detail.

In Fig 6.5(c) the vernier adjustment made has been slightly too much, with the consequence that the x-parallax of the floating mark is now smaller than that of the detail. The mark therefore appears to be below the surface of the detail.

In Fig 6.5(d) the x-parallax of dot and detail are very near correct, but there is now some y-parallax to be seen. This indicates that the axis of the bar has not been maintained in a direction parallel to the detail points. The bar therefore needs to be rotated slightly to eliminate this effect.

Finally, in Fig 6.5(e) the setting of the bar is shown correct within the limits of diagrammatic accuracy. There is no y-parallax as the direction of the bar is now correctly set. The x-parallaxes of dot and detail are also identical and hence the floating mark appears to be in the same plane as the point of road detail. The vernier reading R_l is therefore taken as the required value and can be compared with a similar reading taken elsewhere to some other point of detail. Any difference in reading between two points provides the value of ΔS required for the application of equation (6.8).

6.7 The precision of parallax measurements

From tests carried out to ascertain the stereoscopic acuity of observers and hence their ability to detect small changes in parallax, a surprising low value of 4″ of arc has been derived as an average value. Values were, however, found to fluctuate greatly between 1.5 and 140″ (see also Ref. 0.1, vol. 2).

If we take an observer with an eye base of 65 mm and a least distance of distinct vision of 250 mm then the average angular acuity quoted above provides an x-parallax acuity of just over 5 μm. Figure 6.6 shows one way of deriving this using the value 4″.

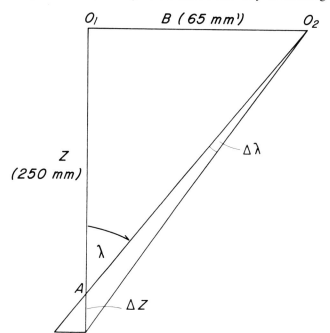

From the diagram we have

$$\tan \lambda = \frac{B}{Z}$$

$$= \frac{\Delta P}{\Delta Z}$$

By differentiation we get

$$\sec^2 \lambda \cdot d\lambda = -\frac{B}{Z^2} \cdot dZ$$

Hence we can write

$$\Delta Z = -\frac{Z^2}{B} \cdot \sec^2 \lambda \cdot \Delta \lambda$$

Therefore

$$\Delta P = -\frac{Z^2}{B} \cdot \sec^2 \lambda \cdot \tan \lambda \cdot \Delta \lambda$$

Figure 6.6. Angular Acuity and x-Parallax

102

If we now substitute the values shown in Fig 6.6 and express ΔP in microns then

$$\Delta P = -\left[\frac{B}{Z}\left(\frac{O_2A}{Z}\right)^2 \frac{Z^2}{B}\right]$$
$$\times \frac{4000}{200\,000}\,\mu m$$

i.e.

$$\Delta P = -\left[\frac{(250)^2 + (65)^2}{250}\right]$$
$$\times \frac{4}{200}\,\mu m$$

i.e.

$$\Delta P = -5.3\ \mu m$$

When taking parallax bar readings the usual procedure is to take the mean of five independent readings and accept the mean of these provided their spread is no greater than 40 μm. Most observers can more readily appreciate a point floating above the surface than one below it. It is therefore preferable to start each measurement with the mark clearly floating above the model surface and slowly adjust the instrument until the mark appears just to touch the surface.

If we refer again to the diagrams of Fig 6.6 we will realise that in each case the eye is comparing the position of the reference mark in relation to the point of detail. In fact, the parallax bar is merely a linear rule so devised that both ends can be viewed simultaneously, i.e. stereoscopically. The precision of any measurement will therefore depend to some extent on the size and quality of the reference marks used but, more importantly in practice, on the nature of the detail under examination. For the best x-parallax determinations we need sharp linear details running in the y direction. Similarly, for the measurement of y-parallax in any model we would require an image pattern showing sharp linear features in the x direction. It should be clearly understood, therefore, that the accuracy of any such stereoscopic measurement depends on the nature of the photo-image patterns present. In extreme cases, such as areas of featureless snow, sand, over water surfaces the process cannot be carried out at all. The accuracies quoted for various instruments and processes all usually assume something near ideal measurement conditions.

6.8 Errors in parallax bar heighting—theoretical considerations

The accuracy of any height values obtained from a parallax bar will not be directly correlated to the precision of the measurements taken unless all the assumptions made in the derivation of equation (6.8) are correct (i.e. vertical photographs taken at a constant flying height). These conditions are rarely encountered in practice and so unacceptably large errors in the derivation of differences in height will usually accrue unless these departures are taken into account. In essence, what we are saying is that the simple transformation formulae used to derive the equation are not satisfactory for general use. We therefore have the alternative of using more correct transformation equations directly as in analytical photogrammetry (see chapter 10) or applying certain corrections to our results based on an understanding of the more correct transformations. At this stage it is in fact more convenient to investigate the latter solution and so we conclude this chapter with some discussion as to the nature of the corrections that can be applied to height values derived from the simple parallax equation (6.8).

In Fig 6.7 we show the effect on the photo co-ordinates of any point – a clue to the presence of a tilt on the photograph. In Fig 6.7(a) we have the ideal situation with the

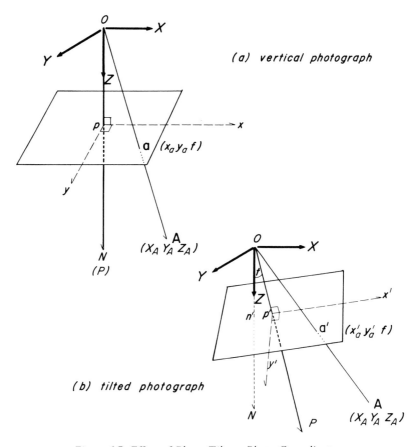

Figure 6.7. Effect of Photo Tilt on Photo Co-ordinates

camera axis vertical, i.e. the z-axis of the photo co-ordinate system coincides with the Z-axis of the ground axial system. The vector **OA** cuts the horizontal photo plane in point a to give the photo co-ordinates (x_a, y_a, f). In Fig 6.7(b) the photograph now has a tilt t and the directions of the photo axes are now shown as primed quantities. In this rotated system, the co-ordinates of the point a′, the point where the vector **OA** now intersects the tilted photo plane, are given by (x'_a, y'_a, f). The effect on the co-ordinates of a point in space owing to a rotation of the axial system can be represented by an orthogonal matrix R (see appendix A). But this alone does not completely describe the new situation because in addition the point of intersection with the tilted photo plane has moved along the vector **OA** from a to a′. This requires the introduction of a scale factor λ whose value is Oa′/Oa at any point and is such that all points once again have a constant z' co-ordinate equal to the focal length f. We can therefore write the following transformation:

$$\begin{pmatrix} x' \\ y' \\ f \end{pmatrix}_a = \lambda_a R \begin{pmatrix} x \\ y \\ f \end{pmatrix}_a \tag{6.9}$$

In passing, we note that

$$\begin{pmatrix} x \\ y \\ f \end{pmatrix}_a = \left(\frac{f}{Z_A} \right) \begin{pmatrix} X \\ Y \\ Z \end{pmatrix}_A \tag{6.10}$$

104

If the tilt of the photograph is within the limits laid down for near-vertical photography (see chapter 2, section 2.9) then the form of the matrix R can take that of the approximate orthogonal matrix derived in appendix A section 5 and so we have

$$\begin{pmatrix} x' \\ y' \\ f \end{pmatrix}_a \simeq \lambda_a \begin{pmatrix} 1 & -\Delta\kappa & \Delta\varphi \\ \Delta\kappa & 1 & -\Delta\omega \\ -\Delta\varphi & \Delta\omega & 1 \end{pmatrix} \begin{pmatrix} x \\ y \\ f \end{pmatrix}_a \tag{6.11}$$

where $\Delta\omega$, $\Delta\varphi$ and $\Delta\kappa$ are the small components of the tilt t about the (X, Y, Z) axes respectively.

From equation (6.11) we see that

$$f = \lambda_a(-x\Delta\varphi + y\Delta\omega + f)$$

hence

$$\lambda_a = \frac{1}{1 - \dfrac{x}{f}\Delta\varphi + \dfrac{y}{f}\Delta\omega}$$

i.e.

$$\lambda_a \simeq 1 + \frac{x}{f}\Delta\varphi - \frac{y}{f}\Delta\omega \tag{6.12}$$

since the quantities involved, other than unity, are very small and we can ignore second-order terms.

If we now substitute the result of equation (6.12) in equation (6.11) we obtain the following by multiplying out

$$x' = \left(1 + \frac{x}{f}\Delta\varphi - \frac{y}{f}\Delta\omega\right)(x - y\Delta\kappa + f\Delta\varphi)$$

$$y' = \left(1 + \frac{x}{f}\Delta\varphi - \frac{y}{f}\Delta\omega\right)(x\Delta\kappa + y - f\Delta\omega)$$

If in the expansion of the above we again ignore the second-order terms we finally arrive at the following results:

$$\left. \begin{aligned} x' &= x - y\Delta\kappa + \left(\frac{f^2 + x^2}{f}\right)\Delta\varphi - \frac{xy}{f}\Delta\omega \\[2mm] y' &= y + x\Delta\kappa + \frac{xy}{f}\cdot\Delta\varphi - \left(\frac{f^2 + y^2}{f}\right)\Delta\omega \end{aligned} \right\} \tag{6.13}$$

Considering only at this time the changes in x co-ordinate as required for our analysis of x-parallax errors we see that equation (6.13) will give a value for the error in parallax introduced by a tilt of one of the photographs. Let this error arise owing to a tilt on the second photograph only and in so doing assume for the moment that the first photograph is a true vertical one. The small elements of rotation now represent the relative tilt of the second photograph with respect to the first. In such a case therefore we have the following relationships applying

$$\Delta\kappa = \kappa_2 - \kappa_1$$

$$\Delta\varphi = \varphi_2 - \varphi_1$$

$$\Delta\omega = \omega_2 - \omega_1$$

where κ_1, φ_1, ω_1, and κ_2, φ_2, ω_2 are the first and second photo tilts respectively.

The parallax error in this case can therefore be expressed as follows:

Error in parallax $= P'_x - P_x$

$$= x_2 - x'_2$$

$$= + y_2 \Delta \kappa - \left(\frac{f^2 + x_2^2}{f} \right) \Delta \varphi + \frac{x_2 y_2}{f} \cdot \Delta \omega$$

Let us now also consider the effects of an inclination of the air base, again relative to the position of the first photograph, by introducing a change in flying height ΔZ between the first and second exposures. For the second photograph all Z values are therefore changed by this amount and so the transformation equation (6.10) can be modified accordingly. For any point (a, A) we have

$$\begin{pmatrix} x'' \\ y'' \\ f \end{pmatrix}_a = \frac{f}{(Z_A + \Delta Z)} \begin{pmatrix} X \\ Y \\ Z + \Delta Z \end{pmatrix}_A \tag{6.14}$$

If ΔZ is small and the inclination of the air base (B) is of the same order of magnitude as the photograph tilts, where $\tan \Phi = \Delta Z / B$, this inclination being relative to the plane of the first photograph, then we have

$$\begin{pmatrix} x''_2 \\ y''_2 \\ f \end{pmatrix}_a = \left(1 - \frac{\Delta Z}{Z_A} \right) \lambda_a R \begin{pmatrix} x \\ y \\ f \end{pmatrix}_a \tag{6.15}$$

Hence for the point (a, A) we have

$$x''_2 = \left(1 - \frac{\Delta Z}{Z_A} \right) \left[x_2 - y_2 \Delta \kappa + \left(\frac{f^2 + x_2^2}{f} \right) \Delta \varphi - \frac{x_2 y_2}{f} \cdot \Delta \omega \right]$$

And if we ignore second-order terms once more this simplifies to

$$x''_2 = x_2 - y_2 \Delta \kappa + \left(\frac{f^2 + x_2^2}{f} \right) \Delta \varphi - \frac{x_2 y_2}{f} \Delta \omega - \frac{\Delta Z}{Z_m} x_2 \tag{6.16}$$

In this expression Z_m becomes the mean of all points, and we make the assumption that the changes in ground elevation are small with respect to the flying height.

Now, we have

$$P''_x = x_1 - x''_2$$

Hence, in this case by rearranging the terms of equation (6.16), we have finally

$$P''_x = P_x - (f \Delta \varphi) + \left(\frac{\Delta H}{Z} \right) x_2 + (\Delta \kappa) y_2 + \left(\frac{\Delta \omega}{f} \right) x_2 y_2 - \left(\frac{\Delta \varphi}{f} \right) x_2^2 \tag{6.17}$$

This equation shows us that parallax bar readings on photographs exposed under the conditions set out above will give rise to an error surface when used in conjunction with equation (6.8). The shape of this error surface, a hyperbolic paraboloid, is given by the terms involving brackets on the right-hand side of equation (6.17) and is one commonly encountered in photogrammetric analysis, e.g.

$$e_H = a_0 + a_1 x + a_2 y + a_3 xy + a_4 x^2 \tag{6.18}$$

106

Finally, we must account for the initial assumption made that the first photograph was a true vertical. If this is not the case in practice then the model we have dealt with so far is one that is tilted with respect to the datum plane by amounts φ_1 and ω_1, i.e. the tilt components of the first photograph. In consequence, for the general case we need only modify the linear terms of equation (6.18). The datum surface will require a further tilt in the x direction due to φ_1 and a further tilt in the y direction due to ω_1. There will also be a change in the value of the constant term. An examination of Fig 6.8 illustrates these points but shows, for clarity, the situation in the xy plane only.

The method of approach used above has been selected because it emphasises some of the important features of the error surface which are as follows:

(a) The relative tilt of the photographs produces the non-linear components of the error surface.

(b) The relative omega tilt $(\omega_2-\omega_1)$ produces the hyperbolic warp of the error surface in the y direction (see Fig 6.9(a)).

(c) The relative phi tilt $(\varphi_1-\varphi_2)$ produces a parabolic bowing of the error surface in the x direction (see Fig 6.9(b)). This is sometimes referred to as the φ-cylinder effect.

(d) The linear term in the x direction is produced by the presence of an inclination of the air base and the φ_1 tilt of the first (the reference) photograph (see Fig 6.9(c)).

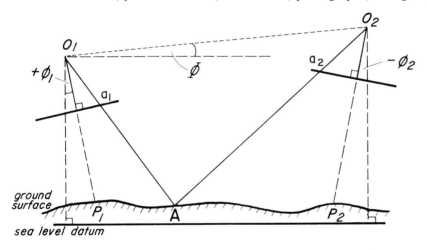

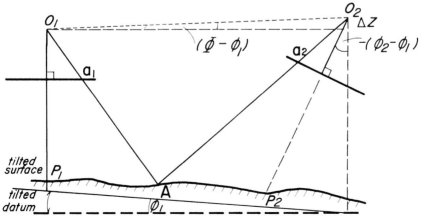

Figure 6.8 (a) and (b). Inclination of the Datum Surface

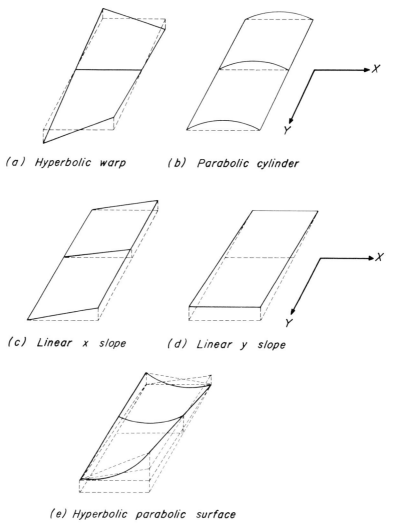

(a) Hyperbolic warp   (b) Parabolic cylinder

(c) Linear x slope   (d) Linear y slope

(e) Hyperbolic parabolic surface

Figure 6.9. Error Surface Produced by Photo Tilts

(e) The linear term in the *y* direction is produced by the presence of an ω_1 tilt of the first (the reference) photograph coupled with any errors in the orientation of the axes of the second photograph.

The importance of appreciating the various effects of the relative and absolute inclinations of the photographs will become more apparent later in the text, for the relative orientation of the photographs is the key factor in the use of analogue plotting instruments. Also, many approximate solution instruments base their outputs on parallax bar measurements corrected, to some extent, by mechanical devices of one sort or another. In such instruments it is the non-linear terms of the equation that call for the most ingenuity on the part of the designer. (Examples of such instruments are the Thompson CPI and the Santoni SMG 5.)

The concept of the error surface is readily understood in practical terms if we examine the following situation. Consider two aerial photographs being used for parallax bar heighting using only the simple parallax equation. Both photographs have tilts and there

108

is some inclination of the air base but all these are within the usual limits laid down for near-vertical photography. The ground surface photographed is in fact perfectly flat. Hence any one point can be taken as a datum point and differences in height can be derived using equation (6.6) for a regular array of points. These spot heights will therefore provide a set of contours that will depict a hyperbolic paraboloid surface, i.e. the error surface. It will be realised that the parameters of this surface are a function of Z, the flying height above the horizontal ground surface. In fact, each Z plane has a set of values for these parameters associated with it. In our derivation of equation (6.16) we had therefore to make a 'flat earth assumption' to the effect that any variations in ground elevation must be such that a mean height value can be taken without the introduction of any significant errors in the correcting surface. Even if the ground is flat, errors may still of course occur if the assumptions made concerning the tilts of the photographs and the air base are incorrect. The use of the approximate orthogonal matrix also gives rise to errors in the corrections in such circumstances. A more detailed treatment of this topic is to be found in Ref 6.5.

6.9 Parallax bar heighting − practical applications

If we accept that in general we will need to apply to the parallax bar heights a correction formula similar in form to equation (6.18), then the practical problem is simply one of evaluating a set of coefficients such as a_0, a_1, a_2, a_3 and a_4. Having obtained these, each height value is then subjected to a correction given by

(Corrected height value) = (parallax bar height value) + e_{Hi}

where

$$e_{Hi} = a_0 + a_1 x_i + a_2 y_i + a_3 x_i y_i + a_4 x_i^2 \qquad (6.19)$$

For this purpose, a set of co-ordinates for each control point is required. As we are only defining the shape of a smooth error surface any system of rectangular co-ordinates on either photograph will suffice. (Although ideally we would prefer to use the undistorted ground values.) The base line is usually taken as the x-axis and for convenience the origin is best situated halfway along the base line. In this way the value of the x^2 term is much reduced in magnitude. Measurements are recorded to the nearest millimetre. If we have a minimum of five control points of known height values then we can solve the five simultaneous equations obtained by substituting five sets of the known quantities in equation (6.19). Having defined the error surface with respect to a rectangular system on one photograph, the correction required to be applied to any other point of that system can then be computed. The use of a desktop computer such as one of the Hewlett–Packard

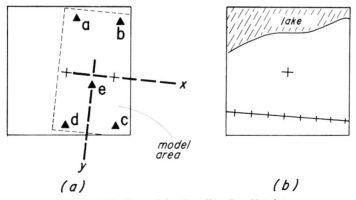

Figure 6.10. Control for Parallax Bar Heighting

9800 or Wang 2200 series is ideal for this purpose. A suitable disposition of control points is shown in Fig 6.10(a).

Three points, such as a, b, c, will serve to define the linear tilt of the surface, while the fourth point d and the fifth point e will monitor the omega warp and phi cylinder effects respectively. If more control points are available, we have the choice of either increasing the complexity of our mathematical model (for example by introducing a y^2 term) or of finding better values for the five original coefficients using a least squares technique.[6.4,6.5]

In the past graphical techniques were often employed in the application of this form of correction. However, with the introduction of the computer the use of such techniques is no longer essential. Details of such a method are described in Ref 6.6. We do, however, note one of the features of these methods here since it provides an example of how a favourable disposition of control can simplify the correcting procedure.

Suppose that four of the central points were provided so that they formed a rectangle with two sides parallel to the base line as in Fig 6.11. We know from our analysis that the height error is linear in the y direction. Hence any number of points can be corrected along

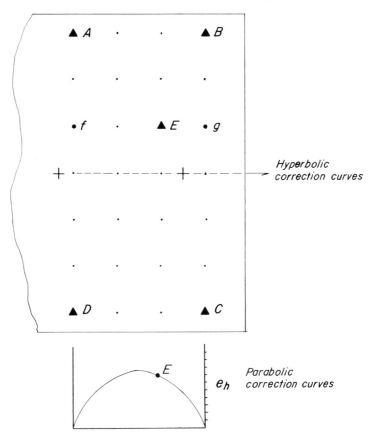

Figure 6.11. *Ideal Control Positions for Parallax Bar Heighting*

the lines AD and BC and a few such points are indicated on the diagram. If we ignore the $\Delta\varphi$ correction for the moment then we can also do a linear interpolation in the x direction. By this means we can provide a network of points as dense as necessary and from these a family of hyperbolic curves can be drawn by interpolation. The error surface represented in this way would be applicable only if $\Delta\varphi$ is zero, i.e. the convergence of the two photographs happened to be zero. In general, this will not be the case and so a fifth point E is necessary. For maximum effectiveness this point should be somewhere on a line normal to the base line and equidistant from AD and BC. Now, we have the magnitude of the total error at this point E and from our hyperbolic correction curves we can find the hyperbolic component of it and subtract this from the observed value. The remainder must therefore be the parabolic component. As indicated, a single simple parabolic correction curve can now be drawn. In the case shown, the error is a positive one indicating the presence of a negative convergence of the two photographs.

Sometimes, the natural disposition of ground features will enable simple graphical techniques to be employed in an effective manner. Figure 6.10(b) is an example of such a situation. The shore line of the lake can provide numerous points of known height provided the water line can be identified. The railway line can also provide a series of points as long as survey data concerning the line are available. Further details of practical methods can be found in Refs 6.4, 0.4 and 0.8.

7: The stereoscopic model

7.1 Introduction

In previous chapters we have investigated how, by quite separate processes, planimetric detail and height information can be obtained from a pair of overlapping aerial photographs. Perhaps the only merit of these methods is the extreme cheapness of the equipment required to carry them out. The results are certainly not of a particularly high quality and the methods can be quite laborious. At the present time, therefore, the standard method of producing accurate topographic mapping at the most economical rate is by the use of analogue stereoscopic plotting instruments. The range of instruments now available is quite large and there are considerable variations in design and capability. However, the majority conform to one set of concepts and it is these that we are to examine in this chapter. A comparative study of the instruments themselves will be found in the next chapter.

Thus far the word 'model' has been used occasionally to indicate little more than the three-dimensional effect experienced by a viewer using a simple stereoscope. From now on, however, the word takes on a more specific meaning. In analogue plotting instruments the requirement is to produce, by some means or other, a true scaled-down model of the ground surface. Measurements are then made on this model surface and so, with a knowledge of the scale factor involved, a contoured map sheet can be drawn. All plotting instruments have in consequence to carry out two basic functions. The first of these is to construct a scaled model of the ground surface with as little distortion as possible. The second is to provide a means by which the operator can view the model and carry out the measuring and recording processes. It goes without saying that the precision of the latter must be in sympathy with the quality of the model forming process. In the most precise analogue instruments for example the measurement system employed is one capable of good metrical standards. Model co-ordinates can be recorded with a precision of the order of 10 μm over the model area. To make such an order of precision worthwhile a model with a very low distortion must be formed.

The analogue model produced by plotting instruments is a direct one. When correctly set up in the laboratory the two projectors, together with the model surface they produce form a situation having a direct correspondence with the two camera stations which provided the photographs and the ground surface. With the exception of the dimensions of the projectors themselves all the remaining parts are at model scale. The projectors are devices to reconstitute directions (vectors) only and so the essential requirement is for them to be geometrically similar to the camera but no more. This analogue modelling process is illustrated diagrammatically in Fig 7.1.

In order to examine this method of mapping we will take as our example a simple form of plotter based on a process of direct optical projection. This type of solution is illustrated in Fig 7.2, which shows a pair of Multiplex projectors in use. In this equipment, reduced diapositives of the photographs are used in the projectors to generate the intersecting light rays that define points on the model surface. When all pairs of rays intersect, the points so formed produce a continuous model surface. The procedure by which this is achieved is called projector orientation and is composed of three distinct and important stages.

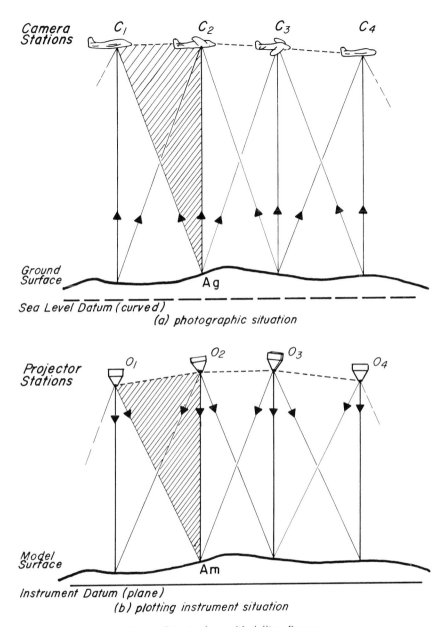

Camera
Stations
C_1 C_2 C_3 C_4

Ground
Surface

Sea Level Datum (curved)

Ag

(a) photographic situation

Projector
Stations
O_1 O_2 O_3 O_4

Model
Surface

Instrument Datum (plane)

Am

(b) plotting instrument situation

Figure 7.1. Analogue Modelling Process

7.2 Interior orientation

This is a requirement of each projector, the object being to reproduce as accurately as possible the geometry of the bundle of rays that entered the camera. (Of course this could be done very well indeed if the camera were itself used as a projector to reproject the developed photograph, but this is not usually a practical proposition.) The geometric requirement is therefore that the projector should faithfully reproduce through the outer node of its lens the directions of the rays of light that originally entered the camera through the outer node of the camera lens. In order to do this, certain dimensions of the projector

113

Figure 7.2.(a). Multiplex

must be the same as, or geometrically similar to, those of the camera. The data required for this purpose are that provided by camera calibration as described in chapter 1.

Referring to Fig 7.3, clearly in order that the two angles may have the same value the following conditions need to be fulfilled:

(i)

$$\tan \theta = \frac{\mathrm{ap}}{f}\bigg| = \frac{\mathrm{a'p'}}{f'} = \tan \theta'$$

i.e.

$$\frac{d}{f} = \frac{d'}{f'}$$

(ii) The principal point (p) of the photograph should fall on the optical axis (p') of the projector.

114

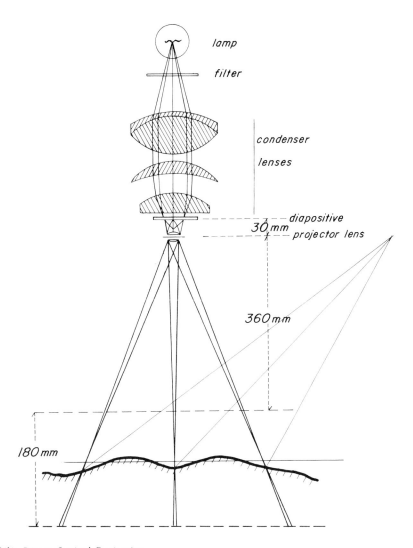

Figure 7.2(b). Direct Optical Projection

Most projectors use full-sized diapositives in order to preserve, as far as possible, the limits of image resolution. The principal distance (f') is therefore made equal to the accepted focal length of the camera lens, usually to within about $\pm 10\ \mu m$. One of the disadvantages of the optical projector arises from the fact that this distance cannot be varied very much, otherwise conditions for the sharp focusing of the model image will not be preserved. When a reduction printer is used to provide smaller diapositives, then the projector principal distance (f') may have a fixed value; small variations in camera focal length being allowed for in the photographic reduction process. For example, in the case of Multiplex equipment the diapositives are reduced to about one-fifth size because a fixed principal distance of 30.00 mm is used for the projectors. When the focal length of the camera is 152.40 mm the reduction factor required is 152/30.0 (i.e. 5.08). When photography of some other focal length (f) is used then the reduction factor required is $f/30.0$.

In order to centre the diapositive correctly in the projector, the projector optical axis is marked by a small circular dot at the centre of the stage plate. Small adjusting screws allow

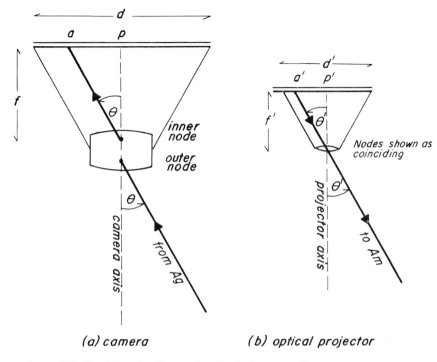

Figure 7.3. Conditions for Correct Interior Orientation of Projectors

the diapositive to be moved on the stage plate until the principal point of the photograph coincides with this dot. In many instruments the plate holders of the projectors can be removed from the instrument. In such a case, the diapositive is accurately located in the holder by means of the four fiducial marks to be found in the corners of the photograph and four equivalent points marked on the stage plate of the diapositive holder. A small illuminated table (i.e. a setting box) and four low-powered magnifiers enable a precision in location of about 10 μm to be achieved. With plotters using mechanical reprojection it is a simple matter to set the principal distance to the focal length of the camera. This can be done in these instruments to a wide range of values. In the case of optical projectors a much more limited range of principal distance settings is possible. This matter is discussed in more detail in the next chapter.

7.3 Procedure for relative orientation

When the two projectors of an instrument have been put into correct interior orientation, generally no model will be produced in the first instance. A model point is generated only when the two appropriate rays (vectors) intersect in the model space. For example, O_1A_m and O_2A_m of Fig 7.1(b) when they intersect define the point A_m. But in the first instance they will not intersect, i.e. they will be skew rays. The projectors must therefore be moved and rotated until this pair of rays, and all other pairs of rays, intersect simultaneously. When this has been achieved, the infinity of points so formed produces the continuous model surface. This process of projector manipulation is called relative orientation and is the essential model forming process. It is the cornerstone of analogue stereo-photogrammetry.

At a first inspection the task of carrying out a relative orientation seems to be a formidable one but in fact once the basic ideas have been grasped the process is quite straightforward. The first essential point was introduced in section 6.8 of the previous chapter, namely that a model will be formed as soon as one projector is placed in its correct position relative to the other. In the first instance, therefore, in order to understand the process we need only consider moving and rotating one of the projectors. The second essential point we need to understand is the concept of Y-parallax and its association with the presence of skew rays. If a pair of rays is skew then they will not define a model point and the observer will see Y-parallax in the model. In theoretical terms, the object of relative orientation is to clear the model of any visible Y-parallax. In practice, for a number of reasons (such as the presence of lens distortion or film shrinkage for example) it might be impossible to clear a model completely, in which case the Y-parallax is reduced to a minimum distribution.

Referring once again to Fig 7.1, in the photographic situation the rays $\mathbf{A}_g\mathbf{C}_1$ and $\mathbf{A}_g\mathbf{C}_2$ cannot be skew — they emanate from the ground point A. In consequence the three points $\mathbf{A}_g\mathbf{C}_1\mathbf{C}_2$ define a plane, the basal plane or epipolar plane. Hence every ground point generates one member of an infinite family of basal planes, all of which contain the air base $\mathbf{C}_1\mathbf{C}_2$. By definition, the X-axis of the local co-ordinate system of this section of the ground surface is the line $\mathbf{C}_1\mathbf{C}_2$. Hence at every height value the two vectors have the same Y co-ordinate, i.e. the Y-parallax is zero. Similarly, when a model has been correctly formed, all pairs of rays such as $\mathbf{O}_1\mathbf{A}_m$ and $\mathbf{O}_2\mathbf{A}_m$ will generate a family of basal planes about the model base line $\mathbf{O}_1\mathbf{O}_2$. No Y-parallax will therefore be visible in the model space either.

In contrast to the above, in Fig 7.4 we show a model situation where there is some Y-parallax present in an imperfectly formed model. In this diagram the ray $\mathbf{O}_1\mathbf{B}'$ is in front of the plane of the paper and the ray $\mathbf{O}_2\mathbf{B}''$ is behind it. The situation in three different height planes is illustrated and clearly shows that only at the air base does the Y-parallax reduce to zero. As the Z value increases so the amount of Y-parallax visible in the model progressively increases. It should be noted that where some residual parallax has to be accepted at some model point then the 'correct' point is taken as that position where the X-parallax is zero. That is the middle of the three Z planes shown in Fig 7.4.

When attempting to clear a model of Y-parallax using only the movements of one of the projectors we realise that there are only six different adjustments that can be made. These are the three translations of the projector (let us say, projector 1) that in effect change the direction of the air base and are usually designated as $\Delta BX_1, \Delta BY_1, \Delta BZ_1$. Then, there are the three rotations of the projector about the three axes which in this case would be designated as $\Delta\kappa_1, \Delta\varphi_1, \Delta\omega_1$. For the purpose of relative orientation, therefore, it would appear that a unit of two projectors has $6°$ freedom. However, on inspection we realise that in fact there are only five effective movements because by definition the X-axis is the air base. The movement ΔBX cannot therefore influence the distribution of Y-parallax. If, then, we can remove the parallax at just five points using the remaining five movements we should then achieve the condition of relative orientation.

In order to understand how to carry out a relative orientation we must first appreciate the effects produced in the model space by the projector movements. These are illustrated in Fig 7.5, where in order to emphasise the effects the movements have been made rather large. They are most readily appreciated if we commence with a model of a flat ground area formed as a result of a perfect relative orientation. The model area is most usefully delineated by the introduction of six standard points, sometimes referred to as the Von Gruber points. These are used a great deal in photogrammetric model analysis, and the various diagrams of Fig 7.5 are typical examples of their use. Only in the first diagram, Fig 7.5(a), have the numbers been shown on the points. Initially all six pairs of rays will intersect to define the six standard points. In the six diagrams of the figure the effects of disturbing projector 1 in each of the six possible ways is illustrated. These effects should always be translated into terms of the X-parallax and the Y-parallax introduced into the model as indicated for points 2, 4 and 6 of Fig 7.5(b). As will be demonstrated in a later

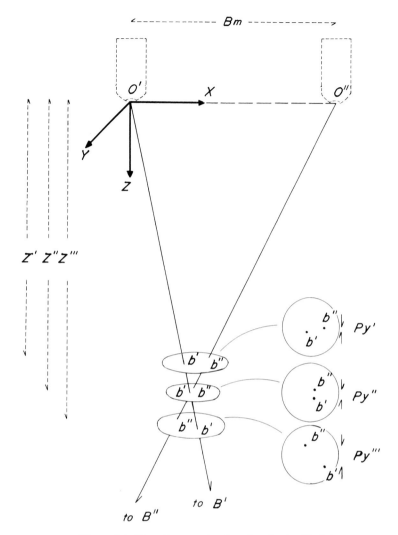

Figure 7.4. *Skew Rays and y-Parallax in the Model*

section, all these effects can be described adequately by two equations of a familiar hyperbolic paraboloid form

$$\left. \begin{array}{l} P_y = b_0 + b_1 X + b_2 Y + b_3 X Y + b_4 X^2 \\[2mm] P_x = c_0 + c_1 X + c_2 Y + c_3 X Y + c_4 X^2 \end{array} \right\} \tag{7.1}$$

A clear understanding of the various effects produced by projector movements is an essential preliminary to any study of stereo-photogrammetry, and a series of useful exercises can be devised for use with an optical plotting instrument to illustrate these. The main points to note are as follows:

(i)   Rotation $\Delta \kappa_1$ produces no P_y at 1 and has a maximum effect at 2, 4 and 6. In a similar way, a rotation $\Delta \kappa_2$ produces no P_y at 2 and has a maximum effect at 1, 3, and 5.

(ii)   Rotation $\Delta \varphi_1$ introduces no P_y along the base line 1, 2 but produces maximum P_y values with opposing signs at points 4 and 6. Similarly, a rotation $\Delta \varphi_2$ produces no effect along the base line and maximum effects at points 3 and 5.

118

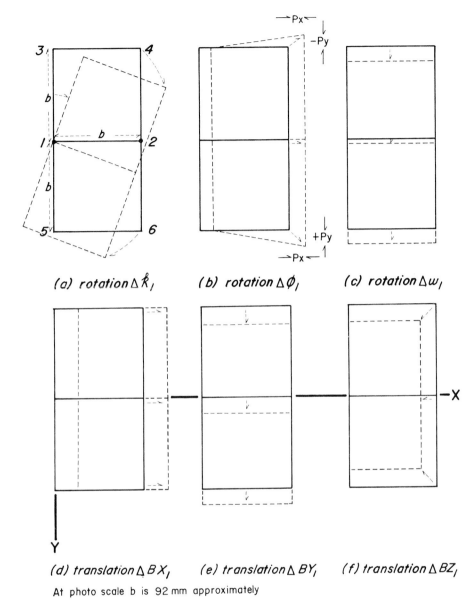

At photo scale b is 92 mm approximately

Figure 7.5. Parallax at the Six Standard Points

(iii) Rotations $\Delta\omega_1$ and $\Delta\omega_2$ have the same effect. All model points will show an introduction of some P_y but the magnitude of the effects increases with the distance from the base line.

(iv) Translations ΔBX_1 and ΔBX_2 introduce no P_y into the model.

(v) Translations ΔBY_1 and ΔBY_2 introduce the same amount of P_y into all model points.

(vi) Translation ΔBZ_1 produces an overall change in the scale of the projection from projector 1. No P_y is introduced into 1 but all other points experience the introduction of P_y in linear proportion to their Y co-ordinate value. A similar effect centred on 2 is produced by ΔBZ_2.

119

The amounts of X-parallax, and therefore the height distortion introduced into the model at the six standard points, may be summarised in a similar manner.

If we now bear in mind the various patterns of Y-parallax introduced into the model by the five relevant movements, the rationale behind the following empirical methods of orientation is clear. If the situation is such that after a coarse preliminary adjustment of the projectors there is some distribution of P_y at the six standard points then by means of a very simple diagram this pattern can usefully be recorded at any stage. Figure 7.6 illustrates a typical example of such a distribution. After some practice the various patterns of parallax introduced by various projector movements become familiar and the relative orientation process becomes easier.

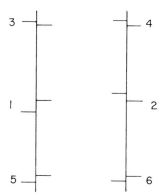

Figure 7.6. Simple Diagram to Show y-Parallax Distribution

7.4 One-projector relative orientation

Using the movements of projector 1 only we proceed as follows:

(i)   At 1 remove Y-parallax with ΔBY_1.

(ii)   At 2 remove Y-parallax with $\Delta \kappa_1$.

Repeat (i) and (ii) until all parallax is removed at these points.

(iii)   At 3, remove parallax with ΔBZ_1.

(iv)   At 4, remove parallax with $\Delta \varphi_1$.

Repeat the above until all parallax has been removed at the four points. This can be done most effectively when the detail points being examined are very sharp with strong delineation in the X direction. In practice, therefore, one selects the most suitable points as near as possible to the standard positions. Only if the points fall exactly on the standard positions and the ground is flat can the process be carried out with little or no iteration, although the sequence of movements has been chosen so that each one has the minimum effect on points cleared previously. Hence the procedure gives a rapid convergence to the point where four of the five points necessary have been cleared of P_y. Unfortunately, the remaining projector movement $\Delta \omega$ we must now use will introduce some P_y into all model points. At the fifth point, therefore, we proceed as follows.

(v)   At 5 (or 6) remove P_y with $\Delta \omega$ and then overcorrect by a calculated amount. For wide angle photography at points 90 mm from the base line this overcorrecting would be about 190%. (See section 7.11 for details of this factor.)

(vi)   Repeat the previous steps (i) to (v) and repeat the whole process until no Y-parallax is visible at any of the five points. With a reasonable estimate of the overcorrection factor, the process should converge quite rapidly. With good photography a skilled operator will complete the process in 15–20 min under normal conditions.

120

The sixth point provides a check on the procedure but it must be realised that lens distortion, instrument imperfections, film shrinkage, etc. will all produce elements of distortion that can produce small amounts of Y-parallax in the model. The sum of these will therefore become apparent at the sixth point. In such a case, skill is required in redistributing this residual parallax over the model so that its effects are minimised.

7.5 A two-projector relative orientation

Many analogue plotting instruments are not provided with all possible projector movements. By reducing the number of movements an instrument can be made more stable, more accurate and perhaps cheaper. Hence many instruments are provided with only a ΔBX translation necessary for changing the scale of the model. In such instruments the base line coincides with the X-axis of the instrument and a relative orientation using the rotations of both projectors is used, i.e.

$$\Delta\kappa_1, \Delta\kappa_2, \Delta\varphi_1, \Delta\varphi_2, \text{ and } \Delta\omega_1 \text{ or } \Delta\omega_2 \text{ (both omegas have the same effect)}$$

In this case the process can be carried out as follows; the reasoning behind the sequence of operations is the same as before:

(i)   At point 1, remove Y-parallax with $\Delta\kappa_2$.

(ii)   At point 2, remove Y-parallax with $\Delta\kappa_1$.

Repeat (i) and (ii) if necessary until these points are clear of parallax.

(iii)   At point 3, remove Y-parallax with $\Delta\varphi_2$.

(iv)   At point 4, remove Y-parallax with $\Delta\varphi_1$.

Repeat these stages (i) to (iv) as necessary until all points are clear.

(v)   At point 5 (or 6) remove the Y-parallax with $\Delta\omega_1$ or $\Delta\omega_2$ and overcorrect by about 190% in the case of wide angle photography.

Repeat the above stages (i) to (v) until all five points are clear. Then finally check at the last point 6.

Many minor variations will be found on the two basic procedures described above and all are designed to speed up the convergence in some way. All are, however, governed by the principles most clearly demonstrated by the above methods. The reader is recommended to work through both these two procedures using simple P_y diagrams of the type shown in Fig 7.6 to indicate the pattern of Y-parallax present after each adjustment.

7.6 The process of absolute orientation

From the above we have seen that using only the clues afforded by a distribution of Y-parallax the two projectors can be adjusted until a three-dimensional model of the terrain is produced. This model is not, however, at any known scale and will be tilted with respect to the datum level surface of the instrument. The absolute orientation process consists therefore of a scaling and levelling up of the model. These two operations should be carried out in the order stated, and to do this we must now have some additional information.

7.6.1 SCALING THE MODEL

The minimum information required for this purpose is the ground length of one line in the model. This could take the form of a line measured on the ground whose terminal points have also been identified on the model surface. More often than not, however, the co-ordinates of the terminal points such as A and C in Fig 7.7 will be provided. These data will give the following:

121

(i)   A scale factor for the model.

(ii)   The orientation of the plot.

(iii) The position of the plot within the co-ordinate frame of reference.

However, using only two control points provides no check on the setting of the model and so at least one further point is required. Four points as in Fig. 7.7 provide a satisfactory and economic use of control points when a series of overlapping models are to be controlled. If the distribution of ground control points is sparse, aerial triangulation can provide the extra points required, as explained later in chapter 9.

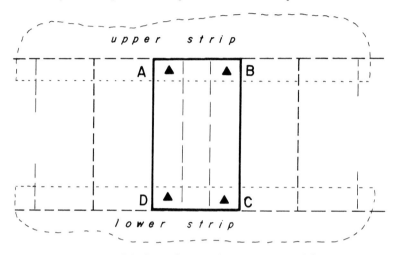

Figure 7.7. Control Points for Scaling a Model

When scaling a model to the pattern of control shown in Fig. 7.7, it is usual to scale on one diagonal (say AC) and then check on the other. If there is a discrepancy in one of these points, that point can be isolated and investigated while the plotting proceeds. The process of scaling can be carried out by direct plotting onto a prepared sheet containing the control points plotted at the required scale. Such a procedure is the normal one for a plotter with a graphical output. For more precise instruments provided with some form of digital readout, the scale factor can be calculated from the ground and model co-ordinates:

$$\text{Scale factor} = \frac{\text{model length}}{\text{ground length}}$$

$$= \frac{\Delta x^2 + \Delta y^2 + \Delta z^2}{\Delta X^2 + \Delta Y^2 + \Delta Z^2} \tag{7.2}$$

Occasionally a model can be scaled using vertical distances derived from spot height values, provided the changes in elevation are sufficiently large. An interesting and rather special case where this technique can be applied is illustrated in Fig 7.8.

If a radar altimeter or airborne profile recorder was used in the aircraft at the time of exposure this too can provide the ground clearance values such as Z_1 and Z_2, which can be used to scale the model. The use of such equipment is described in more detail in chapter 12.

To change the scale of the model we need only change the length of the model air base of the instrument. In instruments with no BY or BZ translations the air base of the model coincides with the X-axis of the instrument. A change in BX is therefore all that is required — see Fig 7.9.

In other instruments, the introduction of projector translations such as ΔBY and ΔBZ

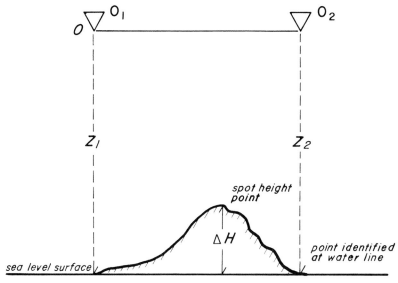

Figure 7.8. Scaling a Model Using Height Values

may mean that the air base is no longer parallel to the X-axis. To change scale, therefore, all base components must be altered in the required ratio. Failure to change the minor components will result in the introduction of some Y-parallax into the model. Figure 7.10(a) shows in plan view a case where a Multiplex type of instrument has a ΔBY component with the result that the air base is no longer parallel to the X-axis of the instrument. Part (b) of the same figure shows the BZ component of the air base. Hence we have

$$\text{Scaling factor required} = \frac{\text{distance required}}{\text{distance plotted}}$$
$$= \lambda$$

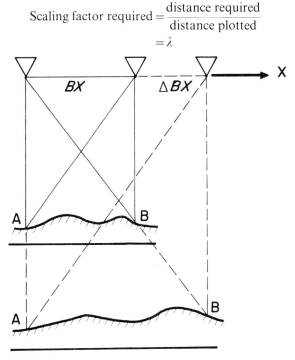

Figure 7.9. Model Scaling Using BX Only

123

where
$$\lambda = \frac{BX + \Delta BX}{BX}$$

$$= \frac{(BY_2 - BY_1) + \Delta BY}{(BY_2 - BY_1)}$$

$$= \frac{(BZ_2 - BZ_1) + \Delta BZ}{(BZ_2 - BZ_1)} \qquad (7.3)$$

where ΔBX, ΔBY and ΔBZ are the changes required in the base components to produce the required model scale.

After scaling, in such a case it would be prudent to check over the model to verify that the relative orientation is still correct.

7.6.2 LEVELLING THE MODEL
For this purpose we need to make the slope of the model surface identical with that of the ground in any two directions. For maximum sensitivity these should be at right angles to one another. Slope information, other than that provided by horizontal water surfaces, is not generally available and so again we use the co-ordinates of identified model points. Figure 7.11(a) shows the minimum distribution of spot heights that can be used for this

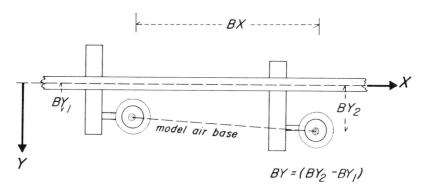

(a) Plan view showing BY components

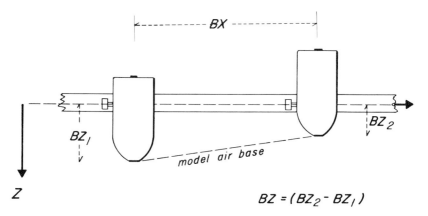

(b) Elevation showing BZ components

Figure 7.10. Base Components BY and BZ in Model Scaling

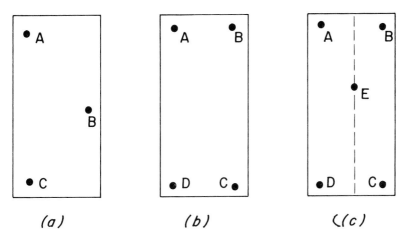

Figure 7.11. Height Control for Levelling a Model

purpose. However, if one uses only three points then no check on the orientation is available, and so in practice at least four points would be required. The distribution shown in Fig 7.11(b) makes levelling up particularly easy for any slope error in the AB (or DC) direction is almost entirely due to a Φ tilt of the model, while any error in the AD (or BC) direction will be the result of error in the Ω tilt. With such a distribution, a few iterations using AB to monitor the Φ slope and AD to monitor the Ω slope will quickly bring the model into the correct position. If more than four points are to be used, theory shows that the optimum position for this point is anywhere along a central line at right angles to the base line, as indicated in Fig 7.11(c). In such a position, any tendency towards the production of a Φ-cylinder bulge in the model surface due to residual errors in the relative orientation will be detected (see section 7.8).

Most frequently, analogue models are levelled up empirically by tilting the model until differences in height in the model surface agree with differences in height on the ground reduced to model scale. Only in special cases, where there are a large number of heighted points that need to be incorporated, will a numerical approach be necessary. For this purpose we take one control point as a local origin (let us say A), and if we let the mis-levelment errors be $\Delta\Phi$ and $\Delta\Omega$ we then see from an inspection of Fig. 7.12 that any error e_Q in the difference in height between datum point A and any other control point Q is given by the following expression:

$$e_Q = \Delta Z_Q - \lambda \Delta H_Q \text{ mm} \qquad (7.4)$$

where λ is the scale factor for the model and ΔH is the true difference in height between Q and point A expressed in the same units as the model co-ordinates. We then have

$$e_Q = \Delta X_Q \tan \Delta\Phi + \Delta Y_Q \tan \Omega \qquad (7.5)$$

Equation (7.5) is therefore an observation equation involving the two unknowns $\Delta\Phi$ and $\Delta\Omega$. To evaluate these we must have at least one more equation, provided by one more control point. The simultaneous solution of the equations then provides the two components of tilt necessary for levelling up the model. If more than three points are available then a least square calculation can be carried out to determine the best values for the unknowns. As an example, the observation equations produced by the distribution of control of Fig 7.11(c) are written out in full, as follows:

$$e_B = \Delta X_B \tan \Delta\Omega + \Delta Y_B \tan \Delta\Omega$$

$$e_C = \Delta X_C \tan \Delta\Phi + \Delta Y_C \tan \Delta\Omega$$

125

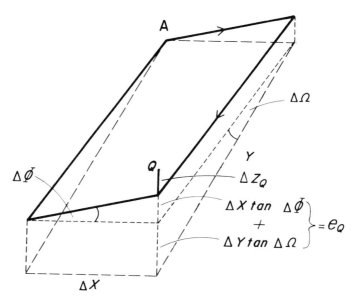

Figure 7.12. Errors in Spot Height Values due to Model Tilt

$$e_D = \Delta X_D \tan \Delta\Phi + \Delta Y_D \tan \Delta\Omega$$

$$e_E = \Delta X_E \tan \Delta\Phi + \Delta Y_E \tan \Delta\Omega$$

i.e.

$$\begin{pmatrix} e_B \\ e_C \\ e_D \\ e_E \end{pmatrix} = \begin{pmatrix} \Delta X_B & \Delta Y_B \\ \Delta X_C & \Delta Y_C \\ \Delta X_D & \Delta Y_D \\ \Delta X_E & \Delta Y_E \end{pmatrix} \begin{pmatrix} \Delta\Phi \\ \Delta\Omega \end{pmatrix} \qquad \Bigg\} \qquad (7.6)$$

In the above equations (7.6) the assumption has been made that $\Delta\Phi$ and $\Delta\Omega$ are small. If for some reason the model had not been previously scaled then the value of e_Q is not known and so equations (7.4) and (7.5) would be combined to form a different observation equation involving the third unknown parameter λ. In this case a minimum of four points would be needed to calculate the three unknowns. It should be realised that the value for λ obtained in this way will not be satisfactory for an overall scaling of the model unless the changes in ground elevation are of sufficient magnitude. To carry out the levelling adjustment ideally, the two projectors should be moved as a rigid unit that is tilted with respect to the datum surface until the correct orientation of the model is achieved. This can be done on a number of direct optical projection instruments; the Kelsh plotter being a good example of this. An alternative to the above is to tilt the datum surface instead, and this too is a commonly used technique. Although in theory, after having produced a good relative orientation, it should thereafter be preserved, some instruments cannot carry out an absolute orientation process without disturbing the setting of the two projectors. Many instruments provided with base components rely on a combination of individual projector movements to produce the effects required for absolute orientation. The introduction of a model tilt $\Delta\Phi$ is the most obvious example of this, as illustrated in Fig 7.13. Here the requirement was to tilt the two projectors as a unit about the Y-axis by an amount $\Delta\Phi$. This same effect is achieved by:

(i)   tilting each projector such that $\Delta\varphi_1 = \Delta\varphi_2 = \Delta\Phi$

and

(ii)   introducing the base component $BZ = BX' \sin \Delta\Phi \approx BX \sin \Delta\Phi$

126

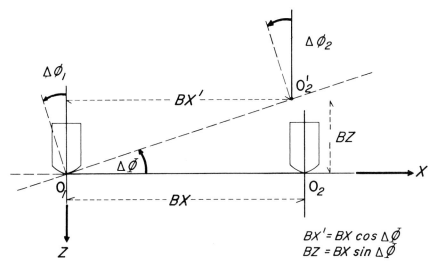

$$BX' = BX \cos \Delta \emptyset$$
$$BZ = BX \sin \Delta \emptyset$$

Figure 7.13. Levelling a Model Using Projector Rotations (φ)

It will also be noticed that the base length should also be reduced to the value $BX \cos \Delta\Phi$. In practice the magnitude of $\Delta\Phi$ will usually be of the order of a few degrees and so this correction would, in fact, be a very small adjustment.

The method of introducing a model tilt $\Delta\Omega$ is illustrated in Fig 7.14. If the air base has no base components then the X-axis of each projector is collinear with the instrument X-axis as shown in Fig 7.14(a). In this case the introduction of a $\Delta\Omega$ model tilt is accomplished, with no complications, by setting each projector such that

$$\Delta\omega_1 = \Delta\omega_2$$
$$= \Delta\Omega$$

However, if the air base is inclined to the machine X-axis as illustrated in Fig 7.14(b) then some small changes in the base components are necessary in order to maintain the relative orientation. From the end-on-view of the X-axis shown in the diagram we see that the base components will need to change very slightly such that the new base components are given by

$$BY' = BY \cos \Delta\Omega$$

and

$$BZ' = BY \sin \Delta\Omega$$

The change in base components with model rotation about a set of instrument axes is neatly dealt with by using the approximate orthogonal matrix derived in appendix A. In the first instance the air base $\mathbf{O_1O_2}$ has base components BX, BY, BZ as indicated in Fig 7.15. If now, in the course of absolute orientation, rotations $\Delta\Omega$, $\Delta\Phi$ and ΔK of the model about the machine axes (XYX) are required, then provided these rotations are of a limited magnitude we have the new base components given by equation (7.7) below:

$$
\left. \begin{pmatrix} BX' \\ BY' \\ BZ' \end{pmatrix} = \begin{pmatrix} 1 & \Delta K & \Delta\Phi \\ \Delta K & 1 & -\Delta\Omega \\ -\Delta\Phi & \Delta\Omega & 1 \end{pmatrix} \begin{pmatrix} BX \\ BY \\ BZ \end{pmatrix} \right\} \quad (7.7)
$$

In graphical work particularly, rotation ΔK about the Z-axis is not usually required, it is easier to rotate the plotting sheet. It should be noted that the signs of the elements appearing in the above matrix are the results of the convention used. Instrument manufacturers do not consistently obey any one set of sign conventions. In consequence these must be found by inspection in each case and the matrix modified accordingly.

127

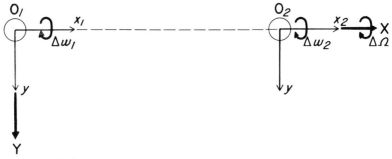

(a) no base component - PLAN view

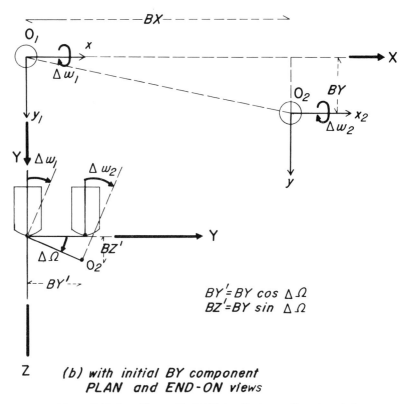

$BY' = BY \cos \Delta\Omega$
$BZ' = BY \sin \Delta\Omega$

(b) with initial BY component
PLAN and END-ON views

Figure 7.14. Levelling a Model Using Projector Rotations (ω)

7.7 Errors in relative orientation

Let us first start with the assumption that we have a model in perfect relative orientation. That is to say, all pairs of rays intersect in model space to define a continuum of model points. The effects of introducing small rotations and translations into the projectors will be examined and the results expressed in terms of distributions of Y-parallax, X-parallax, and height deformations over the model area. The movements are of such a magnitude that they do not destroy the stereo image seen by the operator. The approximate orthogonal matrix can be used to good advantage in this analysis.

In Fig 7.16 we have a typical set of three vectors, defining the base line $\mathbf{O_1O_2}$ and basal plane O_1O_2Q, the ray $\mathbf{O_1q_1Q}$ generated by the left-hand projector, and the ray $\mathbf{O_2q_2Q}$ generated by the second projector intersect in the point Q. At the height Z, therefore, we

128

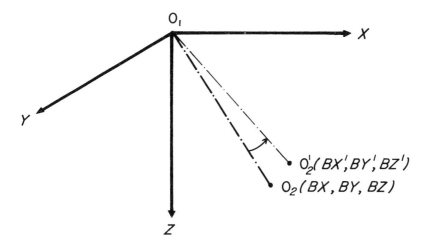

Figure 7.15. Rotation of the Air Base

have no X-parallax and of course no Y-parallax present. A rotation of the left-hand projector will cause the vector $\mathbf{O_1q_1}$ to rotate about the origin of the instrumental axes (XYZ). The new vector $\mathbf{O_1q_1^1Q^1}$ will no longer intersect O_2Q in Q but will, produced if necessary, intersect the same Z plane in Q″. The movement has therefore introduced elements of parallax into the model space and the values of these at the height Z can be quantified as follows.

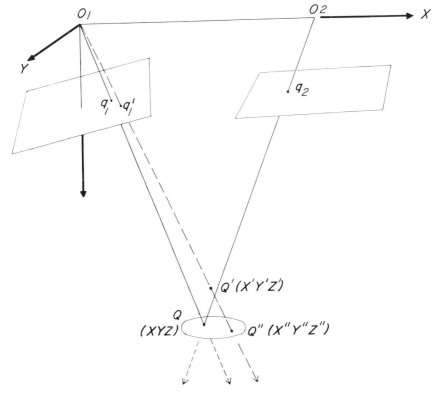

Figure 7.16. Effects of Residual Errors in Relative Orientation

The co-ordinates of the point Q' are given by

$$\begin{pmatrix} X' \\ Y' \\ Z' \end{pmatrix}_Q = R^\mathsf{T} \begin{pmatrix} X \\ Y \\ Z \end{pmatrix}_Q$$

For our purposes the approximately orthogonal matrix will suffice for R^T, hence we have

$$\begin{pmatrix} X' \\ Y' \\ Z' \end{pmatrix}_Q = \begin{pmatrix} 1 & -\Delta\kappa_1 & \Delta\varphi_1 \\ \Delta\kappa_1 & 1 & -\Delta\omega_1 \\ -\Delta\varphi_1 & \Delta\omega_1 & 1 \end{pmatrix} \begin{pmatrix} X \\ Y \\ Z \end{pmatrix}_Q \qquad (7.8)$$

The co-ordinates of Q" are obtained by changing the length of O_1Q' until it intersects the plane Z_Q. This requires the application of a scale factor λ_Q hence

$$\begin{pmatrix} X'' \\ Y'' \\ Z'' \end{pmatrix}_Q = \lambda_Q \begin{pmatrix} 1 & -\Delta\kappa_1 & \Delta\varphi_1 \\ \Delta\kappa_1 & 1 & -\Delta\omega_1 \\ -\Delta\varphi_1 & \Delta\omega_1 & 1 \end{pmatrix} \begin{pmatrix} X \\ Y \\ Z \end{pmatrix}_Q \qquad (7.9)$$

The value of the scale factor comes directly from the condition that

$$Z'' = Z = \lambda_Q(-X\Delta\varphi_1 + Y\Delta\omega_1 + Z)$$

i.e.

$$\lambda_Q = \left[1 + \left(\frac{Y}{Z}\cdot\Delta\omega_1 - \frac{X}{Z}\Delta\varphi_1\right)\right]^{-1} \qquad (7.10)$$

But the term in the internal bracket of equation (7.10) is small, and so using a binomial expansion for the expression on the right-hand side we have

$$\lambda_Q \simeq \left(1 - \frac{Y}{Z}\Delta\omega_1 + \frac{X}{Z}\Delta\varphi_1\right) \qquad (7.11)$$

We can now substitute for the scale factor in equation (7.9) to obtain the following

$$\begin{pmatrix} X'' \\ Y'' \\ Z'' \end{pmatrix}_Q = \left(1 - \frac{Y}{Z}\Delta\omega_1 + \frac{X}{Z}\Delta\varphi_1\right) \begin{pmatrix} 1 & -\Delta\kappa_1 & \Delta\varphi_1 \\ \Delta\kappa_1 & 1 & -\Delta\omega_1 \\ -\Delta\varphi_1 & \Delta\omega_1 & 1 \end{pmatrix} \begin{pmatrix} X \\ Y \\ Z \end{pmatrix} \qquad (7.12)$$

We now carry out the matrix multiplication of the right-hand side and in so doing ignore the second order terms (such as $\Delta\omega \cdot \Delta\kappa$ etc.) for such quantities are very small in this case because the rotations themselves are small. Hence we find

$$\left. \begin{aligned} X'' &= X - Y\Delta\kappa_1 + \left(1 + \frac{X^2}{Z^2}\right)Z\cdot\Delta\varphi_1 - \frac{XY}{Z}\cdot\Delta\omega_1 \\[2mm] Y'' &= Y + X\Delta\kappa_1 + \frac{XY}{Z}\cdot\Delta\varphi_1 - \left(1 + \frac{Y^2}{Z^2}\right)Z\cdot\Delta\omega_1 \\[2mm] Z'' &= Z \end{aligned} \right\} \qquad (7.13)$$

The parallaxes introduced into the model are therefore given by

$$\left. \begin{aligned} P_x &= X'' - X \\ &= -Y\Delta\kappa_1 + \left(1 + \frac{X^2}{Z^2}\right)Z\cdot\Delta\varphi_1 - \frac{XY}{Z}\Delta\omega_1 \\[2mm] P_y &= Y'' - Y \\ &= X\Delta\kappa_1 + \frac{XY}{Z}\Delta\varphi_1 - \left(1 + \frac{Y^2}{Z^2}\right)Z\cdot\Delta\omega_1 \end{aligned} \right\} \qquad (7.14)$$

Of the three projector translations (ΔBX_1, ΔBY_1, ΔBZ_1) we note that the first two introduced into the model directly either X-parallax or Y-parallax respectively. The third element changes the scale of the image produced by the projector, as shown in Fig 7.17. This produces the following changes:

$$\left.\begin{aligned} \Delta X &= -\frac{X}{Z} \cdot \Delta BZ_1 \\[2ex] \Delta Y &= -\frac{Y}{Z} \cdot \Delta BZ_1 \end{aligned}\right\} \tag{7.15}$$

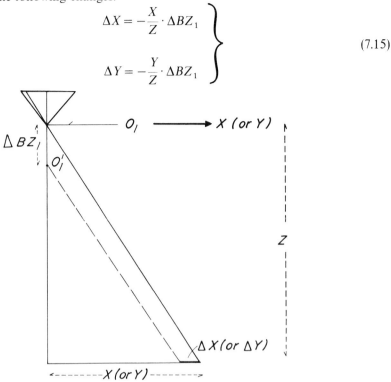

Figure 7.17. The Effect of Projector Translation ΔBZ

In Fig 7.17 the positive movement ΔBZ_1 produces the negative changes $-\Delta X$ and $-\Delta Y$ in the position of point of intersection with the Z plane.

Combining the results of equations (7.14) and (7.15) we now have expressions that describe the effects of all six movements of the left-hand projector.

$$\left.\begin{aligned} P_x &= -Y\Delta\kappa_1 + \left(1 + \frac{X^2}{Z^2}\right)Z\Delta\varphi_1 - \frac{XY}{Z}\Delta\omega_1 + \Delta BX_1 - \frac{X}{Z}\Delta BZ_1 \\[2ex] P_y &= X\Delta\kappa_1 + \frac{XY}{Z}\Delta\varphi_1 - \left(1 + \frac{Y^2}{Z^2}\right)Z \cdot \Delta\omega_1 + \Delta BY_1 - \frac{Y}{Z}\Delta BZ_1 \end{aligned}\right\} \tag{7.16}$$

For the effects of movements of the right-hand projector a similar analysis can be produced. In this case, however, the movements are centred on O_2. The co-ordinates of any point Q must therefore be temporarily quoted using this point as an origin. This only affects the X co-ordinate which becomes $(X - B)$. For the right-hand projector we therefore have a pair of equations similar to equation (7.16) above,

$$\left.\begin{aligned} P_x &= -Y\Delta\kappa_2 + \left(1 + \frac{(X-B)^2}{Z^2}\right)Z\Delta\varphi_2 - \frac{(X-B)Y}{Z}\Delta\omega_2 + \Delta BX_2 - \frac{(X-B)}{Z}\Delta BZ_2 \\[2ex] P_y &= (X-B)\Delta\kappa_2 + \frac{(X-B)Y}{Z}\Delta\varphi_2 - \left(1 + \frac{Y^2}{Z^2}\right)Z \cdot \Delta\omega_2 + \Delta BY_2 - \frac{Y}{Z}\Delta BZ_2 \end{aligned}\right\} \tag{7.17}$$

131

7.8 The use of the X-parallax equations

A distribution of X-parallax over a model surface will be interpreted by the viewer as changes in elevation. The operator will change the Z setting of his instrument at each point until the X-parallax is zero. Hence there is produced a distribution of height distortion over the model area. The relationship between the two quantities is shown in Fig 7.18. By projecting the skew rays on to the XZ plane we produce two similar triangles, from which we see

$$\frac{B}{Z-e_Z} = \frac{P_X}{e_Z}$$

i.e.

$$e_Z = \frac{Z-e_Z}{B} \cdot P_X$$

i.e.

$$e_Z \simeq \frac{Z}{B} P_X$$

$$= -e_H \qquad (7.18)$$

At the 'correct' height setting, where the X-parallax is now zero, there will, of course, remain some Y-parallax. This will introduce an error into the planimetric position also, for this will usually be taken as a position midway between the two points of intersection. This error is quite small in reality although for clarity the diagrams have greatly magnified the effects. For example, a residual Y-parallax of 20 μm might be considered quite large in a first-order plotting instrument.

When a relative orientation is carried out, small residual errors in the projector settings must remain, the values of which will, among other things, depend on the eyesight of the operator. These setting errors produce a distribution of height distortion which for the case of a single projector orientation are given by the following equation derived from equations (7.16) and (7.18):

$$e_H = -\left(\frac{Z^2}{B}\Delta\varphi_1 + \frac{Z}{B}\Delta BX_1\right) + \left(\frac{\Delta BZ_1}{B}\right)X + \left(\frac{Z}{B}\Delta\kappa_1\right)Y + \left(\frac{\Delta\omega_1}{B}\right)XY - \left(\frac{\Delta\varphi_1}{B}\right)X^2 \quad (7.19)$$

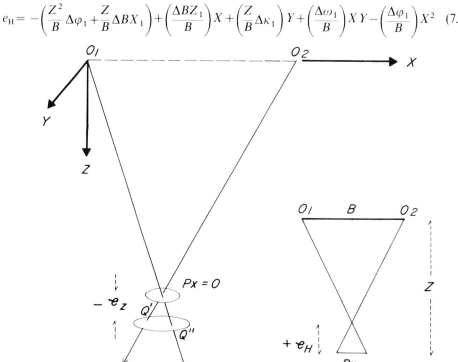

Figure 7.18. Relationship Between x-Parallax and Height Deformation

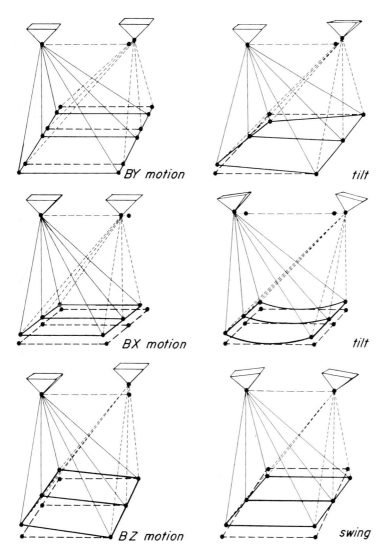

Figure 7.19. Height Deformations Provided by Small Movements of the R.H. Projector

For flat terrain (that is of constant Z value) and a given set of residual errors the equation produced is therefore shown to be of the familiar form of a hyperbolic paraboloid. It should be compared with equation (6.17) which was derived earlier in a similar manner to the above. We note that the error ΔBZ produces a linear slope on the model in the X direction, the $\Delta\kappa$ term produces a linear slope in the Y-direction, $\Delta\omega$ produces the hyperbolic warping of the model and the $\Delta\varphi$ term is responsible for the quadratic curvature in the direction of the air base.

An equation of similar form could also be obtained for the right-hand projector. In this case the substitution of $(X - B)$ in place of X would produce some extra terms. We could combine these two equations to produce a further more complex equation, but again still of the same general form. Such an equation would describe the height distortion introduced into the model by residual errors in both projectors. The diagrams of Fig 7.19 show the effects of movements of the right-hand projector only.

When considering what constitutes an adequate distribution of height control for a model, the diagrams of Fig 7.19 are of interest. We see that with the provision of three points, as in Fig 7.11(a), no warp or curvature of the model would be detected and the model would always be tilted to fit the control exactly. In the case of four points (Fig 7.11(b)) then the model will only fit the control satisfactorily if there is no $\Delta\omega$ warp present. Any $\Delta\varphi$ curvature will, however, remain undetected unless the fifth point is introduced, as in Fig 7.11(c). Of all the elements of relative orientation the φ rotation requires most care since, as we shall see, the Y-parallax test is the least sensitive in the case of this movement.

One further point on the fitting of models to control is the matter of Earth curvature. The datum surface of the ground is, near enough, spherical in shape. On the other hand, the datum surface of the instrument is a plane surface. There must therefore be some discrepancy introduced by this factor, as indicated by Fig 7.20. In the diagram we have taken a model size of 90×180 mm, in which case the semidiagonal length (d) is just over 100 mm long. Now the error due to Earth curvature is given by

$$E_H = \frac{D^2}{2R} \tag{7.20}$$

where D is the distance on the ground and R is the radius of the Earth (taken here at 6360 km), hence

$$D = \frac{d \cdot H}{f}$$

and

$$E_H = \frac{H^2 d^2}{2Rf^2} \tag{7.21}$$

Expressed as a fraction of the flying height we have therefore, for a wide angle camera with a focal length of 150 mm

$$\frac{E_H}{H} = \frac{H}{2R}\left(\frac{d}{f}\right)^2$$

$$= \frac{H}{2 \times 6360} \times \left(\frac{100}{150}\right)^2 \times 10^{-3}$$

$$\frac{E_H}{H} = 0.34 \times 10^{-7} H \quad \text{(where } H \text{ is in metres)}$$

We see therefore that at 1000 m the error is very small, being about 34 ppm of the flying height.

Even at 10 000 m, which is something near to the maximum one might expect for topographical purposes, the error is still not very large. Over the extent of a single model the error can often be neglected unless the plotting instrument has a precision in the measurement of heights of better than $0.3^o/_{oo} H$. (For second order plotting instruments a precision of this order of magnitude is often quoted.) It is worth noting that if a focal length associated with a super-wide camera is introduced into the above calculation then the error term is increased nearly threefold in magnitude.

7.9 The use of the Y-parallax equations

These equations will allow us to solve the relative orientation problem by a direct numerical method provided that model co-ordinates can be recorded by the instrument and the projectors are fitted with linear and angular scales. For example, we can carry out a single projector orientation using only the right-hand projector if, at five or more well-distributed model points, we are able to measure the Y-parallax present at those points.

134

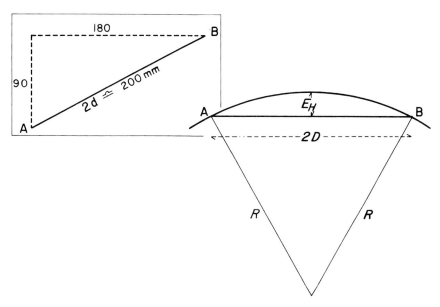

Figure 7.20. The Effect of Earth Curvature

For each point we obtain one observation equation derived from equation (7.17):

$$(P_y)_i = (X_i - B)\Delta\kappa_2 + \frac{(X_i - B)}{Z_i} Y_i\Delta\varphi_1 - \left(1 + \frac{Y_i^2}{Z_i^2}\right)Z_i\Delta\omega_2 + \Delta BY_2 - \left(\frac{Y_i}{Z_i}\right)\Delta BZ_2 \quad (7.22)$$

These equations can readily be solved on a desktop computer to provide values for the adjustments required to complete the relative orientation using the right hand projector only. For a two-projector orientation using only elements of rotation we would obtain a slightly different set of observation equations of the form

$$(P_y)_i = (X_i)\Delta\kappa_1 - (X_i - B)\Delta\kappa_2 + \left(\frac{X_iY_i}{Z_i}\right)\Delta\varphi_1 - \frac{(X_i - B)Y_i}{Z_i}\Delta\varphi_2 -$$
$$\left(1 + \frac{Y_i^2}{Z_i^2}\right)Z_i(\Delta\omega_1 - \Delta\omega_2) \quad (7.23)$$

Because the omega terms do not involve the X co-ordinate, only the difference in omega settings is relevant to the relative orientation process. There are therefore in effect only five unknowns and the equations can be solved with five or more points as before.

If suitable image patterns are to be found at or near the six standard points, most observers would probably choose to carry out an empirical orientation. However, if for some reason some of the points are unusable then a numerical technique is likely to provide the best solution in the least time. This topic is discussed in rather more detail in the last section of this chapter.

7.10 The semi-numerical method for relative orientation

In empirical orientations the setting that gives most trouble is that of omega, for this involves the use of an estimated overcorrection factor. In order to speed up the process a seminumerical technique is often used. In this context it is a most useful exercise to examine the behaviour of the observation equations at the six standard points. We therefore select, as just one example, equation (7.23) concerned with the two projection

135

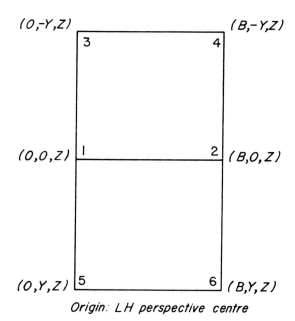

$$(0,-Y,Z) \qquad\qquad (B,-Y,Z)$$

3   4

$$(0,0,Z) \qquad\qquad (B,0,Z)$$

1   2

$$(0,Y,Z) \qquad\qquad (B,Y,Z)$$

5   6

Origin: LH perspective centre

Figure 7.21. The Co-ordinates of the Six Standard Points

method of orientation. Noting the co-ordinates of the standard points (see Fig 7.21) we can write out the Y-parallax occurring at each point due to the various projector misorientations, as set out in Table 7.1.

	$\Delta\kappa_1$	$\Delta\kappa_2$	$\Delta\varphi_1$	$\Delta\varphi_2$	$\Delta\omega = \Delta\omega_1 - \Delta\omega_2$
P_1	0	B	0	0	Z
P_2	B	0	0	0	Z
P_3	0	B	0	$-\dfrac{BY}{Z}$	$-\left(\dfrac{Y^2+Z_2}{Z}\right)$
P_4	B	0	$-\dfrac{BY}{Z}$	0	$-\left(\dfrac{Y^2+Z^2}{Z}\right)$
P_5	0	B	0	$+\dfrac{BY}{Z}$	$-\left(\dfrac{Y^2+Z^2}{Z}\right)$
P_6	B	0	$+\dfrac{BY}{Z}$	0	$-\left(\dfrac{Y^2+Z^2}{Z}\right)$

Table 7.1

It should be noted in passing that if we take a model formed with wide angle photography then, at photo scale, the above coefficients have the following maximum values approximately. If

$$B=92 \text{ mm} \qquad Y_{max}=85 \text{ mm} \qquad Z=152 \text{ mm}$$

then

$$\frac{BY}{Z}=51 \quad \text{and} \quad \left(\frac{Y^2+Z^2}{Z}\right)=199$$

The φ settings are therefore the least sensitive and require the most care. From an inspection of Table 7.1 we can derive the following useful relationships.

(i)
$$P_6 - P_4 = \frac{2BY}{Z} \Delta\varphi_1$$

i.e.
$$\Delta\varphi_1 = \frac{Z}{2BY}(P_6 - P_4)$$

(ii)
$$P_5 - P_3 = \frac{2BY}{Z} \Delta\varphi_2$$

i.e.
$$\Delta\varphi_2 = \frac{Z}{2BY}(P_5 - P_3)$$

(iii)
$$P_5 + P_3 - 2P_1 = P_6 + P_4 - 2P_2$$
$$= \frac{2Y^2}{Z} \Delta\omega$$

i.e.
$$\Delta\omega = \frac{Z}{2Y^2}(P_5 + P_3 - 2P_1)$$
$$= \frac{Z}{2Y^2}(P_6 + P_4 - 2P_2)$$

Combining the two expressions we have the best value for omega given as

$$\Delta\omega = \frac{Z}{4Y^2}[\Sigma P - 3(P_1 + P_2)] \qquad (7.24)$$

where
$$\Sigma P = (P_1 + P_2 + P_3 + P_4 + P_5 + P_6)$$

(iv)
$$P_1 = B \cdot \Delta\kappa_2 + Z\Delta\omega$$

$\therefore$
$$\Delta\kappa_2 = \frac{1}{B}(P_1 - Z\Delta\omega)$$

Likewise
$$\Delta\kappa_1 = \frac{1}{B}(P_2 - Z\Delta\omega)$$

In both these cases, $\Delta\omega$ is determined from the earlier expression.

If we now go back and examine once again the empirical methods of carrying out the orientation process the justification for these methods, and diagrams such as those found in Fig 7.5 become clear. We note particularly that the first two steps using $\Delta\kappa_1$ and $\Delta\kappa_2$ (or $\Delta\kappa$ and ΔBZ) do not in fact correctly base line the model unless the difference in the ω tilts is zero. There is therefore an argument in favour of correcting for this error first, by observing the Y-parallax at all six points, computing the $\Delta\omega$ value and then setting this correction on either of the projectors. If this is done with sufficient accuracy then the first four stages of the empirical routine should then lead to a correct relative orientation. Such a technique would be termed a seminumerical method. If the $\Delta\omega$ setting is calculated at some later stage in the empirical procedure, the result is the same although the calculation is made a little easier. For example, after the first four steps we should have

$$P_1 = P_2 = P_3 = P_4 = 0$$

Hence we are left with

$$\Delta\omega = \frac{Z}{4Y^2}(P_6 + P_5)$$

7.11 The overcorrection factor

We can now appreciate the theory behind the overcorrection process for in this we are simply making allowance for the fact that part of the $\Delta\omega$ displacement was taken up in the first two steps of 'base lining' and was also affected by the $\Delta\varphi$ corrections. The various diagrams in Fig 7.22 illustrate this. In Fig 7.22(a) we show a model free of Y-parallax into which was introduced a $\Delta\omega_2$ rotation of the right-hand projector as indicated by (b). If we accept this situation and now carry out the empirical process of (say) a two-projector

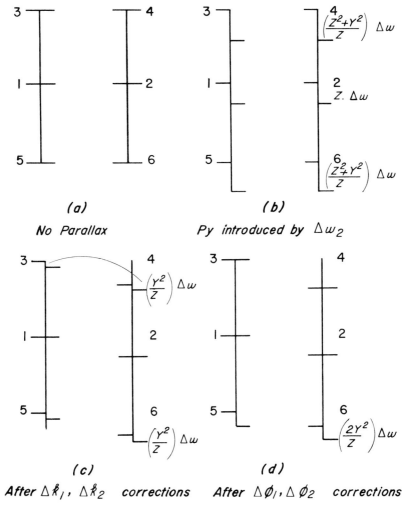

Figure 7.22. The Over-Correction Factor

138

orientation then the results of this are shown in diagrams (c) and (d). In this latter diagram we see that the parallax present at point 5 (and 6) is given by

$$P'_5 = 2\left[\frac{Y^2}{Z}\right]\Delta\omega$$

In fact the correct value should be

$$P_5 = \left[\frac{Y^2 + Z^2}{Z}\right]\Delta\omega$$

If we let m be the overcorrection factor required, then we have

$$mP'_5 = P_5$$

i.e.

$$m = \frac{Y^2 + Z^2}{2Y^2}$$

i.e.

$$m = \tfrac{1}{2}\left[1 + \frac{Z^2}{Y^2}\right] \tag{7.25}$$

When we come to the fifth stage in the orientation process, we record the parallax present at this fifth point, multiply it by the factor m and introduce this as the correction. If we now repeat the first four stages of the procedure, the parallax at the fifth point must now reduce to zero, if the overcorrection introduced was perfectly correct.

As an example, let us take wide angle photography with a 25% lateral overlap and wing points situated nicely in the centre of this. We therefore have, at photo scale, the following,

$$\frac{Y^2 + Z^2}{Z} = \frac{y^2 + f^2}{f}$$

$$= \frac{(85)^2 + (152)^2}{152}$$

$$= 199.5 \text{ mm}$$

$\therefore$

$$m = \tfrac{1}{2}\left[1 + \frac{Z^2}{Y^2}\right]$$

$$= \tfrac{1}{2}\left[1 + \left(\frac{152}{85}\right)^2\right]$$

$$= 2.10$$

In Fig 7.22(d) the parallax observed at points 5 and 6 is therefore

$$P_5 = P_6$$

$$= \left(\frac{2y^2}{f}\right)\Delta\omega$$

$$= 95.06 \times \Delta\omega \text{ mm}$$

For convenience, let the value of $\Delta\omega$ be such that $P_5 = P_6 = 95.0 \ \mu$m. That is to say, $\Delta\omega$ is 0.0001 rad, very nearly.

139

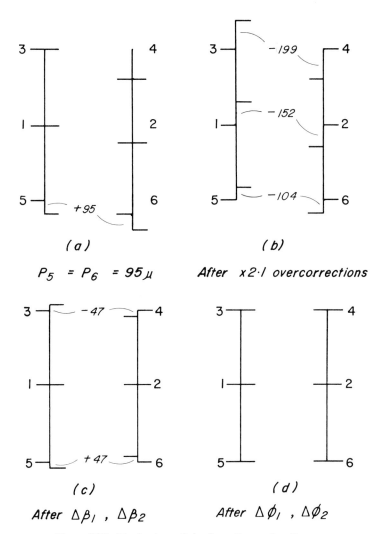

Figure 7.23. Mechanism of the Over-Correction Factor

Figure 7.23 shows the mechanism of the overcorrecting factor at work. In diagram (b) an overcorrection has been applied with the aid of $\Delta\omega_2$ so that

$$P'_5 = P'_6$$
$$= -(2.10 \times 95 - 95)$$
$$= -104.5 \ \mu m$$

and

$$P'_3 = P'_4$$
$$= \left(\frac{y^2 + f^2}{f}\right) \Delta\omega$$
$$= -199.3 \ \mu m$$
$$P'_1 = P'_2$$
$$= f \cdot \Delta\omega$$
$$= -151.8 \ \mu m$$

140

We can now clear the Y-parallax on the baseline using $\Delta\kappa_1$, $\Delta\kappa_2$ with the results shown in part (c) of Fig 7.23, where

$$P_1'' = P_2''$$
$$= 0$$
$$P_3'' = P_4''$$
$$= -199.3 + 151.8$$
$$= -47.5 \ \mu m$$
$$P_5'' = P_6''$$
$$= -104.3 + 151.8$$
$$= +47.5 \ \mu m$$

The use of $\Delta\varphi_1$ and $\Delta\varphi_2$ will now clear the model of the remaining parallax and the situation of Fig 7.22(a) re-established.

7.12 Methods of measuring Y-parallax

With instruments provided with BY base components equipped with accurate linear scales the distribution of Y-parallax values over the model points can be determined directly. Take any point as reference, usually point 1. Clear the model of parallax at that point by using the BY translation of either projector and take note of the scale reading, say BY_1. Now move to the other points in turn and at each clear the model of parallax at that point as before and then note each reading. The Y-parallax at each point is simply the difference between the scale reading at the point and the first reading, i.e.

$$\Delta P_i = BY_i - BY_1$$

In cases where the instrument is not fitted with base components, then either of the omega projector rotations can be used in a similar manner. In some literature, these differences in omega values have been termed lambda-parallax (P_λ). This circular measure can be converted into the more usual linear quantities. From Fig 7.24 we see that

$$P_y = Z \sec^2 \omega \cdot P_\lambda$$

i.e.
$$P_y = \left[\frac{Y^2 + Z^2}{Z} \right] \cdot P_\lambda \tag{7.26}$$

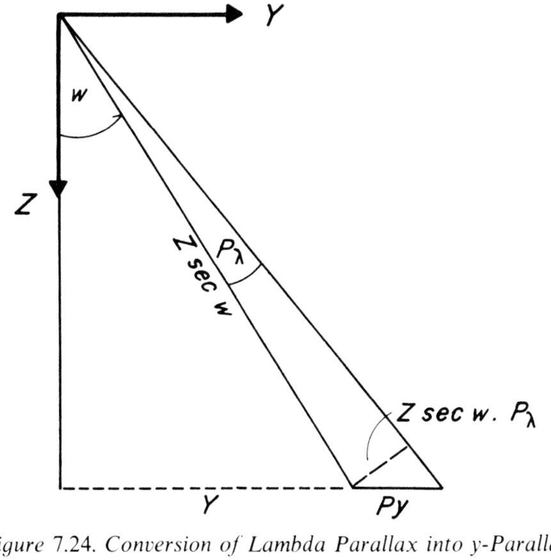

Figure 7.24. Conversion of Lambda Parallax into y-Parallax

On the base line, and using wide angle photography, we note that 1 centigrad of λ-parallax is indicative of approximately 24 μm of Y-parallax, at photo scale. Commonly the working model scale will be twice photo scale (or greater) in which case one unit will signify 12 μm (or less).

7.13 Incomplete models and failure cases

The difficulty that arises when it is not possible to use y-parallax measurements taken at, or in the vicinity of, the six standard points was mentioned earlier in this chapter. This may arise, for example because of cloud patches covering those parts of the photo detail. (It is for this reason that some specifications for photography reject photography with cloud covering the principal points.) Another cause of difficulty can be photo imagery of an unsatisfactory nature. A strong pattern of detail in the X-direction is essential for good P_y perception and similarly sharp, well-defined images in the Y-direction are necessary for satisfactory height determinations. Natural features do not always provide such ideal targets. In the case of water surfaces there is no usable pattern. In the case of sand, snow and similar features the pattern of detail can be non-existent or too poor in texture to be of satisfactory use.

As mentioned above, one possible solution is to use a full numerical relative orientation choosing the points as far apart as possible. If only one or two standard points are not usable an appreciation of the results of the analysis carried out earlier in this chapter can often be put to good use.

As an example, in Fig 7.25(a) one principal point is shown obscured by cloud. In this case it would probably be wise to carry out the $\Delta\omega$ correction first by observing the Y-parallax at the points 2, 4 and 6. The base lining would then be carried out very carefully using the shortened base $1'2$. We would now proceed (in a two-projector orientation) to eliminate the parallax at 3 with $\Delta\varphi_2$ and at 4 with $\Delta\varphi_1$. If all corrections had been applied properly then there would now be no parallax visible at 5 and 6. However, the shortened base line might have proved a little insensitive. In fact, some residual parallax might remain there with the consequence that parallax errors of the same magnitude would be introduced into points 3 and 5, as shown diagrammatically in Fig 7.25(b). On clearing 3 with $\Delta\varphi_2$ as shown, double this amount would therefore appear in 5 but not in 6. We could therefore amend our $\Delta\kappa_2$ setting by eliminating half of this residual. On the other hand, equal errors at 5 and 6 would suggest that the $\Delta\omega$ setting was less than perfect. A better value could be obtained by using parallax observations at $1'$, $2'$, $5'$ and 2, 4, 6.

Finally, when an acceptable model has been achieved, if the cloud cover has resulted in the $\Delta\kappa_2$ setting being somewhat below standard one result of this will be a model with a tilt in the Y direction. This will be corrected for automatically in the absolute orientation process.

Quite often it will be the presence of water that prevents the use of certain points. In such cases the water level can be used sometimes to provide useful height information and the correct model can be produced in effect with the aid of both X-parallax and Y-parallax information.

Figure 7.25(c) illustrates another problem where both base points are missing. In this case we might proceed in the following way. First make the Y-parallax at 3 and 5 equal and opposite in value using $\Delta\kappa_2$. Then carry out the same process at points 4 and 6. Now remove the parallax at these four points using $\Delta\varphi_2$ and $\Delta\varphi_1$. The process is then repeated until all four points are clear of parallax. If we can now identify four points along the edge of the water, as shown in the diagram, we can assume these to be points of equal height. Any misorientation in respect of the $\Delta\omega$ setting will therefore show up as a warp in these points. Either $\Delta\omega_1$ or $\Delta\omega_2$ is therefore adjusted until the water surface in the model becomes flat – it will not, of course, be horizontal until all settings are finally correct.

Figure 7.25(d) illustrates another example of the use of a water surface datum. In this

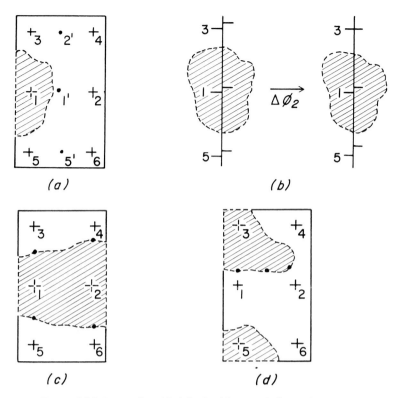

Figure 7.25. Incomplete Models. Problems in Relative Orientation

case the missing points 3 and 5 prevent the usual $\Delta\varphi_2$ adjustment. However, if three points at water level can be identified as shown in the diagram, then any curvature of the datum surface observed in the model after the elimination of parallax at points 1, 2, 4, and 6 should be corrected for using $\Delta\varphi_2$.

In both cases where we have used the water surfaces as a datum level we can use an equation such as (7.19) to provide some idea of the sensitivity of these processes. In the case of Fig 7.25(c) we were making use of the ω coefficient only, in the form

$$e_H = \left(\frac{XY}{B}\right) \cdot \Delta\omega$$

Using a model at photo scale we have therefore

$$X = B$$
$$= 92 \text{ mm}$$

and

$$Y = 85 \text{ mm}$$

Hence the magnitude of any warp present is given by

$$\text{Warp value} = \frac{2 \times 85 \times B}{B} \cdot \Delta\omega$$

$$= 170 \, \Delta\omega$$

This is a good value compared with the size of the coefficients used under normal conditions. In the case of Fig 7.25(d) we are making use of the φ coefficient

$$e_H = \left(\frac{X^2}{B}\right)\Delta\varphi$$

Here
$$X = B/2$$

and
$$e_H = 23\,\Delta\varphi$$

This low value for the coefficient indicates a poor sensitivity for this technique. Hence great care and discretion should be exercised in its use.

It is now perhaps profitable to consider as a theoretical exercise the completion of a relative orientation using X-parallax data only. Using six points distributed as in Fig 7.26 and appreciating the nature of the height deformations illustrated in Fig 7.19, we could proceed along the following lines:

(i)   As the scale of the model is unknown, all the points used must be at the same level in order that various slopes in the model can be compared. At all times remove any Y-parallax using BY.

(ii) Along line AF make the slope correct (i.e. zero) using $\Delta\kappa_1$.

(iii) Along line CD make the slope correct using $\Delta\omega_1$.

(iv) Check that slopes AF and CD are now equal and zero and that slopes AC and FD are also equal.

(v)   Now make slopes AC and FD zero using ΔZ_1.

(vi) Now examine heights at the auxiliary points B and E. Any curvature of the datum surface should be corrected using $\Delta\varphi_1$.

It is not suggested that the above should be considered as a practical method of carrying out a relative orientation. However, in difficult cases some parts of such a procedure might be utilised to complete or check an orientation.

7.14. Failure cases for relative orientation
The analysis of errors in stereophotogrammetry as developed in this chapter follows the generally accepted technique of using a flat earth assumption where necessary in order to

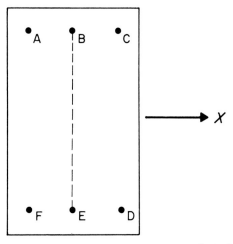

Figure 7.26. Relative Orientation Using Height Values

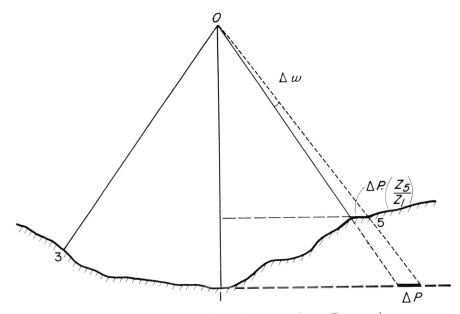

Figure 7.27. Problems in Relative Orientation due to Topography

simplify the development. That is to say the Z co-ordinates over the model surface are assumed to be constant. In the majority of cases the changes in terrain height are small compared with the magnitude of the flying height. The assumption is therefore a reasonable one. However, the combination of a low flying height and mountainous terrain can introduce some difficulty. For example, in the estimation of the overcorrection factor large changes in elevation between the centre line and the wing points can have an influence. Figure 7.27 illustrates the nature of this problem when flying along the axis of a steep-sided valley. In such a case the amount of overcorrection required is less than the theory would suggest. Such effects therefore slow down the relative orientation process, for the procedure will not give such a rapid convergence towards the correct settings unless the effects of changes in elevation are taken into account.

This lack of sensitivity relating changes in Y-parallax value to model point position can be a serious factor when the ground surface conforms to, or is similar to, the critical surface. When the model points fall on such a surface the basis of the method of orientation breaks down. If we take a parallax equation such as (7.16)

$$P_y = (X)\Delta\kappa + \left(\frac{XY}{Z}\right)\Delta\varphi - \left(\frac{Y^2 + Z^2}{Z}\right)\Delta\omega + \Delta BY - \left(\frac{Y}{Z}\right)\Delta BZ$$

for any selected point the bracketed quantities are the fixed coefficients and the projector movements are the unknown variables. When these latter are all zero then the left-hand side is also zero. In moving from point to point during the relative orientation process, the value of the left-hand side is expected to vary. If, however, it does not then the procedure will not work and this indicates that the model points we have been able to select all lie on a critical surface such that P_y is always zero.

Hence by rearranging the parts of the equation we have,

$$0 = Y^2 + Z^2 - \left[\frac{\Delta BZ}{\Delta\omega}\right]Y + \left[\frac{\Delta BY}{\Delta\omega}\right]Z + \left[\frac{\Delta\varphi}{\Delta\omega}\right]XY + \left[\frac{\Delta\kappa}{\Delta\omega}\right]XZ \qquad (7.27)$$

If the projector settings are now regarded as the constant coefficients then this equation describes the critical surface. Fortunately the ground configuration will rarely appro-

145

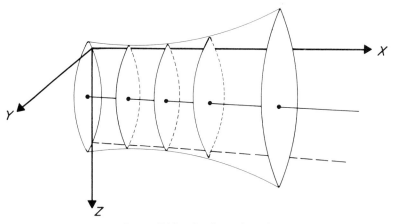

Figure 7.28. The Critical Surface

ximate to this surface but when it does the normal relative orientation procedures will not work. Further, if the model surface is similar in shape to that of the critical surface the convergence of the process is affected. The shape of this critical surface therefore requires some investigation. Consider a series of cross sections in various YZ planes commencing with a first one where $X = 0$. The equation then reduces to that of a circle:

$$0 = Y^2 + Z^2 - \left[\frac{\Delta BZ}{\Delta \omega}\right] Y + \left[\frac{\Delta BY}{\Delta \omega}\right] Z$$

where the radius R is given by

$$R^2 = \left(\frac{\Delta BZ}{2\Delta \omega}\right)^2 + \left(\frac{\Delta BY}{2\Delta \omega}\right)^2$$

and the co-ordinates of the centre are

$$(Y_0; Z_0) = \left(-\frac{\Delta BZ}{2\Delta \omega}; \frac{\Delta BY}{2\Delta \omega}\right)$$

The origin of the co-ordinates, i.e. the camera station, lies on the circumference of this circle. When the projector settings are such that this circle passes through the points on the model surface we can appreciate why the relative orientation procedure fails to work.

If we take any other YZ plane such that $X = X_i$ the equation still reduces to a circle with the base line on its circumference but the radius of the circle and the co-ordinates change in value:

$$R^2 = \left(\frac{-\Delta BZ + X_i \Delta \varphi}{2\Delta \omega}\right)^2 + \left(\frac{\Delta BY + X_i \Delta \kappa}{2\Delta \omega}\right)^2$$

$$(Y_i; Z_i) = \left(\frac{-\Delta BZ + X_i \Delta \varphi}{2\Delta \omega}; \frac{\Delta BY + X_i \Delta \kappa}{Z\Delta \omega}\right)$$

The locus of the circle centre is therefore a straight line.

If we now take a section in the XZ plane with $Y = O$ we find

$$O = Z^2 + \left(\frac{\Delta BY}{\Delta \omega}\right) Z + \left(\frac{\Delta \kappa}{\Delta \omega}\right) XZ \tag{7.28}$$

146

We therefore have the two straight lines

$$Z = O \text{ (i.e. the base line)}$$

and

$$O = Z + \left(\frac{\Delta\kappa}{\Delta\omega}\right) X + \left(\frac{\Delta BY}{\Delta\omega}\right)$$

These lines are generators of the surface which is hyperboloid in character. From the above we can sketch the general shape of the surface as indicated in Fig 7.28.

In the above we have used the one projector parallax equation. A similar result will follow if the equation for a two-projector orientation is used. The coefficients will involve rotation elements only in this case.

7.15 The effects of errors in interior orientation
When using near-vertical aerial photography, the small setting errors that must arise when carrying out the interior orientation process must give rise to imperfections in the bundles of rays generated. These in turn will produce small deformations in the model produced. There are two possible sources of error. The diapositives may be incorrectly placed in the plate carrier so that the principal point of the photograph does not fall on the axis of the projector. The principal distance setting of the projector may not be correct. (In affine methods, of course, this latter is the intended state of affairs.) The effects of any lens distortion can be regarded as a situation where the principal distance is incorrectly set for some particular zone of the lens.

7.15.1 THE EFFECTS OF DIAPOSITIVE MISCENTRING
In this analysis we will assume that the photography is vertical and that the small additional errors introduced by the presence of small tilts can be ignored.

Figure 7.29(a) shows the effect of displacing the left-hand diapositive a small positive distance (δx) from its correct position in the projector. Any such displacement shifts the projected rays and produces the effects shown in the diagram. The displacements ΔX shown in the datum plane of the figure are of equal value because these depend on the Z co-ordinate value:

$$\Delta X_i = \frac{Z_i}{f} \cdot \delta x$$

In the higher plane containing the points A, B, the value of the displacement is less, being $2/3\Delta X$ for the circumstances depicted in the diagram. The ray displacements produce the image displacements that in turn produce the distorted surface A′D′C′B′. A close inspection of this will show that the surface now has a slight upward tilt in the X direction and also that the line D′C′ is at a slightly smaller scale than the line A′B′. Also, the line A′D′ is at a slightly smaller scale than the line B′C′. If we were to displace the left-hand projector by an amount $-\Delta X$ then the overall scale of the model would be increased and the points D′ and C′ would then coincide with D and C respectively. If we displaced the projector by two-thirds of the amount, then the points A′ and B′ would coincide with A and B. In part (b) of Fig 7.29 the displacement has been selected to give the best mean fit, although the residual errors indicated are much enlarged for reasons of clarity. In practice, with a less drastic change in ground elevation a satisfactory model could be produced. The effects of a small diapositive shift δy can be deduced in a similar manner to the above.

An equal and like displacement of both diapositives will result in the introduction of no x- or y-parallax into the model but the model will show the effect of a shear as indicated in Fig 7.30. The lines A′D′ and B′C′ are no longer vertical.

147

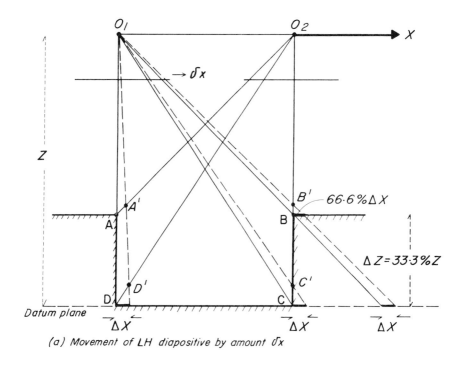

(a) Movement of LH diapositive by amount δx

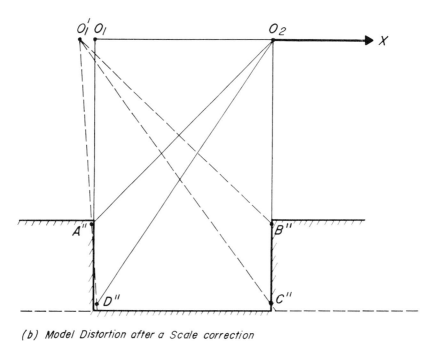

(b) Model Distortion after a Scale correction

Figure 7.29. Effects of Errors in Interior Orientation: Miscentring

148

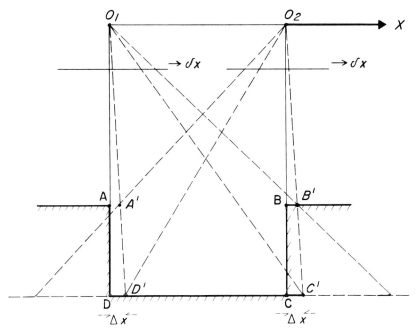

Figure 7.30. Effects of Errors in Interior Orientation: Miscentring of Two Projectors

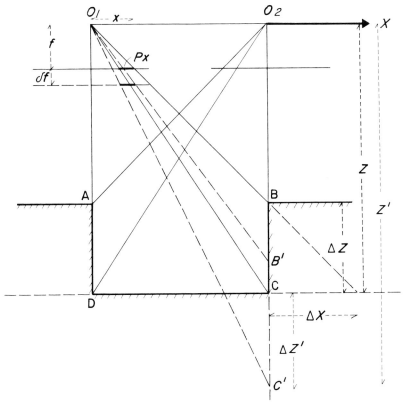

Figure 7.31. Effects of Errors in Interior Orientation: Principal Distance Setting

7.15.2 THE EFFECTS OF ERRORS IN THE PRINCIPAL DISTANCE

Model deformations arising from errors in the settings of the principal distances are most easily appreciated if we consider vertical photographs taken from the same elevation. In Fig 7.31 the effect of an error in only the left-hand projector is illustrated; the value being ($f + \delta f$) instead of the correct value f. This will introduce both X- and Y-parallax into the model at all points in proportion to the (X, Z) and (Y, Z) co-ordinates respectively. If the ground surface is horizontal then the pattern of displacements in the X-direction introduces an erroneous slope on the model surface in that direction. The magnitude of this slope is proportional to the value of Z.

If both projectors are given the same error in principal distance then the model will be free of any parallax but the combination of the effects from the two projectors will cause a vertical displacement of the model and a change in the vertical scale of the model, the horizontal scale being back to the original value. Figure 7.32 shows the nature of the effects.

Making use of the dimensions shown in Figs 7.31 and 7.32 we see the following. From Fig 7.31 we have

$$\frac{\Delta X}{\Delta Z} = \frac{x}{f} \quad \text{and} \quad \frac{\Delta X}{P_x} = \frac{Z}{f}$$

hence

$$\Delta Z = \frac{Z P_x}{x}$$

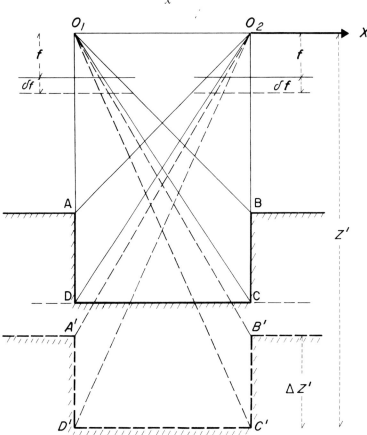

Figure 7.32. Effects of Errors in Interior Orientation: Principal Distance Settings

Similarly, for the dimensions shown in Fig 7.32 we have

$$\Delta Z' = \frac{Z'P_x}{x}$$

Hence the change in vertical scale is given by

$$\frac{\Delta Z'}{\Delta Z} = \frac{Z'}{Z}$$

$$= \frac{f + \delta f}{f}$$

The change in vertical scale is therefore proportional to the error in principal distance setting. In practice, such an error would arise if for some reason or other the correct focal length of the camera was not used.

8: Analogue plotting instruments

8.1 Introduction

In this chapter we will examine a representative range of photogrammetric instruments based on analogue solutions. As mentioned earlier, the vast majority of instruments are of this type, although analytical plotters are now becoming available. In order to be able to select the correct instrument for a particular mapping requirement an overall appreciation of the characteristics and capabilities of analogue instruments is most essential. A knowledge of these instruments is also necessary to any study of analogue aerial triangulation because the main design features of an instrument will determine how, if at all, it can be used for this purpose.

In order to make an effective study of the range of equipment available some systematic method of approach is necessary. And in order to do this we will select a form of classification based on the methods used to represent the camera-ground point vector. Each ground point together with its photo image and the camera node defines a vector in ground space. All photogrammetric plotters using a rigorous solution attempt to reproduce such vectors correctly positioned in the model space with respect to all other vectors of the bundle. This reproduction can take one of three forms which are as follows.

(a) A ray of light along the whole of the path length (e.g. Multiplex type instruments, Zeiss Stereoplanigraph C8, Kern PG1).

(b) A metal space rod (e.g. the Wild range of instruments. Kern PG2, Zeiss Planitop).

(c) Part ray of light and part space rod (e.g. Thompson-Watts Mark 2, Poivilliers Stereotopograph B, Nistri Photostereograph β).

Loosely speaking, the instruments of group (a) are often referred to as optical instruments, those of group (b) as mechanical instruments and those of the third group as optical-mechanical machines. Most instruments fall into one of these three categories. However, there is one further group of plotters that will be mentioned in this chapter which does not readily fall into the above form of classification and these are the approximate solution plotters. In these, no attempt is made to reproduce accurately the bundles of vectors. The models produced are therefore distorted due to the fact that any photo tilts and air base inclination have been ignored initially. Corrections, in accordance with the theoretical considerations outlined in earlier chapters, are then applied to the measurements taken from such models.

There is also one further category of instruments that will not be discussed in this chapter. This is the group of instruments known as affine plotters. Here again a distorted model is produced. In this case, however, this is due to the fact that in the interior orientation process, the principal distance is not made equal to the focal length of the camera. The consequences of this situation were described in the final section of the previous chapter. As these instruments are quite complex and also somewhat rare they will not be the subject of any further discussion in this chapter.

8.2 Instruments using optical projection

In this category there are two main types of instrument; those in which the operator views the stereoscopic image using transmitted light and those in which he sees the image by the reflection of light from an opaque screen or platen. The latter are by far the most simple to appreciate and so this type of instrument will be studied first.

In instruments of this latter kind each projector has a light source that illuminates the diapositive centred in the instrument according to the requirements for inner orientation. Each projector lens projects a magnified image onto the opaque platen which is then viewed directly by the operator. The platen is usually a small round table that can be moved within the model space, usually by hand. It has at its centre a small illuminated mark and vertically below this is the pencil point that records the table position. In use, the operator views the two projected images and the illuminated mark simultaneously. In order to experience a stereoscopic effect it must be arranged so that the left eye sees only the projected image from the left-hand projector, while the right eye sees only the image produced by the right-hand projector. (If the arrangement is crossed over then a pseudoscopic model would result.) When the operator has carried out a correct relative orientation of the projectors and has set the table to the correct height then a single illuminated mark will appear to rest on the surface of the model. In this situation there is therefore no Y-parallax or X-parallax present in the model at that point. The planimetric position of the model point can be plotted by the pencil while the height of the platen above the instrument datum surface can be measured with a screw-thread micrometer. The general appearance of such an instrument is shown in Fig 8.1. The two projectors are

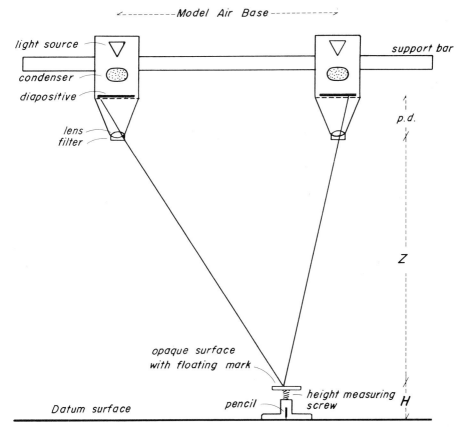

Figure 8.1. *Diagram of Instrument Using Optical Projection*

153

held on a support bar and often have all six projector movements (not shown in the diagram). These movements can be quite large since there are only the usual constraints of mechanical construction to limit them. They are not restricted in any way by any connections or linkages with the viewing system. Their simplicity and the ability to accept a wide range of photo tilts and base inclinations are therefore two important virtues of such instruments. On the other hand, a serious disadvantage lies in the rather poor quality of images produced. The image seen by reflection from an opaque surface cannot be of high quality and added to this are the difficulties of maintaining sharp focus and adequate illumination. The difficulty of focusing arises from the fact that the lens to diapositive distance must be set accurately at some fixed distance according to the requirements of interior orientation. Therefore the laws of geometric optics dictate that a sharp image will be produced in one plane only, the position of which depends on the focal length of the projector lens. In these instruments, therefore, the focal length of the lens is chosen to give the sharpest definition at the mean working distance. In order to provide a workable depth of focus the aperture of the lens must be stopped down to a low value. This greatly restricts the amount of light passing through the lens and this can only be compensated for by using an intense and efficient form of diapositive illumination. The geometric relationships are shown in Fig 8.2.

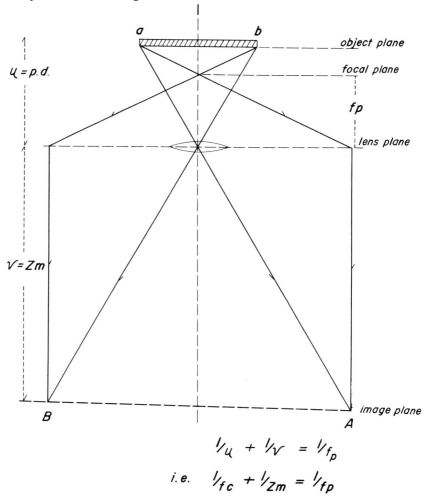

$$\frac{1}{u} + \frac{1}{v} = \frac{1}{f_p}$$

i.e. $\frac{1}{f_c} + \frac{1}{Zm} = \frac{1}{f_p}$

Figure 8.2. *Interior Orientation. Geometric Considerations for Sharp Focusing*

154

As examples, if we take the case of a camera with a focal length of 152.40 mm then in the case of Multiplex and Kelsh instruments we might have the following results shown in Table 8.1, where all dimensions are given in millimetres.

	Multiplex (mm)	Kelsh (mm)
Format size	230/5	230
u	$\dfrac{152.40}{5}$, i.e. 30.48	152.40
Z_m	360	760
f_p	28.10	126.94

Table 8.1

It will be noticed from the table that the light emerging from these projectors is not parallel but convergent. In this respect, therefore, these projectors differ from the camera where the light entering the camera lens was parallel.

There are a number of methods by which the two projected images can be viewed separately by the two eyes. These are as follows:

(a) The two images can be separated by using a difference in wavelength; by using a different colour for the light projected by each projector. This is achieved by fitting each projector with an optical filter that only allows light of a restricted waveband to be transmitted. The operator wears spectacles fitted with two matching filters. The two wavebands used most frequently are one in the blue and one at the red part of the visible spectrum. This technique is commonly referred to as the anaglyph method and its use cuts down dramatically the amount of light being allowed through the projectors. The process will only result in a complete separation if the two filter bands do not overlap at all. In practice they do a little and so faint ghost images are sometimes seen. Figure 8.3 illustrates this and also provides some indication of the amount of light absorbed in the process.

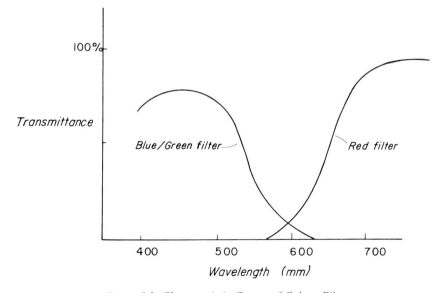

Figure 8.3. Characteristic Curves of Colour Filters

155

(b)   The two images can be separated in time by using mechanical or electromechanical means. In its simplest form, each projector is fitted with a revolving shutter that allows light to pass for just under one-half of each revolution. The two shutters rotate at synchronous speed and are set 180° out-of-phase. By this means light from only one of the projectors is illuminating the platen at any one time. A similar pair of shutters operating in unison with those of the projectors are fitted to the pair of spectacles used by the operator. In this way, the left eye views only the projected image from the left-hand projector and the right sees only the image produced by the right-hand projector.

A device of this nature has been fitted to some types of Williamson Multiplex equipment. In this case just under 50% of the available light is utilised. A similar method is used in the Kern P-1 instrument. Here each light source is a high intensity mercury vapour discharge lamp. Cold cathode lamps of this nature produce light of alternating intensity, in phase with the voltage applied to them. In this case, therefore, a 100 Hz voltage is supplied to each lamp 180° out-of-phase with each other. By this means, only one lamp is operating at any one time during the cycle of $\frac{1}{100}$th second. The operator views the platen through eyepieces fitted with mechanical shutters rotating at 100 revolutions per second and in phase with the appropriate light sources. One advantage of this instrument is its ability to switch instantly from stereo to pseudo-stereo viewing by an electrical switch. If the floating mark is not quite set onto the model surface, operating the switch will cause the mark to jump vertically. Better height settings are therefore possible with this device; the best reading being that at which no vertical movement can be detected.

(c)   The two images can be separated spatially. This requires two platens separated by a constant distance and means that the two rays of a pair do not in fact intersect. The two images are viewed through a pair of low-powered binoculars with the left eye viewing the left-hand platen and the right eye viewing the right platen. With this method all the light passing through the lenses is used. The two platens and the binoculars in moving through the model space must be held so that the two floating marks and the eye base of the binoculars are always parallel to the X-axis of the instrument. The Zeiss (Jena) Topoflex uses this technique.

Adequate illumination is a problem with most instruments of this category and, with the exception of the PG1 and the Topoflex, all should be used under dark-room conditions. Various attempts at obtaining a more satisfactory light intensity can be seen within the range of instruments. Among the most commonly used are the following:

(i)   Using reduced diapositives the amount of light required to illuminate the whole of the format area is much reduced. Even so, air cooling of the lamp housing is often required. The Williamson Multiplex equipment is of this type. In the Bausch and Lomb Balplex instruments, diapositives of 110×110 mm are illuminated with the aid of ellipsoidal reflectors.

(ii)   Using guide rods attached to the plotting table, the light is directed only onto the working parts of the diapositives. In this manner full-sized diapositives can be used.

The Kelsh instrument and the Nistri Photocartograph VI are instruments of this type.

(iii) The use of cold cathode mercury vapour lamps is a more expensive light source but a higher intensity of illumination can be achieved without the generation of heat. The PG1 uses these together with selective illumination.

8.2.1 METHODS FOR INTERIOR ORIENTATION
One of the disadvantages of instruments using optical projection is their limited accommodation with respect to the focal length of the taking camera. The diapositive-lens principal distance cannot be altered very much without changing the position of the plane

156

of best focus by a significant amount. In some instruments such as the Kelsh plotter and the Nistri photocartograph VI, for example, the principal distance can, however, be changed within the range 152 ± 3 mm. In the case of instruments using diapositives of reduced size, small changes in camera focal length can be accommodated in the reduction printer. If C is the fixed principal distance of the projector then the reduction required by the printer is given by

$$\text{Reduction factor} = \frac{C}{f_c}$$

$$= \frac{\text{principal distance of projector}}{\text{focal length of camera}}$$

The application of this reduction factor will of course produce a diapositive of the correct nominal size only if the camera negative is of standard size and the reduction factor is that associated with the nominal focal length. To complete the interior orientation process by placing the principal point of the diapositive on the optical axis of the projector, reference marks are either engraved on the stage plate at the principal point position or in the four-corner positions. In some optical instruments (for example the Balplex) the diapositive holders can be removed from the projectors and are interchangeable. In such a case a setting box is used to centre the diapositive in its holder.

8.2.2 METHODS FOR RELATIVE ORIENTATION

The floating mark used is usually an illuminated mark of small diameter. Sometimes this is interchangeable with an opaque black cross that is preferred by some operators when carrying out relative orientation. Because of their simpler construction, most optical instruments have projectors with ample linear movements in all three directions (BX, BY and BZ) and large tilt settings of at least 10 grad for the three rotations (κ, φ, ω). The linear movements are most often arranged so that BX is primary, with BY and BZ being secondary and tertiary respectively. For the rotations, κ is almost invariably a tertiary movement while, most commonly, φ and ω are arranged to be primary and secondary respectively. The significance of these two statements is illustrated in Fig 8.4. An unusual and most significant variation from the norm can be found in the Nistri Photomultiplex

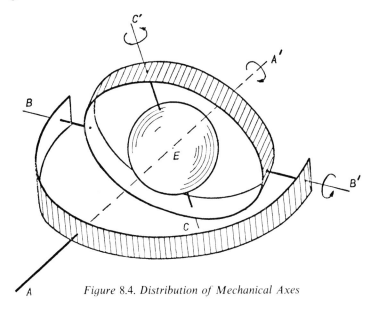

Figure 8.4. Distribution of Mechanical Axes

157

DIII instrument. In this case the κ rotations of the projectors are the primary ones and this feature facilitates one particular method of carrying out aerial triangulation as described in chapter 9.

Because of their full complement of projector movements, both single and double projector relative orientations can be used. In some instruments, the φ tilts possible are such that convergent photography can be used. In others, the ω tilt settings can be made sufficiently large for oblique photographs to be accommodated.

8.2.3 METHODS FOR ABSOLUTE ORIENTATION

In many of these instruments the projector guide-rails can be tilted in both the common omega (Ω) and common phi (Φ) directions with respect to the instrument datum plane. The range of settings possible is of the order of 5–10 grad. Unusually, in the case of the Williamson three-projector equipment, the projector support is maintained in a horizontal position and datum surface is tilted with respect to it using four levelling screws.

To scale the model, the separation of the projectors is adjusted by using the BX movement. If the two projectors have different BY and BZ settings on them then the scaling will involve small adjustments of these components also in order to maintain the relative orientation. The minimum scale possible is limited by the minimum possible separation of the projectors. The maximum model size is limited by the maximum Z range of the plotter, but working well away from the optimum Z value will mean some deterioration in image quality. Usually we would wish to work with the maximum model size possible but when choosing a model scale it is prudent to check that the maximum Z value present over the model surface can be accommodated.

8.2.4 CORRECTIONS FOR CAMERA LENS DISTORTION

If the lenses of the projectors have the same distortion characteristics as the camera lens then no model deformation will arise from this source. If the lenses of the projector are distortion free but the distortion curve of the reduction printer lens matches that of the camera then once again compensation will have been achieved. Such methods are based on a concept first put forward by Ignazio Porro but now generally known as the Porro–Koppe principle. In practical terms the method is not as attractive as might appear at first sight for although lenses might be constructed to provide reasonable compensation for one particular camera lens (or group of lenses) different characteristics will be required for different types of camera lens. The application of this principle is probably of the most value when the lens distortions present are large. With modern camera lenses the distortion values have been reduced to very low values. So much so, that the term 'distortion free' is sometimes used. In this situation the Porro–Koppe concept has lost much of its practical significance.

In chapter 1 we demonstrated that radial lens distortion could be described in terms of small displacements of the inner nodal point of the lens, that is to say in small changes in the focal length of the lens. In some instruments we see therefore a method of compensation carried out by introducing the required changes in principal distance with the aid of eccentric cams. This method is used in the Kelsh plotter but was devised and first used by Santoni in some of his instruments. In the Kelsh instrument the guide rods linking the plotting table to the swinging light sources are used to rotate eccentric cams that in turn convey to the projector lenses the appropriate changes in principal distance (see Fig 8.5).

8.2.5 CAPABILITIES AND USES

The precision of the instruments of this category is limited to some extent by problems of inadequate illumination and image quality. If reduced diapositives are used then this must also reduce the precision of measurement possible.

158

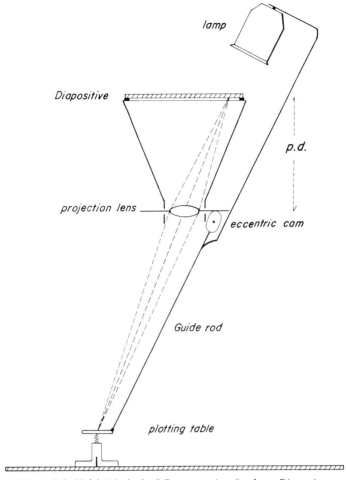

lamp

Diapositive

p.d.

projection lens

eccentric cam

Guide rod

plotting table

Figure 8.5. Kelsh Method of Compensation for Lens Distortion

The size of the floating mark used in an instrument is some indication of the capabilities of the instrument. We find that in the case of direct optical projection instruments the diameter of the mark is about twice that used in a second order plotting instrument and about four times that used in a stereocomparator. In optical plotters the floating mark has an apparent diameter of the order of 80 μm at photo scale. The instruments are therefore best suited to the direct plotting of detail and contours. Some instruments, particularly those using full-sized diapositives, sometimes employ a polar pantograph to give a variable magnification between model scale and plotting scale.

Topographical mapping plotting at scales of $\frac{1}{10\,000}$ and smaller are appropriate to this category of instruments. They also find applications in the plotting of exploratory and inventory surveys. Because of their lack of complexity they are also much used for training purposes.

Some instruments of the category have a unique feature that cannot be found in more complex and expensive instruments. This is the facility of being able to set up more than two projectors at a time. With equipment using reduced diapositives and small units as many as eight projectors can be set up along the bar. Using such equipment it is possible to produce a continuous strip model. Such a facility can be valuable, for example, in road alignment and route surveys. The facility also lends itself to a particular method of carrying out aerial triangulation described in the next chapter.

In general, when assessing the performance of any plotting instrument there are two main factors that we must consider. Firstly, there is the planimetry and here there are often three scale factors to consider — photograph, model and plotting table scales. For economic reasons the enlargement between photo scale and the required plot scale should be as large as possible with the photo scale itself as small as possible. In order to obtain the maximum information from any given set of photographs, the quality of the image seen by the operator should be as high as possible. This calls for an instrument with a very good optical performance if the maximum possible resolution of detail is to be achieved. The direct optical projection instruments are therefore at a disadvantage from this point of view.

If the model point position is not recorded directly, then any mechanical or electromechanical linkage between the two points must be of high precision. In this respect, for example, pantographs of any type are not particularly satisfactory. It is sometimes stated that a plotting instrument should be capable of producing a satisfactory output at $\times 5$ photo scale or more. The larger optical plotters (such as the Kelsh and Topoflex for example) have the capability of producing a satisfactory graphical output at this order of magnification; the plotting accuracy being of the order of ± 0.2 mm. On the other hand, first order plotting instruments of other types are capable of recording model co-ordinates with an accuracy of the order of $\pm 15\ \mu$m. This suggests that a satisfactory graphical output can be obtained with an overall magnification of about $\times 12$ using such equipment.

The second factor we need to consider is the ability of the instrument to produce accurate spot height values. This is most commonly formulated as the mean height error produced by the instrument expressed in parts per thousand of the flying height (i.e. $^o/_{oo}\ Z$). For a direct optical plotter a precision of $0.25\ ^o/_{oo}\ Z$ would be considered good while a first order instrument would achieve a value of something in the order of $0.05\ ^o/_{oo}\ Z$. (We note, however, that such results are only obtained in practice under the most favourable conditions.) This factor will of course dictate the minimum contour·interval that can be confidently produced by the instrument. If we accept the criterion that no contours should be in error by more than 50% of the contour interval, then the spot height errors must be no greater than a quarter of the contour interval at most. In cases of large-scale mapping where a contour accuracy in excess of the topographical criterion might well be expected (as in construction work for example), it would then be prudent to arrange matters so that the contour interval is five to ten times greater than the spot height precision of the instrument.

In practical situations the vertical co-ordinate often has a significance greater than the planimetry and this is reflected in mapping specifications. The contouring capability of an instrument is therefore, more often than not, a deciding factor when considering the suitability of an instrument for carrying out a particular mapping operation with a given set of photographs. Or, looking at it the other way round, this is the factor that dictates the maximum flying height that can be specified using a particular type of plotting instrument. This is particularly true at the larger topographical scales and· when mapping for engineering requirements. An attempt to convey the overall capability of an instrument is found in the use of the C-factor, which is defined as follows:

$$C = \frac{\text{maximum flying height}}{\text{minimum contour interval}}$$

With good photography, an instrument in perfect adjustment and a competent operator a certain precision in spot height determinations can be found by experiment. From this the minimum contour interval that can be satisfactorily produced can be deduced. The meaning of the word 'satisfactory', however, is somewhat indefinite and various mapping agencies and different mapping requirements suggest different definitions of the word. The factor is therefore best used for purposes of comparison. As an example of the use of the C-factor, the manual of the Kelsh plotter states that under average conditions of commercial mapping, competent operators should be able to maintain a factor of about 1200 with the instrument and should rarely fall below 950 with any usable photography. If we therefore take a figure of 1000 and assume that the spot height precision (e_Z) will be not less than one-fifth of the contour interval (I) then we have

$$1000 = \frac{Z_{max}}{I}$$

$$= \frac{Z_m}{5e_Z}$$

Hence $$e_Z = \pm 0.2 \, ^o/_{oo} \, Z$$

On this scale of values, a third-order plotter using optical projection and reduced size diapositives might have a C factor of about 600. A large first-order plotting instrument would have a value of the order of 2000.

At the smaller topographical scales (1 : 50 000 and less) the circumstances change for in such work the minimum photo scale that can be usefully employed depends on the ability of the instrument to resolve the photographic detail and so enable the operator to interpret this correctly. For normal mapping purposes a photo scale of about 1/80 000 is often considered to be about the minimum by many mapping agencies.

Apart from training, one other task for which the direct optical projection instruments seem to be particularly useful is the activity of map revision. This is especially true when the revision areas are only parts of models and the use of more expensive equipment might well be considered uneconomic. For such purposes, it is an advantage if the Z motion is assigned to the projectors and not the platen. In this way the map sheet undergoing revision can be laid flat on the datum table surface and the projectors raised or lowered in unison to register changes in model surface elevation. The Zeiss (Oberkochen) DPI and the Gamble Stereoplotter are examples of such instruments. For the purposes of revision, the use of a single projector should not be overlooked. In such a situation the optical projector is being used as an 'automatic' rectifier. However, the image is only automatically in focus if the tilt is small and the image plane falls within the zone of sharp focus produced by the use of a small lens aperture. The Balplex wide angle projectors have an unusual facility in this connection, for in these instruments the Scheimpflug condition can be maintained by tilting the projector lens about an axis through its outer node. This facility is made use of when using highly convergent or oblique photography.

8.3 Optical instruments using auxiliary telescopes

In order to improve the image quality obtained by optical projection, the use of auxiliary telescopes or goniometers is essential. They do this in two ways. Firstly, more of the light passing through the exit pupil of the projector lens is collected and eventually reaches the operator's eye. As the diameter of the telescope objective can be quite large there are no serious problems of illumination, low wattage bulbs being quite adequate for the purpose. Secondly, the image seen by the operator is one formed by transmitted light. There are no images formed by reflection from opaque screens. The images can therefore be magnified much more and to a degree dependent only on the quality of the diapositives themselves. The use of binocular viewing eliminates the need to use the instrument in a darkened room and also, of course, automatically provides the separation of the two photo images necessary for stereoscopic viewing. At some stage in the optical trains (and as close as possible to the diapositive plane) floating marks are superimposed on the two photo images. These marks are now much smaller and an apparent diameter of the order of 40 μm at photo scale is quite common.

The introduction of these additional optics does, however, mean that the projectors and the viewing system are now interconnected. For scanning the diapositives, the telescopes must be equipped with a system of mechanical axes. The instruments are therefore much more complicated in design, and in consequence some versatility is inevitably lost. In order to ensure an adequate utilisation of the enhanced image quality, a high standard of workmanship is required in the construction of the scanning and measurement components.

8.3.1 THE ZEISS STEREOPLANIGRAPH

One of the first instruments to successfully overcome the design problems inherent in such instruments was the Zeiss Stereoplanigraph. This instrument was available in the early 1930s as the C4 model and production was only finally discontinued in the mid-1970s with the C8. There are, therefore, many of these instruments in use throughout the world. Because of this and the general significance of many of its design features, it is instructive to take this instrument as an example and comment on its more important characteristics. The instrument is in fact one of the most versatile instruments ever produced; being suitable for the restitution of terrestrial, near-vertical, oblique and convergent photographs. A schematic diagram of the instrument is shown in Fig 8.6.

The optical projectors of the instrument take standard-sized diapositives which are illuminated by low voltage lamps. A number of different lens cones can be fitted to accommodate a variety of focal lengths. Only small adjustments of ± 7.5 mm are possible

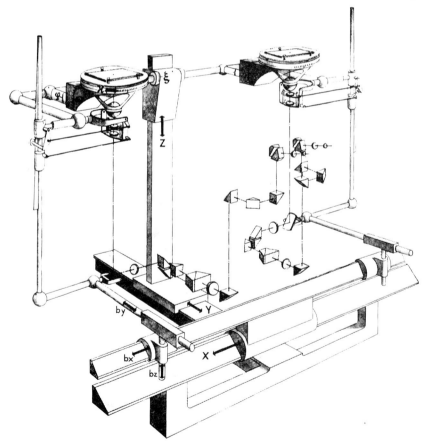

Figure 8.6. Zeiss Stereoplanigraph Universal Instrument

to the nominal focal length of the projector lens fitted. With the setting at the nominal focal length value the emergent light is parallel. (At other settings the light is either very slightly divergent or convergent.) At any particular pointing of the telescope the light is brought into focus by the telescope optics in a plane at right angles to the axis of the telescope that contains the centre of the mirror M. At the centre of each mirror is a small black dot which forms the floating mark. The mirrors are suspended in universal (cardanic) mountings such that the centre of the mirror is also the centre of rotation. This is important, for if on rotation of the mirror the floating mark moved with respect to the projected image then spurious parallaxes would be introduced into the system. The mirrors are constrained to rotate so that when scanning the model space the light from the projector is always directed along the X direction into the lens L_2 of the fixed section of the viewing optics. Figure 8.7 illustrates the process. Rotation of the guide rod R about an axis in the X direction causes a direct rotation of the mirror about its primary axes xx'. Rotation of the guide rods in the XZ plane causes a half-angle rotation of the mirrors about its secondary axes through y. The mirror rotation about xx' also causes a corresponding rotation of the reflected image. This is compensated for by the rotating Dove (or Amici) prism D. A half-angle rotation of this compensates fully for the rotation produced by the mirror.

The Bauersfeld telescope is a pancratic arrangement composed of two lenses of equal power, one being concave and the other convex. The unit revolves about the outer node of the camera lens which also coincides with the second nodal point of the Bauersfeld unit. The focal length of the combination and the positions of the two nodal planes are given by the three formulae set out below. The separation of the two lenses is d and their focal lengths are $\pm f$.

$$\text{Focal length of the combination} = \frac{f_1 f_2}{f_1 + f_2 - d}$$

$$= \frac{f^2}{d}$$

$$= \bar{f}$$

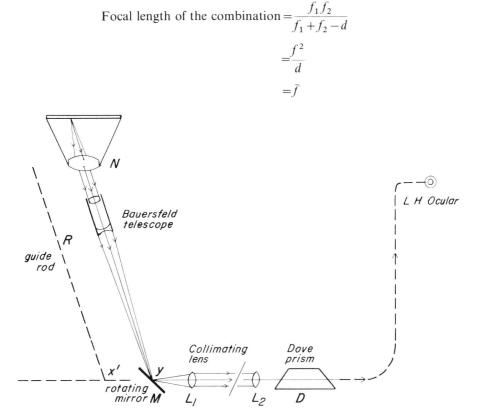

Figure 8.7. Optical Projection Process of the Stereoplanigraph

$$\text{Distance of first node from first lens} = \frac{\bar{f}\, d}{f_2}$$

$$= -f = S_1$$

$$\text{Distance of second node from second lens} = -\frac{f\, d}{f_1}$$

$$= -f = S_2$$

These results are shown in diagrammatic form in Fig 8.8. Now the concave lens is fixed in the unit mounting so that the outer nodal point of the camera coincides with the second node N'. The convex lens slides in the mounting and the distance d is always arranged to be such that a sharp image is formed at the mirror M. This adjustment is made with the aid of a hyperbolic cam and is such that as the length of the guide arm R varies when scanning the model, the convex lens moves the appropriate amount. It is important to ensure that movement of this lens does not introduce any image displacement with respect to the floating mark, otherwise unwanted parallaxes will be introduced. If the light emerging from the projector lens is parallel the point about which the telescope rotates is not critical. If, however, this condition is not exactly fulfilled then the rotation must take place about the outer node if unwanted parallaxes are to be avoided. A note on the geometric optics of compound lenses is to be found in section 1.10.

It will be appreciated that with instruments of this category the operator is viewing each diapositive through a lens similar to the camera lens and tilted to the correct angle. The images seen by the operator are therefore rectified and any changes of size and shape introduced by camera tilts have been removed. Also the operator is viewing in epipolar planes. Hence the conditions of viewing are those most satisfactory for stereoscopic work.

8.3.2 THE ZEISS PARALLELOGRAM
In an instrument such as the Stereoplanigraph the pair of rays defining a model point cannot physically intersect in the model space. Hence, in order to maintain the correct spatial relationships, the concept of the Zeiss parallelogram was developed. Instead of being required to intersect at a point the two rays must now intersect with a line of fixed length that is free to move in the model space but is constrained to remain parallel to the

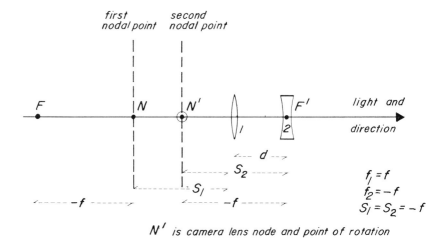

Figure 8.8. Bauersfeld Telescope Arrangement

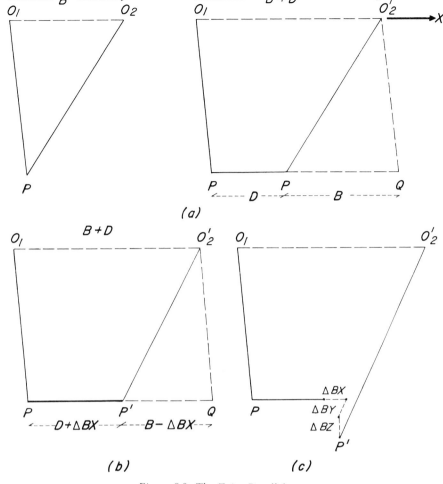

Figure 8.9. The Zeiss Parallelogram

X-axis of the instrument. Figure 8.9(a) illustrates the concept. The model point can be any point on the line PP' and in the figure the points $O_1 O_2' PQ$ form the parallelogram. With such an arrangement the projector separation in the instrument can be a fixed quantity and any projector movement required in the X direction (for changing the scale of the model for example) can be simulated by introducing equal but opposite movements at the other ends of the space rods. This is illustrated in Fig 8.9(b), where in order to reduce the scale of the model a reduction in base length to $(B-\Delta BX)$ was required. This was achieved by increasing the length of the line PP' to $(D+\Delta BX)$. In a similar way ΔBY and ΔBZ movements of the projectors can be simulated by the introduction of equal but opposite base components at the other ends of the space rods as illustrated in Fig 8.9(c), where for the sake of clarity only base components of the right-hand projector have been shown. Figure 8.6 shows how the components are introduced in the stereoplanigraph at both ends of the base by displacing the two mirrors. To scan the model in the X-direction, the base bar has a movement in the X-direction but always remains parallel to the X-axis. The model space is scanned in the Y and Z-directions in a somewhat unusual manner, by using a primary Y and secondary Z movement of the projector unit. All the diagrams so far have shown what is known as the 'base in' position where the projector separation $O_1 O_2$ is greater than the base length PP'. However, in this instrument the 'base out' position can

165

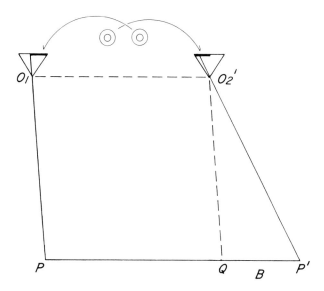

Figure 8.10. The 'Base-Out' Position Using a Zeiss Parallelogram

also be utilised and this situation is illustrated in Fig 8.10 for the stereoplanigraph. In this case the base length is greater than the projector separation, the model is formed by placing the photo overlap on the inside parts of the stage plates and the viewing system is switched so that the left eye views the right-hand photograph and vice versa. (Otherwise the operator would see a pseudoscopic model.)

The incorporation of some form of Zeiss parallelogram in an instrument obviates the need for providing projector translations. The instrument can therefore be rather less complex and more stable. Also as we will see in the next chapter the capability to move from the base-in to the base-out position allows aerial triangulation to be carried out on the machine.

In a single chapter there is not sufficient space to discuss further details of this instrument and its versatility. The reader must therefore be referred to the information provided by the manufacturer and other sources. The *American Manual of Photogrammetry* volume 2[0.1] is a particularly valuable source of information on many analogue plotters.

8.4 Optical instruments using goniometers

A number of instruments of this type have been designed since the method was first used in 1865 in the photogoniometer designed by Porro. Among the instruments of this type successfully produced are the Poivilliers Stereotopograph B (1937), the Nistri Photostereograph β instruments (1952), and the Thompson–Watts plotters (from 1953). Such instruments fall into group (c) of the classification introduced in section 8.1 where the ray paths are defined partly by optical and partly by mechanical means. Figure 8.11 shows in a diagrammatic way the differences in image geometry produced by the three different methods of optical projection and normal viewing.

Part (a) of Fig 8.11 shows the case of the direct optical projection. Here the photo images are projected on to a horizontal surface. After orientation both images are of the same size and shape, for all the effects of camera tilts and air base inclination have been removed everywhere.

Part (b) of Fig 8.11 shows the case of projection using telescopes of variable power. Again the effects of camera tilts and air base inclination have been rectified. However, the

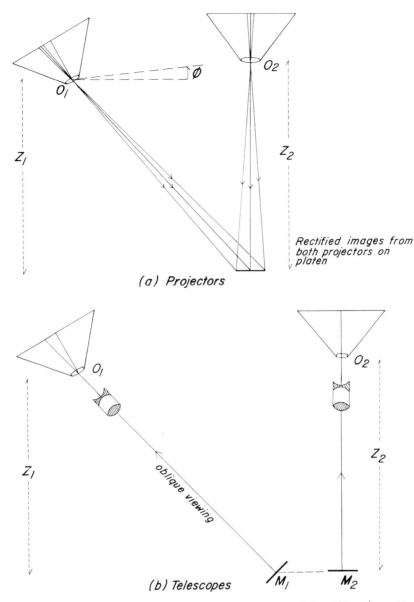

O_1

ϕ

O_2

Z_1

Z_2

Rectified images from
both projectors on
platen

(a) Projectors

O_1

O_2

Z_1

oblique viewing

Z_2

(b) Telescopes M_1 M_2

Figure 8.11. (a), (b). *Four Methods of Viewing the Photographs and their Effects* (cont'd on p. 168)

method by which the observer views the images is somewhat different from Fig 8.11(a). In this case the observer's viewpoints are along the optical axes of the telescopes and so the observer's eyes are normal to these directions. We notice therefore that the left-hand viewpoint is an oblique one while that of the right hand is very similar to that of part (a) of the diagram. In the Y direction (not shown), of course, both telescopes will have the same inclined viewpoint. This viewing situation results in an apparent curvature of the datum surface where the ground appears to fall away from the observer. This is an interesting characteristic of instruments using telescopes or goniometers and one that distinguishes them from mechanical instruments where the photographs are invariably viewed normally.

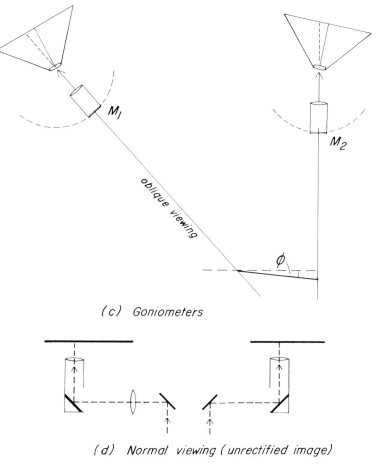

(c) Goniometers

(d) Normal viewing (unrectified image)

Figure 8.11. (c), (d). (cont'd.)

In part (c) of Fig 8.11 we see that the effects of camera tilts are rectified but that inclination of air base is not: In addition, we notice that the two images formed in the focal planes of the goniometer lenses (we assume the emergent light is parallel) must be of varying sizes, according to the projector lens–diapositive distance. At all angles the object distance remains at the constant value of f_g. Although the brain can accommodate changes of image size of up to about 15%, it is clear that some form of compensation is required for comfortable viewing. All modern instruments of this type therefore have some form of pancratic system.

There is one further general point we need to note concerning telescopes and goniometers. These must be provided with axes of rotation in order to scan the diapositives. Or, if they are not, then the projectors themselves must be capable of rotation for the same purpose. In either case, it is necessary to take care in the selection of primary and secondary axes. In the case of goniometers it is necessary for the centre of rotation to coincide with the outer node of the projector lens. In the case of both telescopes and goniometers it is necessary for the X-axis to be primary and the Y-axis to be secondary. For good stereoscopic fusion it is necessary always for lines in the X-direction to be parallel to one another and the air base. As will be seen from an inspection of Fig 8.12 this requires a primary X-axis. Away from the centre, the lines in the Y-direction will not appear to be perpendicular to the X-direction.

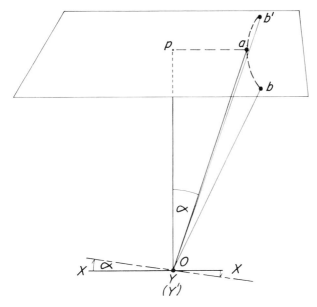

Figure 8.12. The Axes of Rotation of a Goniometer

In Fig 8.12 the X-axis is the primary one. Starting from the vertical position Op a rotation α about the secondary axis YY' will carry the pointing to point a in the photo plane. A rotation about the primary axis XX will cause the line Oa to generate a cone with apex O. The intersection of the surface of this cone produces the hyperbolic line bab'. At any point the direction of the Y line will appear to be tangential to this curve. A rotation about the YY'-axis generates a plane surface which intersects the photo plane to give a straight line parallel to the X-axis. The effect is usually only noticeable when carrying out the calibration of an instrument using grid plates.

8.4.1 THE THOMPSON–WATTS MK2 PLOTTING INSTRUMENT

As an example of an instrument using goniometers and optical-mechanical projection we will examine briefly the main design characteristics of the Thompson–Watts instrument. This machine first became available in the mid-1950s; the mark 2 version appearing about eight years later. The instrument was designed about two basic principles that were first used by H. G. Fourcade in his Stereogoniometer of 1926. In the instrument we find therefore that the model surface is constructed about the air base, this being the primary axis of the instrument. The second feature of the instrument is that the model forming processes (i.e. interior, relative and absolute orientations) is quite distinct and separate from the model measuring and plotting processes. To see how this is achieved we need to examine Fig 8.13, which shows the general arrangement of the machine. The instrument is described fully by Thompson.[8.8] In this chapter we are concerned only with the main features of the design.

The two projectors are held on a beam which represents the air base. The outer nodal point of each projector lens lies on the axis of the beam. The projectors have no translations but individual rotations $(\Delta\varphi_1, \Delta\varphi_2, \Delta\kappa_1, \Delta\kappa_2)$ are provided, the axes of which are at right angles to the beam and are centred on the nodal points. The beam carrying the projectors is split at the central point so that one-half of the beam can rotate axially with respect to the other. In this way a $\Delta\omega$ rotation can be introduced. These rotations constitute the minimum possible for carrying out a relative orientation using both projectors. The principal distance is set using four micrometer screw gauges that hold the

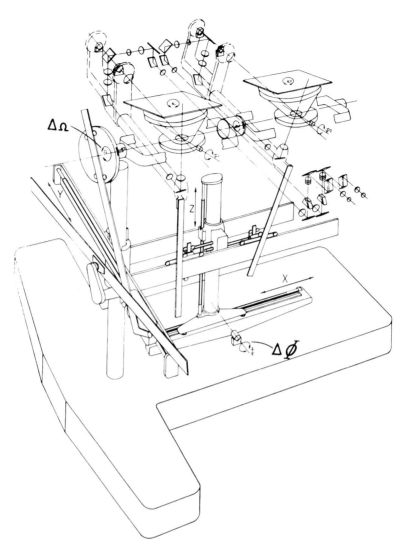

Figure 8.13. *Thompson–Watts Plotter Model* 2

diapositive holders normal to the lens axes and at a nominal distance of 152.40 mm ±3 mm. One of the aims in the design was to keep the number of critical movements to a minimum. In this way, the machine would then be as stable and precise as possible.

The diapositives are viewed with the aid of goniometers that rotate only about Y-axes passing through the outer nodes of the projector lenses. Scanning in the Y-direction is achieved by the unusual method of rotating the projectors about the main beam. In this way, lines in the X-direction remain parallel to the X-axis. Each goniometer brings the emergent parallel (or very nearly parallel) light from the projector into focus in the focal plane of the goniometer objective. Each floating mark is located on the optical axis in this focal plane and so the measuring marks are impressed on the photo images at a very early point in the optical train. This means that after this point, no spurious parallaxes can be introduced into the images by some accidental relative movement between mark and detail.

170

Beyond the goniometers, the directions of the vectors are defined by the machined inner edges of the two lineals (4, 4) as shown in more detail in Fig 8.14. The lineals do not intersect but are separated by two spacing bars (9, 9). The separation is adjusted to introduce a change in model scale, i.e. the instrument has a form of Zeiss parallelogram that introduces ΔBX changes only. (This is a most satisfactory way of changing scale without moving the projectors.)

The images projected into the focal planes of the goniometer lens vary in size according to the angle of inclination. These images are transmitted to the eyepieces but in the process they must be subjected to varying magnifications so that the two images seen by the operator are the same size. Figure 8.15 indicates how this situation arises. When applying rotations to the goniometer the measuring mark can be regarded as a point moving on the surface of a circle of radius R. In the focal plane of the goniometer, the degree of magnification at any position is given by

$$M_i = \frac{R}{Oa_i}$$

The increase in magnification required must therefore be equal or proportional to the following factor:

$$\text{Increased magnification required} = \frac{OA_i}{R}$$

$$= \frac{Z \sec \theta_i}{R}$$

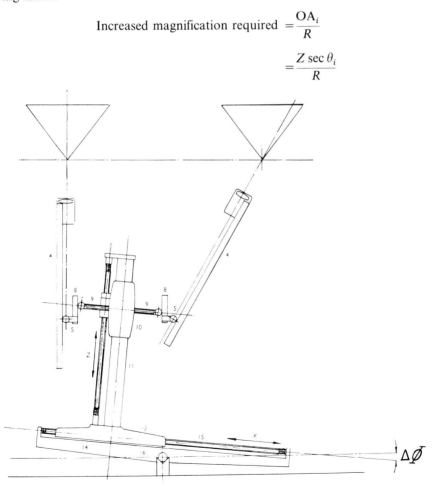

Figure 8.14. *Optical-Mechanical Projection in the Thompson–Watts Instrument*

171

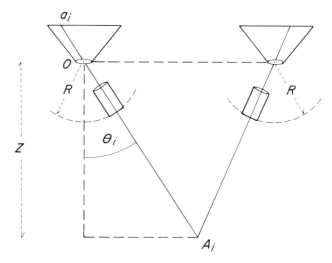

Figure 8.15. The Need for Pancratic Optics

This applies to both projectors and so the pancratic unit of each optical train is required to introduce a variable magnification proportional to the secant of the angle of the goniometer. In the instrument, rotations of the goniometers are picked up and arranged to introduce the required changes in magnification using a simple mechanical analogue device and a movable negative lens. It will be realised that the rotations of the projectors will also produce changes in image size for the same reason. In this case, however, the two rotations have the same value and so there is no differential change in image size. A simple modification to the pancratic units enables the size of the images seen to be changed overall within a range of $\times 5$ to $\times 15$ magnification. The most important consequence of the scanning system adopted for this instrument is that the operator always views the model in the epipolar (or basal) plane which remains in the fixed vertical plane defined by the rotations of the goniometers. In scanning in the Y-direction the model rotates about the air base. The vertical distances measured by movements of the Z bridge are therefore not the Z ranges required but in general something rather greater. The half-angle link mechanism of the instrument is the analogue device used to correct for this effect. Figure 8.16 illustrates this point. From this diagram we see that ranges (D) measured in the basal plane O_1O_2A are greater than the required Z co-ordinate at all points away from the base line of the model. Inspection shows that

$$Z = D - Y \tan \frac{\lambda}{2}$$

The correction $Y \tan \lambda/2$ is automatically introduced for all values of Y by the half-angle link mechanism.

For the purposes of carrying out an absolute orientation on this machine we note the following points. If the photographs are vertical then the half-angle lineal (20) in Fig 8.16 is vertical for points on the base line. However, if the photographs have a common tilt $\Delta\Omega$ then this element of absolute orientation is introduced by rotating the two projectors about the beam axis. An inclination of the air base $\Delta\Phi$ is introduced by tilting the model datum surface the appropriate amount. This is effected in the instrument by giving an inclination $\Delta\Phi$ to the X-axis, as indicated in Fig 8.14.

8.5 Instruments using mechanical projection

In many ways the projectors of an instrument using optical projection closely resemble the cameras. However, in the case of instruments using a mechanical solution to the problem

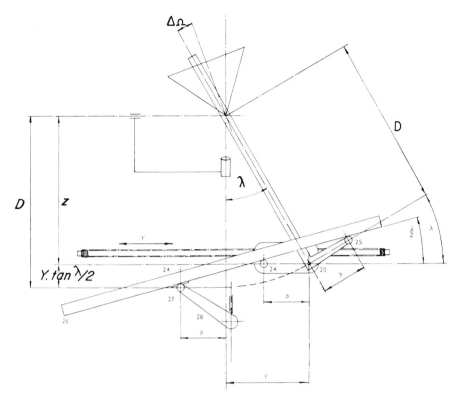

Figure 8.16. Half-Angle Link Mechanism

of restitution, the projectors are quite different. They, in fact, translate into mechanical terms the mathematical model of central perspective projection that is so often associated with camera geometry. It is this fact that gives rise to many of the basic differences between the optical and mechanical instruments.

In mechanical-type instruments the directions of the rays of light are represented by space rods or lineals along the whole of their length from projection planes to model point. The outer nodal points of the camera lens are represented by a mechanical point defined by the intersection of a set of axes about which each space rod can rotate (i.e. the centre of a universal or cardan joint). The situation is illustrated in Fig 8.17. Here points such as a_1 and a_2 in the projector planes 1 and 2 are projected into model space through the centres O_1 and O_2 by space rods a_1O_1A and a_2O_2A. If the two projection planes are correctly orientated about the two centres O_1 and O_2, then the first photo image point associated with a_1 will be identical with that associated with a_2 on the second photograph. In the more straightforward instruments the various photo tilts are introduced directly onto the two projection planes and the principal distances are made equal to the nominal focal length of the camera lens. In this case we therefore have a simple case of mechanical rectification of each photograph. Mechanical complications arise in coupling the two photo image points with their corresponding points in the projection planes, for it is not practicable to make the photo planes coincide with the projection planes. An inspection of the instruments of this category will show a variety of methods used to overcome this difficulty.

Whatever method is used to scan the photographs they all have some points in common. The photo images are viewed by low-powered microscopes held normal to the photo planes. The images seen by the operator are therefore unrectified and the viewing situation is similar to that of a simple stereoscope. If the two photographs have unlike tilts

173

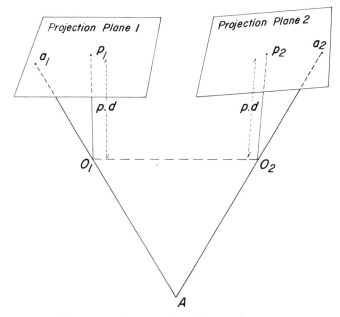

Figure 8.17. Concept of Mechanical Projection

of a magnitude in excess of about 10° then the differences in the size and shape of the images may be such that an operator cannot obtain a strong stereoscopic impression. Another consequence of this method of viewing is that the observations do not take place in the basal plane, except along the base line. However, in practice these points are of little consequence with near-vertical aerial photographs where the average tilt in recent years seems to have been a little over 1.5°. Again, because of the normal viewing, the plane of reference appears to the observer as a horizontal plane and this is claimed by some to be an advantage when contouring.

At the present time, instruments using mechanical projection are by far the most popular type. In a single chapter only a brief introduction can be made to this category of instrument. There are two approaches to the solution using either space rods or lineals in two planes. As an example of the use of space rods we will take the Wild A8 instrument.

8.5.1 THE USE OF SPACE RODS — THE WILD A8 AUTOGRAPH

This instrument was introduced in the early 1950s and has undoubtedly been one of the most successful photogrammetric restitution instruments so far produced. The design of the instrument is relatively simple and provides an excellent example of mechanical restitution.

In Fig 8.18 the two projection planes are swept out by the centres of the two cardan joints a_1 and a_2, for these points are constrained to move in these planes by the lazy tongs arrangements that are pivoted about their central points l_1 and l_2. At the remote ends of these tongs are held the two microscopes m_1 and m_2. The floating marks are introduced, as opaque black dots apparently 60 μm in diameter, onto the images formed in the first focal planes of these microscope objectives. The scanning optics move on swinging girder arrangements that ensure that for all orientations of the projectors and all viewing positions of the diapositives the images are directed into the fixed parts of the optical viewing system. Dove prisms d_1 and d_2 are included in the optical trains and these can be operated manually to eliminate image rotations introduced when comparatively large tilts are encountered during the process of relative orientation. The operator sees a $\times 6$ magnified image of the diapositives. These are adequately illuminated by low wattage

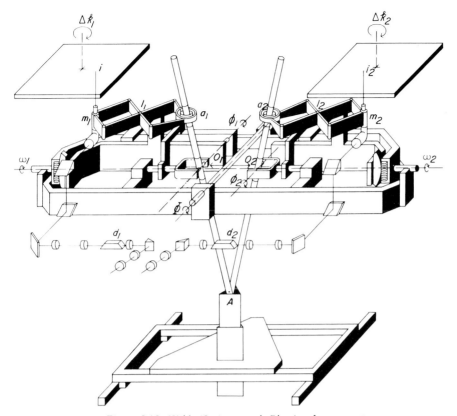

Figure 8.18. Wild A8 Autograph Plotting Instrument

bulbs, the brightness of which can be adjusted by individual rheostats. (Uniform brightness of the two images promotes a higher accuracy in the stereoscopic measurement.)

For the purposes of interior orientation the principal distance of the projectors can be set over a wide range. Unlike optical projectors there is much less restriction on the range of values that can be accommodated. By means of a cranking handle the perpendicular distance between projection plane and projection centre can be varied continuously within the range 98 to 215 mm. This therefore is one of the advantages claimed for mechanical projection instruments. As is customary, the diapositive holders can be removed from the instrument and the photos accurately centred in the holders using a setting box and low powered magnifiers.

Each projector unit carries out a form of mechanical rectification between projection plane and model plane in a most direct and obvious way; the projection plane being tilted to accord with the values of the camera tilt. Relative orientation is carried out therefore by means of a primary ω, secondary φ, and a tertiary κ rotation of each projector about its perspective centre. In making these rotations it should be noted that the projection plane, microscope plane, diapositive plane and all parts associated with them must rotate as a single unit about the centre. The tertiary κ rotation is carried out by rotating the diapositive in its carrier. It does not therefore rotate directly about the perspective centre. The range of φ and ω movements possible on each projector is limited to 5 and 6 grad respectively. Because the instrument has no Zeiss parallelogram and the projectors can move only in the X direction, a two-projector method of relative orientation is all that is possible. The scaling of the model is achieved by a BX screw that moves both projectors by equal and opposite amounts in the X direction. The model tilt required for absolute

175

orientation is introduced as two components. The Φ rotation comes directly from a rotation of the air base and projector units about a centrally placed Y-axis. The Ω component is introduced by setting individual projector rotations such that

$$\Omega = \omega_1 = \omega_2$$

Any model point A is defined in model space by the direct intersection of the two space rods. This point of intersection is moved within model space with the aid of an YXZ scanning system formed by three precision screw threads held mutually at right angles to one another. These threads have a pitch of 1 mm and dial readings can be read off to 1/100th part of one revolution. In this way model XYZ co-ordinates can be read off to 10 μm. As an addition, optical shaft encoders can be fitted which feed into a digitising unit such as the Ectomat EK8. From this a nine-digit Nixie tube display is produced also reading to 10 μm. The output from the Ectomat can be used to provide punched tape, magnetic tape and teletype outputs, suitable for use with a wide variety of computer programmes. Where a graphical output is needed the X, Y shafts of the instrument can be connected mechanically through a gear box to a large plotting table. The range of gears provides a model to plot scale ratio of between 1:1 and 1:4.

The instrument can be classed as a first-order precision plotting instrument capable of handling near-vertical photographs of standard format within the focal range values 98 to 215 mm. The instrument is therefore much used for large-scale plotting of detail required for topographical and engineering purposes. The digital output can be used for the production of digital ground models and for earthwork estimations. Although not primarily designed as an instrument for aerial triangulation, analogue aerial triangulation can be carried out most successfully on the instrument when inclinometers are used. The digital output also makes the instrument highly suitable for independent model aerial triangulation, as described in the next chapter.

The A8 Autograph is typical of the Wild range of plotters, all of which utilise mechanical projection and straight space rods. After orientation, therefore, the projectors take on the same tilts as the camera had at the two air stations. The B range of instruments are plotting instruments designed for medium and small-scale mapping from wide and super-wide angle photographs. These instruments use freehand movement for XY scanning and an illuminated scale for Z readings. They employ a one-way projection method, that is to say, the diapositives are held in the positive planes on the same side of the projection centres as the model plane. To some extent this limits the range of magnification between photo and model scales (about $\times 1.5 - \times 2.2$). Plotting can be carried out directly or by means of a polar pantograph.

The A10 is the universal instrument capable of the highest precision. It can be used with both aerial and terrestrial photographs as the footwheel associated with the Z movement can be interchanged with the Y movement for terrestrial purposes. Focal distance values of 85–308 mm can be set directly. The projection system uses double-arm space rods, i.e. a negative projection plane is used. The instrument is fitted with a full Zeiss parallelogram. This instrument would be used where results of the highest precision are required. In photogrammetric mapping it would therefore be used for high quality aerial triangulation, and for this purpose it would be used in conjunction with the Ectomat EK8 co-ordinate printer. As with all Wild instruments, lens distortion is corrected for using aspheric plates. Figure 8.19 shows the manner in which such plates introduce image displacements calculated to compensate for distortion.

In 1976, the new Aviomap range of Wild instruments was introduced, the stereo plotters being designated AM, AM-H and AM-U. All are based on space rod projection from negative projection planes. These three instruments cover the range of requirements previously filled by the B8/B9, A8, A10 instruments. In order to reduce the frictional effects great use is made of air bearings. All the instruments now have fixed perspective centres but only the universal (U) instrument has a full Zeiss parallelogram. For plotting purposes

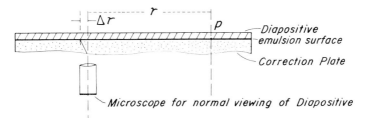

Figure 8.19. Compensation for Lens Distortion by means of Aspheric Plates

the instrument can be connected to the electric plotting table Aviotab TA which has a number of useful cartographic facilities. The AMU instrument can be used in a number of modes that facilitate plotting, the production of profiles and digital ground models, and the production of orthophotos.

There are many other types of instruments using space rods at present available for at the moment this form of solution is most favoured by the manufacturers. For example, Zeiss (Oberkochen) and Santoni both produce a full range of instruments falling within this category. From the design point of view some features of these instruments are of particular interest but such considerations unfortunately lie beyond the scope of this chapter.

8.5.2 THE USE OF LINEALS IN MECHANICAL PROJECTION
When we consider the possibilities for mechanical rectifications the use of a cranked space rod is an attractive proposition as the method allows the photograph plane to be held in a fixed, usually horizontal position. The viewing system can therefore be made quite simple. The Kern PG2 and the Zeiss (Jena) Topocart both adopt this solution. However, there are difficulties in introducing a spatial crank into a rod and so both of these instruments adopt methods using lineals working in the XY and XZ planes. In the case of the PG2 the two planes used are in fact the XZ and YZ planes of the instrument. In the Topocart all projection planes are folded into the XY plane for the sake of compactness.

To appreciate these particular methods of solution we must first examine Fig 8.20. In the left-hand diagram we show a photograph with a φ-tilt only. This will produce a pattern of scale changes over the photograph such that the scale along any parallel is constant value, with the value of the scale along any particular parallel being a linear function of its X co-ordinate (see section 3.3). From Fig 8.20(a) we note that the scale along a parallel through any point (a, A) is given by

$$S_a = \frac{Oa}{OA}$$

$$= \frac{f \sec (\alpha + \varphi)}{Z \sec \alpha}$$

$\therefore$
$$S_a = \frac{On_2}{Z}$$

$$= \frac{On_2}{f} \cdot S$$

where
$$S = \frac{f}{Z}$$

From Fig 8.20(b) we note that

$$S_a = \frac{On_2}{Op'} \cdot S$$

177

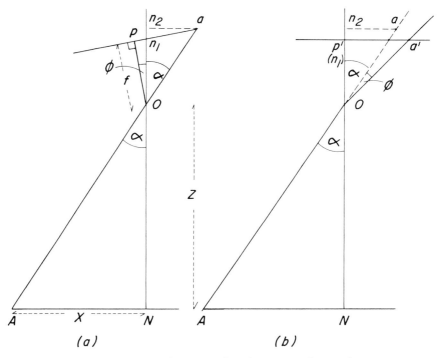

Figure 8.20. *Cranked Space Rod to Compensate for φ-Tilt*

The scale of any small element (δx) in the neighbourhood of the parallel through a, and at right angles to it, is given by equation (3.5), i.e.

$$\frac{\delta x}{\delta X} = S_a \cos \varphi \, (= S'_a)$$

In part (b) of Fig 8.20 the camera (projector) geometry has been rotated through an angle φ about O so that the photo plane is again horizontal. In order to do this we have had to introduce the crank angle φ into the lineal AOa at O. By this means, the rectification of scale changes in the direction of the line of greatest slope (in this case the X-direction) can be carried out. The method is used in both the Kern and Zeiss instruments. In the case of the PG2 there are just two such lineals. That of the left-hand projector is cranked in a YZ plane by an amount $\Delta\omega_L$ (where $\Delta\omega_L = \omega_2 - \omega_1$) while the other one, associated with the right-hand projector, is cranked in the XZ plane by an amount $\Delta\varphi_R$ (where $\Delta\varphi_R = \varphi_2 - \varphi_1$). The diagrams of Fig 8.21 show the projections of the right-hand space rod into the YZ and XZ planes. In the course of relative orientation the right-hand lineal is cranked in the XZ plane by the amount $\Delta\varphi_R$ and the left-hand lineal is cranked in the YZ plane by the amount $\Delta\omega_L$. To maintain the correct scale along the photo parallels the principal distances must be varied continuously for the YZ projection of the right-hand space rod and for the XZ projection of the left-hand space rod in accordance with the requirements of equation (3.3). For the YZ plane of the right-hand projector, we note from an examination of Fig 8.20(a)

$$On_2 = ZS_a$$

$$= \frac{f \sec (\alpha + \Delta\varphi_R)}{\sec \alpha}$$

$$= C''_y$$

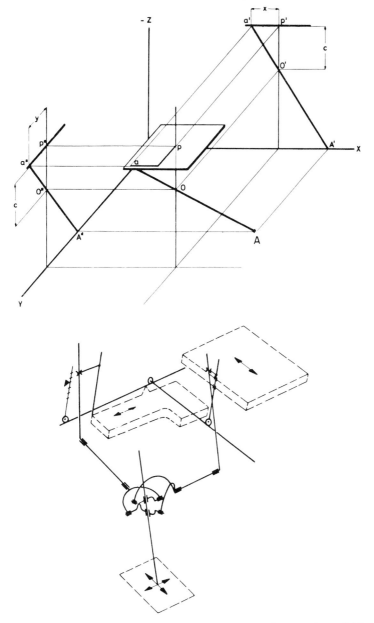

Figure 8.21. Projection of R.H. Space Rod into Two Planes in Kern PG2

In the instrument the angle picked up and used by the analogue corrector for the projection distance is in fact $(\alpha + \varphi)$, hence we will write

$$C_y'' = \frac{f \sec \alpha'}{\sec (\alpha' - \Delta\varphi_R)}$$

$\therefore \qquad C_y'' = [f \cos \Delta\varphi_R] + \sin \Delta\varphi_R [f \tan \alpha']$

The analogue corrector takes the form of an inclined bar the first setting of which introduces the constant correction given to the principal distance in accordance with the

179

term in the first bracket. The inclination of the bar is set at the angle $\Delta\varphi_R$ and in scanning the model in the X-direction the term in the second bracket is introduced (i.e. p′a′ of Fig 8.20(b)). In fact, the corrector applies a correction of p′a′ tan $\Delta\varphi_R$ but as the maximum value of $\Delta\varphi_R$ is ±6 grad the error introduced can be neglected.

In a similar manner we have for the left-hand projector and the XZ plane

$$\text{On}_2 = ZS_a = \frac{f \sec \beta'}{\sec (\beta' - \Delta\omega_L)} = C'_x$$

i.e.
$$C'_x = [f \cos \Delta\omega_L] + \sin \Delta\omega_L[f \tan \beta']$$

To carry out a relative orientation of the PG2 model the elements κ_1, κ_2, $\Delta\varphi_R$, $\Delta\omega_L$ and $\Delta\Phi$ are used. The element $\Delta\Phi$ is introduced as an inclination of a short base separating the two space rods in the object space. On the completion of this process the model has an inclination φ_L and ω_R. This is removed in absolute orientation by tilting the datum plane of the instrument with respect to the projectors. The scale of the model is adjusted by changing the length of the base. Figure 8.22 gives a general view of the instrument that illustrates these points.

In the case of the Zeiss Topocart there are four cranked lineals to compensate for the four elements of photo tilt φ_1, ω_1, φ_2, ω_2. All these lineals work in flat XY planes, one above the other. In order to understand the nature of this interesting solution we will examine the means by which the scale corrections for the φ, tilt of the left-hand photograph are carried out. For this case, the line of greatest slope for a φ tilt is the X-direction and so the two diagrams (a) and (b) of Fig 8.20 apply once more. The required scale corrections in the X-direction are introduced directly by the cranking of the lineal

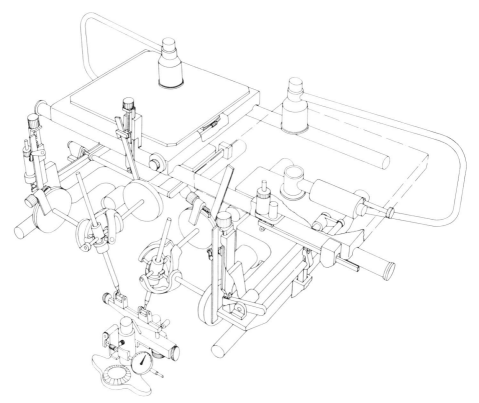

Figure 8.22. Kern PG2 Plotting Instrument

through the angle φ_1 about the projection centre, the point a_1 being connected to the X slide of the horizontal photo carriage. At O there are two lineals representing the lines AOa and Oa'. The required angular cranking φ_1 is introduced by rotating arm Oa' with respect to Oa. On each of these two arms is a pin (1 and 2) positioned at a fixed and equal distance c from O, as shown in Fig 8.23. The sliding joints at points 1 and 2 move the parallel rails 11' and 22' in accordance with the magnitude of the angle α introduced by scanning the photograph in the X direction. The perpendicular distance of rail 11' from O is therefore $c \cos \alpha$ and that of rail 22' is $c \cos(\alpha + \varphi_1)$. At the second centre of rotation O' we have two more lineals associated with scanning in the Y direction. The lineal BO'1' represents the uncorrected direction of the space rod in the YZ plane. This intersects the rail 11' to define the point 1'. From this point, by mechanical means, a perpendicular is dropped from 1' to intersect the second rail to define the point 2'. This point is on the lineal O'2' which now has, in consequence, the small angular separation ε with the main lineal BO'1'. Each of the two arms has a perpendicular arm attached to it to provide the directions O'b and O'b'. The intersection of the arm O'b' with a rail set at a distance f from O' locates the corrected image point b'. In this way the correct displacement Y is given to the Y slide of the photo holder. From an inspection of the right-hand side of the diagram we note that

$$\tan \beta = \frac{S}{c \cos \alpha} \quad \text{and} \quad \tan(\beta + \varepsilon) = \frac{S}{c \cos(\alpha + \varphi_1)}$$

$\therefore$

$$\frac{\tan(\beta + \varepsilon)}{\tan \beta} = \frac{\cos \alpha}{\cos(\alpha + \varphi_1)}$$

Also

$$y' = n'b$$

$$= f \tan \beta$$

and

$$y = n'b'$$

$$= f \tan(\beta + \varepsilon)$$

$\therefore$

$$\frac{y}{y'} = \frac{\tan(\beta + \varepsilon)}{\tan \beta}$$

$$= \frac{\cos \alpha}{\cos(\alpha + \varphi_1)}$$

$\therefore$

$$y = y' \frac{\cos \alpha}{\cos(\alpha + \varphi_1)}$$

This of course is the correct value, for we have

$$S_a = \frac{y}{Y}$$

$$= \frac{y'}{Y} \cdot \frac{\cos \alpha}{\cos(\alpha + \varphi_1)}$$

$$= S \cdot \frac{\cos \alpha}{\cos(\alpha + \varphi_1)}$$

$\therefore$

$$S_a = \frac{f}{Z} \cdot \frac{\sec(\alpha + \varphi_1)}{\sec \alpha}$$

as required.

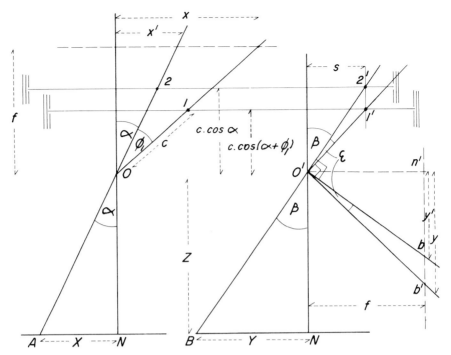

Figure 8.23. Mechanical Rectification in the Zeiss Topocart Instrument

For the sake of clarity, Fig 8.23 shows only the φ_1 correction process. The ω_1 correction would be applied by cranking lineal BO'1' by the amount ω_1. (The diagram therefore, in fact, indicates a zero value on the arm O'1'.) Two pins, 3' and 4', on the arms O'2' and O'3' at a constant distance c from O' serve to locate two further parallel rails 3'3 and 4'4. The intersection of rail 3'3 with O2 locates the point 3. A perpendicular from this point 3 onto the rail 4'4 locates the point 4. This in turn defines the direction of the third lineal O4 which has a small rotation δ with respect to O2. The x co-ordinate corrected for both φ_1 and ω_1 tilts is therefore given by

$$x = f \tan (\alpha + \varphi_1 + \delta)$$

Figure 8.24, taken from the Zeiss instruction manual, illustrates the combined mechanisms to produce ε and δ corrections.

8.5.3 APPROXIMATE SOLUTION INSTRUMENTS
The last category of instrument we mention in this chapter is that known as the approximate solution plotter. These machines do not attempt to reconstruct a correct surface in the model space. They usually accept the model produced by normal viewing of the photographs and apply corrections to any measurements made in accordance with the geometric properties investigated in chapter 3. The accuracy achieved does not usually approach that of any of the instruments mentioned earlier in the chapter. They are, however, cheaper, and satisfactory results can certainly be produced at small and medium scales by the two instruments that will be discussed in this chapter, the Santoni Stereo-micrometer and the Thompson CP1. Instruments that combine the use of paper prints with inferior optics are best disregarded.

A justification for using this type of instrument can be made out for at least two situations. A mapping organisation with an obligation to produce small-scale mapping of an extensive region in the shortest possible time might choose to use these instruments; the argument being that their lower cost will enable more plotting instruments to be

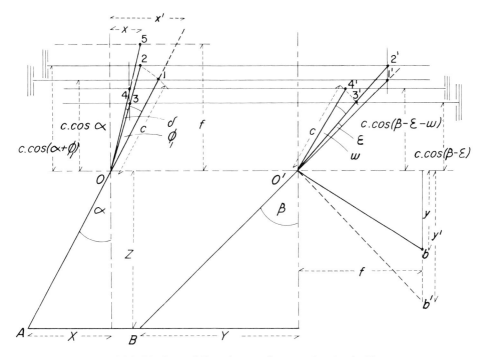

Figure 8.24. Mechanical Rectification for φ, and ω, in the Topocart

purchased and so the task can be completed in a shorter space of time, provided of course that the time taken to plot each model is not significantly longer than that using more expensive equipment. In another situation, an organisation using maps but not in any way a map producer might decide to expend a limited amount of money on an instrument of this type that can be used from time to time to map limited areas. It should, however, be noted that, in general, more skill and more control is required on each model to produce a product comparable with that produced by a better instrument.

The Santoni Stereomicrometer was first put on the market some years ago and the present model available is the mark V. The instrument is one of the most successful of this category. It uses glass diapositives and has excellent optics. An examination of its working principles provides a useful and instructive exercise. Figure 8.25 shows the general arrangement of the instrument and, without going too deeply into the details of its construction, we can note the following points. The planimetry is taken from the right-hand photograph. Scanning the model with the stereoscope causes the 'space rod' arrangement to reproduce the detail of this photograph on the plotting table. Examination shows that this arrangement is in fact a space rod in two halves, which can be cranked in the ω direction. The diapositives are held in a fixed horizontal plane, but the plotting table can be tilted in the φ direction. In this way a mechanical rectification of the photograph is realised. Figure 8.26 illustrates the method.

The two halves of the space rod are Oa and O′A. When plotting from a vertical photograph the two rods would be parallel and the plane of the plotting table would be horizontal. To introduce an ω correction the arm O′A can be rotated at O′ about the longitudinal axis of the main support beam. The length of the projection distance Oa is set to f, the nominal focal length of the photography. A change of scale (between photo and plot) is set by altering the value of Z. In order to compensate for the changes in scale brought about by changes in ground elevation, this instrument uses an ingenious technique whereby measured changes in ground elevation automatically introduce small

183

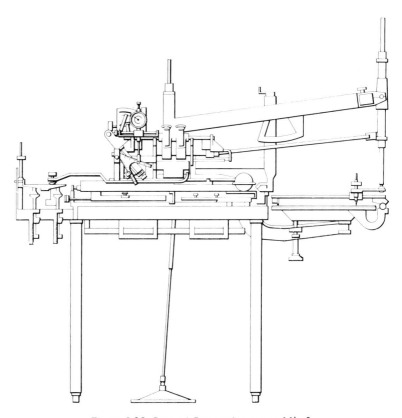

Figure 8.25. *Santoni Stereomicrometer Mk. 5*

changes Δf into the principal distance f which then produce the required changes in plot scale. The device is therefore a mechanical differential rectifier. (Such devices are usually of an optical nature, e.g. orthophotoscopes – see chapter 5.) In order to appreciate the method of differential scale correction we need to examine the method used for measuring height values.

In essence, the instrument measures X-parallaxes using a normal stereoscope, the floating marks being introduced within the viewing optics. In the elimination of parallax

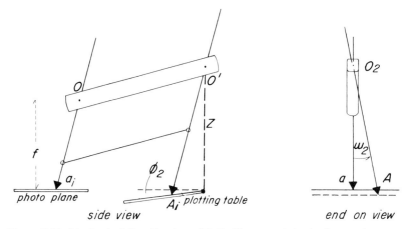

Figure 8.26. *Mechanical Rectification of R.H. Photograph in the Stereomicrometer*

the left-hand mirror moves in the X-direction, and this movement provides a direct measurement of the X-parallax differences. These values will not yield accurate changes in ground elevation unless the photo tilts and the air-to-base inclination are all zero. A mechanical correction device, a Cavalcanti surface, is therefore used to provide a correction to the parallax measurements. This surface is an approximate mechanical realisation of the hyperbolic error surface derived in chapter 6 (see equation (6.18)). As the sensor moves across the surface, the arm movements are transmitted to the left-hand mirror of the stereoscope. Small rotations of this mirror, about an axis in the Y direction, modify the X-parallax measurement by the required amount. The parallax measurements are not displayed on a horizontal scale, as one might expect, but on a vertical one that makes use of the expression

$$\Delta P_x = \Delta f . \tan \theta$$

$$= \frac{\Delta f . b}{f}$$

The significance of this expression can be seen from an examination of Fig 8.27.

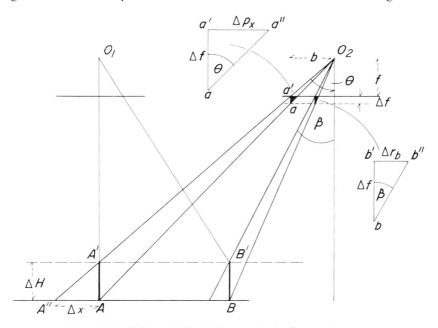

Figure 8.27. Differential Scale Changes in the Stereomicrometer

Using $\Delta O_2 AA'$ we see that

$$\frac{\Delta H}{\Delta f} = \frac{\Delta X}{\Delta P_x} \quad \text{and} \quad \tan \theta = \frac{\Delta X}{\Delta H}$$

Hence

$$\Delta f = \Delta P_x . \cot \theta$$

Since the point A is also the nadir point of the left-hand photograph we have

$$\tan \theta = \frac{b}{f}$$

An estimate of the value of the quantity b is found by measuring the photo air bases of both photographs. In the instrument, the geometry of the triangle $aa'a''$ is reproduced

185

mechanically to form the connection between the ΔP_x measurements required for height determinations and the differential scale change required for correcting the planimetric scale of the right-hand photograph. The required inclination θ of an arm is calculated from a measurement of the two photo air bases and a knowledge of the focal length of the camera lens. Thereafter, movements of the left-hand floating mark $\Delta P_x(=a'a'')$ are produced by changes $\Delta f(=aa')$ in the principal distance of the left-hand space rod. This provides a correction for height displacement at any point such as B, for we have

$$\Delta r_b = \Delta f . \tan \beta$$

If we consider the treatment of the observations carried out by the instrument we also note that the main areas of approximation are as follows. In the case of the planimetry, the mechanical rectification for tilt is correct within the limits of mechanical precision of the instrument. The correctness of the differential change in scale is, however, directly dependent on the accuracy of the parallax heighting and the value of θ used. Both of these depend on the assumption that the changes in ground elevation are comparatively small. As was noted in an earlier chapter, a particular hyperbolic paraboloid surface is only correct for one particular ground plane; different planes requiring different values for the parameters of the equation. In addition the basis of the derivation of the surface shape was one of small camera tilts and inclination of the air base. It will be appreciated, therefore, that significant errors can occur when the tilts are in excess of a few degrees and the changes in ground elevation are more than about 10% of the flying height. One further point that may be noted is the fact that in the later models of the instrument the Cavalcanti surface cannot be given a parabolic deformation. Significant errors across the centre of the model can be introduced if the differential ω tilt of the two photographs is in excess of about 4°.

To orientate a model using this instrument patterns of ground control similar to those shown in Fig 8.28 are used. In part (a) of the diagram, three planimetric points across the base line will provide for the effects of scale and ω tilt, the fourth point being used to rectify the effects of the φ tilt. Part (b) of the diagram shows a distribution of height control. The two centre points will detect any residual error due to the absence of a parabolic correction on the surface. If this is significant, the model can be plotted in two halves as indicated.

The Thompson CP1 is one of the latest approximate solution instruments to be produced, being introduced in 1971. In this instrument both photographs are rectified independently for the image displacements brought about by camera tilts. The optical-mechanical mechanisms designed to do this are to some extent approximate, for they assume the magnitudes of these tilts to be no more than those associated with near vertical

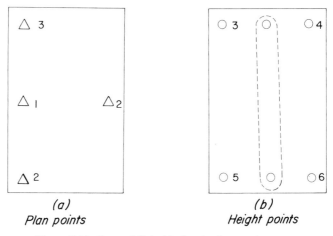

(a)
Plan points

(b)
Height points

Figure 8.28. Control Suitable for the Stereomicrometer

186

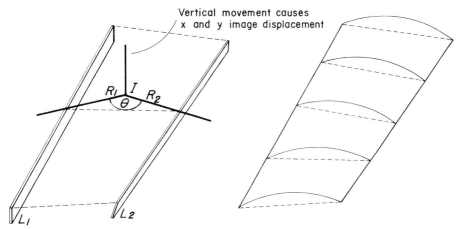

Figure 8.29. *Corrector Surface Generator in Thompson CPI Plotter*

photography. In such cases the parallaxes (x and y) introduced into each photograph can be described by first order corrections similar in form to those of equations (7.1) and (7.16). Four identical mechanical correctors of the form shown in Fig 8.29 are used to introduce compensatory x and y image displacements into the viewing optics. Each photograph is rectified in this way for x and y displacements due to tilt.

The two lineals L_1 and L_2 can be tilted about a common transverse axis and so the two upper surfaces define a warped surface. The two intersecting rods R_1 and R_2, set at some angle of θ, rest on the two lineals. Movement of the rods in the x–y directions causes the intersection point I to trace out a hyperbolic paraboloid surface. The vertical movement of the point I is linked to a mirror in the optical train that either causes the image to move in the x or y direction.

With the aid of four of the above correctors X, Y-parallax is eliminated from the model. The remaining X-parallax is assumed to be entirely due to changes in ground elevation. Here a simple analogue device is used to solve the standard parallax equation (6.8) and provide height values. The X-parallax measurement mechanism also brings about a change in scale in the mechanical projection of the right-hand photograph. One of the advantages of this method of approximation is that, in this case, the errors in plan and height are no longer a function of the changes in elevation of points. A detailed description of the instrument and its capabilities are given in Refs 8.1 and 8.2.

8.6 The checking and calibration of analogue plotting instruments
From time to time, a user will want to check the performance of an analogue plotting instrument in order to verify that it is functioning correctly and within the expected tolerances. The checks that can be carried out easily by the operator include the following:

(i)   The zero positions of all linear and angular scales can be checked. These would include angular scales of individual projectors that give values for $(\kappa_1, \varphi_1, \omega_1)$ and $(\kappa_2, \varphi_2, \omega_2)$ and scales showing model tilts (Φ, Ω). The linear scales include the principal distance settings (f) of individual projectors and base settings (BX_1, BY_1, BZ_1) and (BX_2, BY_2, BZ_2).

(ii)   The ability of each projector to carry out an accurate projective transformation of the form given below within limits that can be determined.

$$X - X_0 = \left(\frac{Z - Z_0}{f}\right)\frac{a_{11}x + a_{12}y + a_{13}}{a_{31}x + a_{32}y + a_{33}}$$

$$Y - Y_0 = \left(\frac{Z - Z_0}{f}\right)\frac{a_{21}x + a_{22}y + a_{33}}{a_{31}x + a_{32}y + a_{33}}$$

187

where (X, Y) are instrument co-ordinates obtained by projecting into the plane (Z) the photo co-ordinates (x, y) using a projection distance (f).

(iii) The ability of the instrument to generate an undistorted model of the surface, within limits that can be evaluated.

(iv) In addition to the above, the instrument co-ordinates of the perspective centre of each projector (X_0, Y_0, Z_0) may be required for such purposes as aerial triangulation by independent models (see section 9.7).

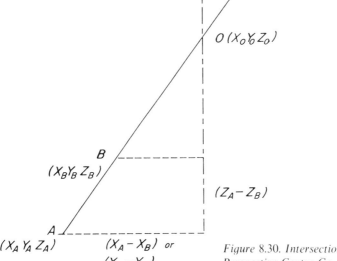

Figure 8.30. *Intersection Method of Finding Perspective Centre Co-ordinates*

The majority of the above tests are carried out with the aid of accurate reseaux plates placed in the photoholders of the projectors. The test (iv) may, however, be carried out using photographs containing points of well-defined detail. As the results of this test are of use in some of the calculations concerned with the others, we will consider this one first. The method is essentially one of space intersection and is fully described in Ref 8.3. If we examine Fig 8.30 we can write the following equations for the line ABO

$$\frac{X_B - X_A}{Z_B - Z_A} = \frac{X_A - X_O}{Z_A - Z_O} \quad \text{and} \quad \frac{Y_B - Y_A}{Z_B - Z_O} = \frac{Y_A - Y_O}{Z_A - Z_O}$$

These equations may be rearranged to give the following:

$$\begin{pmatrix} -(Z_A - Z_B) & 0 & X_A - X_B \\ 0 & -(Z_A - Z_B) & (Y_A - Y_B) \end{pmatrix} \begin{pmatrix} X_0 \\ Y_0 \\ Z_0 \end{pmatrix} = \begin{pmatrix} X_A & -X_B \\ Y_A & -Y_B \end{pmatrix} \begin{pmatrix} Z_B \\ Z_A \end{pmatrix}$$

Given a number of sets of observations to pairs of points such as A and B, the values of the unknowns (X_0, Y_0, Z_0) can be evaluated using a least squares method.

In order to check the principal distance of a projector, a reseau plate must be used. This is placed in the photo holder and the projector tilts set to their zero values, it being assumed that in doing this the plate is then horizontal (i.e. parallel to the datum plane of the instrument) to within $1°$. Observations are now made to reseaux points (1 2), (2 3), (3 2) and (2 1) at two Z levels $(Z_{max}$ and $Z_{min})$ as indicated in Figs 8.31 and 8.32. For pairs of points such as (2 1) and (2 3) the following relationship applies; the distance d being known to a high degree of accuracy.

188

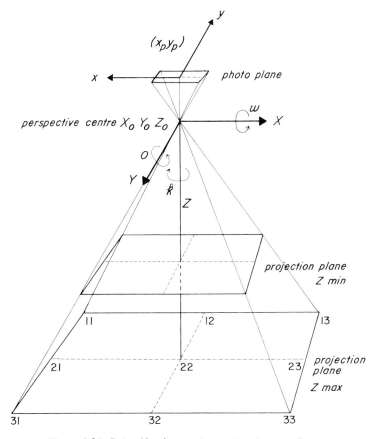

Figure 8.31. Point Numbering System for Reseaux Points

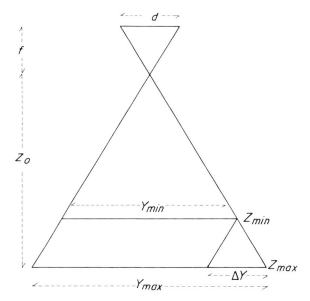

Figure 8.32. Determination of Principal Distance

For points (2 1), (2 3) we have

$$f = d\frac{Z_{max} - Z_{min}}{X_{max} - X_{min}}$$

$$= d \cdot \frac{\Delta Z}{\Delta X}$$

For points (1 2) (3 2) we have

$$f = d \cdot \frac{\Delta Z}{\Delta Y}$$

We can also obtain a value of Z_0 from

$$Z_0 = X_{max} \cdot \frac{\Delta Z}{\Delta X}$$

$$= Y_{max} \cdot \frac{\Delta Z}{\Delta Y}$$

In order to check a projector, a monogrid test involving a regular array of reseaux points is used. In many instruments this array will be confined to the 60% of the format involved in the model-making process. The projector is levelled up as accurately as possible and a series of $(X_i Y_i)$ co-ordinates read off in some datum plane Z in a systematic manner designed to minimise any backlash in the instrument measuring system and systematic errors on the part of the observer.[8.4] To process these observations it is convenient to take the reseau cross at (or very close to) the principal point position as the origin of both photograph and instrument co-ordinates. Knowing the principal distance (f) and the projection distance (Z), the instrument co-ordinates recorded can be scaled down to those of the reseau system and compared directly with them. The discrepancies observed arise not only from errors (v_x, v_y) in the transformation process carried out by the projector but some errors are also introduced by slight errors in the orientation of the projector. Because of their small magnitude, first order expressions of the type derived in equation (7.14) are sufficient to describe these. Hence we have the following

$$X'_i = \frac{f}{Z_0} X_i \quad \text{and} \quad Y'_i = \frac{f}{Z_0} Y_i$$

$$X'_i - x_i = v_x + \delta x \quad \text{and} \quad Y'_i - y_i = v_y + \delta y$$

where

$$\delta x = -y\delta\kappa + \frac{x^2}{f}\delta\varphi - \frac{xy}{f}\cdot\delta\omega$$

$$\delta y = x\delta\kappa + \frac{xy}{f}\delta\varphi - \frac{y^2}{f}\cdot\delta\omega$$

The residuals of the expression given below are therefore assumed to be due to imperfections in the transformation process carried out.

$$\begin{pmatrix} v_x \\ v_y \end{pmatrix} = \begin{pmatrix} X'_i - x_i \\ Y'_i - y_i \end{pmatrix} - \begin{pmatrix} -y & \dfrac{x^2}{f} & -\dfrac{xy}{f} \\ x & \dfrac{xy}{f} & -\dfrac{y^2}{f} \end{pmatrix} \begin{pmatrix} \delta\kappa \\ \delta\varphi \\ \delta\omega \end{pmatrix}$$

Before the advent of the desktop computer it was common practice to attempt to set the projector in its correct orientation with zero rotations by observing the distribution of errors in the instrument co-ordinates and adjusting the projector accordingly. The errors finally observed after carrying out this procedure could then be regarded as those due entirely to errors in transformation. However, given a computer with the ability to normalise and solve a limited number of observation equations, the analytical procedure outlined above should give a rather better result in less time.

It should be noted that the process of calibrating a projector is exactly the same in principle as that of calibrating a camera. The spatial resection procedure described in section 10.3 can therefore be applied to this problem. If we accept the validity of the interior orientation then the residual errors can be evaluated using the collinearity equations directly to obtain the unknowns ($\delta\kappa$, $\delta\varphi$, $\delta\omega$, δX_0, δY_0, δZ_0). If required, a further three unknowns can be introduced (δx_p, δy_p, δf) to account for residual errors in the interior orientation; δx_p, δy_p being small errors in the assumed position of the principal point. A paper by Dowman[8.5] gives details of testing procedures and quotes results obtained for typical calibrations.

In order to check the overall performance of an analogue plotting instrument a stereo-grid test should be carried out. This involves the use of a grid plate in both projectors. The interior orientation is carried out carefully and this is followed by a very careful relative orientation. For a rapid check of the instrument, the model produced may be made horizontal. If this is done correctly then any variations in Z values can be regarded as imperfections in the model generated. Positional errors, which are likely to be of a smaller magnitude, can be regarded in the same manner. In order to check the height distortion more thoroughly it is desirable to observe spot heights at each reseau point on a model inclined at some small angle $\Delta\Phi$, $\Delta\Omega$ to the datum surface. By this means variations are introduced into the Z readings and these will tend to reduce any unconscious biasing of results by the observer. The effects of the model slopes are readily subtracted from the height values. The slopes $\Delta\Phi$ and $\Delta\Omega$ can be determined graphically by finding the X and Y slopes for every section of spot heights along these two directions and taking the means as the best available values. Alternatively, they can be computed from a set of simple observation equations in the following manner. Take the point (1 1) as the origin of planimetric co-ordinates and reduce all observations accordingly. If the grid interval is a distance d then the (i, j)th point has reduced co-ordinates (id, jd), hence we have

$$\delta Z = (id)\tan\Phi + (jd)\tan\Omega$$
$$Z_i - Z_1 = v_z + \delta Z$$

i.e.
$$v_z = (Z_i - Z_1) - (id)\tan\Phi - (jd)\tan\Omega$$

Where i and j take on integer values only. The set of observation equations is normalised in the usual way and then solved to give values for Φ and Ω and the residual height error.

9: Analogue aerial triangulation

9.1 Introduction

In chapter 7 we described the control requirements necessary for bringing the model into absolute orientation. Usually this control is provided in the form of the ground co-ordinates of identified photo points, although the number of such points can be greatly reduced in certain circumstances by making use of the data provided by auxiliary instruments (see chapter 12). If the area to be mapped photogrammetrically is small and covered by a limited number of models, then the provision of full ground control on each model is a reasonable possibility. This is particularly true when using large-scale photography where the ground area covered by each model is quite small. In such a case ground survey techniques can provide the necessary control points in a most economical manner. However, where large numbers of models are required and these form continuous strips of models, and perhaps the strips themselves overlap to form blocks of models, then perhaps the prospect of supplying ground control for each and every model becomes a daunting task. In these cases the technique of aerial triangulation is used in order to reduce greatly the number of ground control points required to orientate each model of the strip or block.

9.2 The problem of absolute orientation

Consider the situation where the first model formed from a strip of photographs has been put into absolute orientation and we wish to carry forward this orientation to the next model without recourse to further external control information. Then we have the situation illustrated in the two parts of Fig 9.1.

The second model can be formed and brought into relative orientation using photographs 2 and 3. Since the fore-and-aft overlapping of the photographs is not less than 55% we also have a usable overlapping of models M1 and M2. The information contained in this common overlap of the two model surfaces can therefore be used to contribute to the absolute orientation of the second model. Clearly there is enough information here to scale the model and orientate it in the Ω direction. The Φ tilt cannot be satisfactorily carried forward from this data; to attempt it, would involve a precarious extrapolation. To complete the orientation we therefore look to the other factor these two models have in common, the common photograph 2. It will be realised that after the absolute orientation of the first model, the position and orientation of this photograph are known. Therefore, if after a relative orientation, the second model is rotated until the aspect of the common photograph is correct and the scale of the model is changed until the nadir distance is correct (i.e. $O_2N_2 = O_2N_2'$) then we must have achieved absolute orientation. We see that the two factors provide more than enough information to carry forward the accepted orientation of the first model. In fact, the only element for which there is no check is the Φ_2 setting. All methods of aerial triangulation, analogue and analytical, must employ the above concepts in some form or other. The analogue methods to be described in the following sections of this chapter illustrate how this can be done mechanically in a number of ways. It will be realised that the strong link of models along

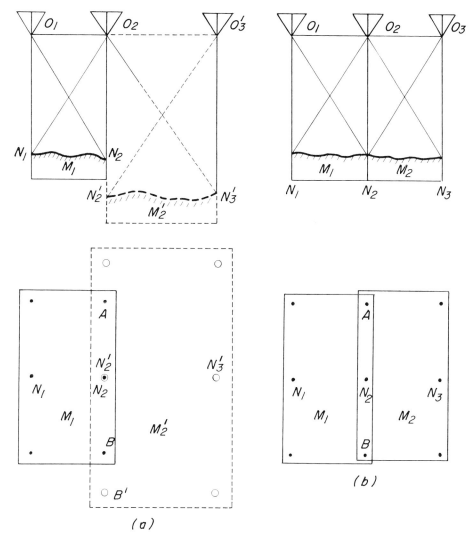

Figure 9.1. *Absolute Orientation of Adjacent Models on Long Bar*

the strip arises from the 60% fore-and-aft overlap. A weaker link is also provided between models by the 25% lateral overlap of strips. Of course none of the processes is error free and so errors in model co-ordinates must slowly accumulate as we proceed along the strip. The nature of this propagation will be considered after a review of the methods used.

9.3 Long bar Multiplex method
The value of this method is that it illustrates in the most direct way the principles introduced above. The first two projectors are placed at one end of the long bar and a relative orientation and then an absolute orientation carried out using control. A third projector is now introduced and placed on the bar at about the correct base length; a one-projector orientation using only the elements of that projector is now carried out. In no way should the positions of the first two projectors be altered with respect to the instrument datum surface. To scale the second model we note the height reading of a point on the base line of the first model in the vicinity of the second nadir point. The air base of

the second model is then changed, by moving the third projector until the height reading of this point is identical in both models. Figure 9.1(b) illustrates the method with a base line point N_2. The common overlap now provides two checks. First, the heights of the two wing points (A, B) should now agree, showing that the Ω tilt is the same in both models and, secondly, the planimetric positions of the two wing points should be identical showing that the scale and direction of the second model are also correct.

Further models may be connected up in a similar way until the end of the bar is reached. Usually about eight models can be joined in this way on the long bar multiplex-type of equipment. We see therefore that a strip model is formed by this type of instrument – a rather unusual but sometimes useful characteristic. The method is essentially a graphical procedure with the planimetric positions of all the points being plotted after every connection. Only the heights of the points are recorded numerically.

We note that each model connection depends primarily on an ability to carry out an excellent relative orientation. Any residual errors in the orientation of a projector are carried forward as errors in the absolute orientation of the next model. However, the heights of the wing points do provide an adequate check on the Ω tilt and the scaling of the model. The Φ orientation does, however, depend entirely on the relative orientation, and unfortunately the y-parallax criterion is least effective in the case of this element of orientation. A few simple calculations can be used to illustrate this point. If we assume the instrument has a spot height capability of something of the order of $0.3\ °/_{oo}\ Z$, then we have for the scaling procedure using the height of a centre line point a precision in scaling of this same order of magnitude. For example, if at photo scale, the wing points have Y co-ordinates of ± 90 mm this requires an ability to scale, using the planimetric positions of the tie points, to a precision given by

$$\text{Scale error between tie points} = \frac{3 \times 180}{10\,000}\ \text{mm}$$

$$= 0.05\ \text{mm (approx) at photo scale}$$

The detection of such an error in plotting by a graphical inspection is barely possible, unless the model scale is at least twice photo scale.

Again, if we consider the possibility of detecting an error in the Ω-tilt of the model using the heights of the tie points, we have, using the same Y range and a Z value equal to the focal length of a wide angle camera (152 mm),

$$\text{Minimum } \Delta\Omega \text{ error detectable} \simeq \tan^{-1}\left[\sqrt{2} \times \frac{3}{10\,000} \times \frac{152}{180}\right]$$

i.e. $\quad\quad\quad\quad\quad\quad\quad\quad\quad\quad \Delta\Omega \simeq 72'' \text{ of arc}$

Now, the ability to spot height to $0.3\ °/_{oo}\ Z$ suggests an ability to detect changes in x-parallax to values given by

$$\Delta P_x = \frac{B}{Z} \cdot 0.3\ °/_{oo}\ Z \text{ (see equation (6.8))}$$

i.e. $\quad\quad\quad\quad\quad \Delta P_x = \frac{92 \times 3 \times 1000}{10\,000}\ \mu\text{m at photo scale}$

i.e. $\quad\quad\quad\quad\quad \Delta P_x \simeq 28\ \mu\text{m at photo scale}$

If we now assume a similar acuity in detecting differences in y-parallax values (ΔP_y), again at photo scale, we have

$$\Delta P_y = \left(\frac{y^2 + z^2}{z}\right)\Delta\omega \text{ (see section 7.10)}$$

i.e.
$$\frac{28}{1000} \simeq \frac{(90)^2 + (152)^2}{152} . \Delta\omega$$

$$= 205 \; \Delta\omega$$

hence
$$\Delta\omega = \frac{28 \times 200\,000}{205 \times 1000} \;'' \; \text{of arc}$$

i.e.
$$\Delta\omega = 28''$$

This would suggest that the relative orientation process yields an Ω orientation rather better than that produced from the observation of heights of tie points. It does, however, require the assumption that the y-parallax acuity is comparable to that of the x-parallax.

9.4 The classical method using a Zeiss parallelogram

The method merits the description of 'classical' because for many years it was regarded as the most satisfactory method of carrying out aerial triangulation. To use this method it was necessary for the instrument to have a Zeiss parallelogram arrangement such that base-in and base-out viewing positions were possible (see section 8.3.2) and also that a one-projector method of relative orientation could be carried out. The technique used allows the photograph common to the two models to remain undisturbed in the instrument during the connecting process. In this way a strip model is formed using only two projectors. The diagrams of Fig 9.2 illustrate the main steps in the technique.

If we start with the base in the inside position as in Fig 9.2(a) then, with diapositives in the negative planes, the common areas of the two photographs are placed in the outside positions. The first model is oriented into either an absolute or a convenient orientation. Photograph 1 is then replaced by photograph 3 and the common overlap now falls on the inside sections of the photographs. The viewing optics are therefore switched over so that the left eye now views the right-hand photograph and vice versa, otherwise the operator will experience a pseudoscopic effect. The B_x base component of the Zeiss parallelogram is then changed to give the base-out position. Now, a one-projector relative orientation is carried out using only movements associated with the left-hand projector. The model is scaled using small variations in the base length and the checks carried out as before using the co-ordinates of the common tie points. In this way the second model is connected to the first with the correct orientation. The process can now be repeated by replacing photograph 2 by 4, switching the optics again and moving from the base-out to the base-in position and so on. In this way a homogeneous set of model co-ordinates can be obtained for all the tie points of the strip. In this method, the model co-ordinates of the six tie points are recorded after each connection. Some method of automatically recording these co-ordinate values is therefore highly desirable.

For many years, universal machines such as the Zeiss Stereoplanigraph and the Wild A7 Autograph were used to carry out strip aerial triangulation in this way. As the precision of such instruments is very high, the errors in orientation carried forward from model to model were correspondingly low. The elimination of parallaxes can be carried out with a precision of about 10 μm, while spot height precision is of the order of $0.07^o/_{oo}$ Z. One factor that obviously limits the accuracy of this technique is the ability of a projector to maintain the same projection characteristics in the base-in and base-out positions. The considerable change in instrument geometry involved requires a very high precision in manufacture if stability and accuracy are to be maintained. In recent years no machines have been designed with the base-in and base-out facility since other methods of aerial triangulation have been developed that make this requirement obsolete.

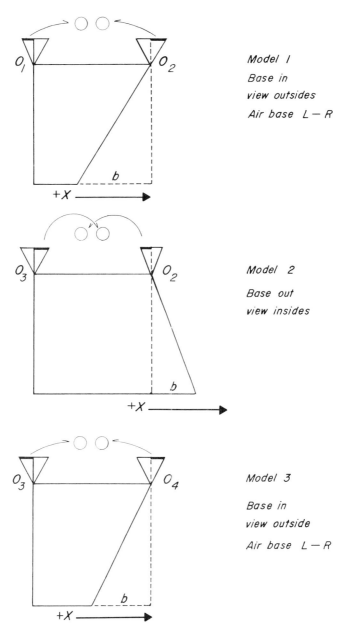

Figure 9.2. Aerial Triangulation by means of the Zeiss Parallelogram

9.5 The Nistri Photomultiplex

This method is based on an idea that is quite simple but it is not one that has been taken up. It would appear that the Nistri Photomultiplex was the only instrument to be designed for this method. The instrument using direct optical projectors and anaglyph viewing was constructed so that the Z-axis of each projector was the primary axis. With such an arrangement it means that rotation of the projector in azimuth (i.e. kappa) does not affect the phi and omega tilts. Hence, after carrying out the orientation of model 1, the right-hand projector is rotated through approximately 180° so that the base line now runs in the

196

opposite direction. Photograph 3 is now placed in the left-hand projector and a relative and absolute orientation carried out using only the movements of that projector. On completion, this projector is rotated through approximately 180°, photograph 2 is replaced by photograph 4 in the right-hand projector and so on. The positive direction of the air base changes direction by about 180 at each connection.

In the vast majority of projectors, kappa rotations take place about tertiary axes and these are sometimes displaced away from the main centres of rotation. Only in the case of direct optical projection is it comparatively simple to arrange for kappa rotations to be primary.

9.6 The use of inclinometers

In the three methods described so far, the tilt of the common projector has, by some means or other, remained undisturbed with respect to the instrument datum surface during the connecting process. A number of other methods have been devised by which the tilt of the common photograph is recorded in some way and then reintroduced after moving the photograph from the right-hand to the left-hand projector (assuming of course the positive X-direction is from left to right).

One obvious method that might appear to be satisfactory is one making use of the graduated dials provided for the rotational elements of the projectors of the larger instruments. The elements of orientation could be read off the right-hand projector and then reset on the left-hand projector when the photograph is moved across. In practice, however, it is found that the transfer cannot be made with sufficient accuracy. The readings cannot be taken to the required precision and the dials would need to be calibrated in order to allow for any zero errors.

A most straightforward and satisfactory technique for accurately transferring the tilt values has been devised using inclinometers which can record directly the φ and ω tilt of the photograph in its plate holder. This device consists of two sensitive spirit bubbles held at right angles to one another, both being capable of being levelled using two foot screws. Figure 9.3 shows a diagram of the unit.

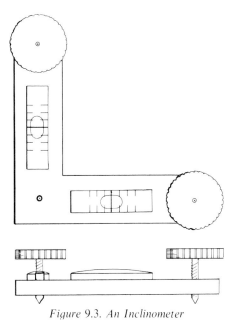

Figure 9.3. An Inclinometer

197

After the orientation of a model, the inclinometer is placed on the right-hand plate holder in its locating grooves. The two bubbles are then levelled up using the finely threaded levelling screws. The inclinometer is then carefully removed and placed in a safe place. The next model is now set up after moving the right-hand photograph, in its plate holder, over to the left-hand side. The new photograph is then placed in the right-hand side. If the instrument cannot be used to carry out a one-projector relative orientation then this must be done using both projectors. The inclinometer is then replaced on the left-hand plate holder and the model rotated, using the elements of absolute orientation, until the two bubbles are in the middle of their runs, indicating that they are once again in levelled positions. The plate holder and photograph must now have their original tilt within certain limits. The correct scale of the model can be introduced as previously using the height of a point on the base line and/or the planimetric co-ordinates of the two wing points. The height readings of the wing points will provide a check on the Ω orientation. In this method the heights of all tie points of the strip are recorded while the planimetric positions can either be recorded or plotted directly at the scale of the strip model.

In this method we make the basic assumption that the projective properties of each projector are identical. If they are not, then the slight differences will introduce systematic errors into the co-ordinates of the strip model as we progress along the strip. Usually $20''$ bubbles are considered sufficiently accurate for most purposes. More sensitive bubbles have sometimes been suggested, especially for the Φ-tilt determination, but these can take a long time to settle. If we repeat some of the arithmetic of section 9.3 we note that a precision of $\pm 0.1^{o}/_{oo}$ Z for the instrument capability suggests a precision of about $\pm 24''$ in the Ω-tilt setting using the heights of the two wing points. A bubble sensitivity of the order suggested above would therefore seem to be appropriate even for a first order plotter with the precision quoted above.

The method can be adapted for use on any instrument where the projectors take up their correct positions in space after relative orientation. Tests carried out using a Wild A8 showed that the results obtained in strip triangulation were comparable with those obtained using the classical method on a Zeiss Stereoplanigraph or Wild A7 instrument.[9.1]

9.7 The semi-analogue method using independent models

The last method we describe in this chapter is not strictly a pure analogue method. The models are formed as accurately as possible using an analogue plotting instrument but no attempt is made to join the models mechanically. Instead, the information available is used to carry out an analytical joining up of the models. This can be done on one of the larger dedicated desk top computers (such as the Hewlett–Packard 9830 or Wang 2200) or a large computer. Data from any instrument giving model co-ordinates can be used, some form of automatic reading of the co-ordinates being highly desirable. For the most sophisticated applications, punched or magnetic tape feeding directly into the computer is used. At the present moment, this hybrid method is considered to be one of the best ways of carrying out aerial triangulation and is being used extensively.

The analogue instrument is required to produce a model of the ground surface with as little distortion as possible. It is also required to be as stable as possible geometrically. Therefore no attempt is made to carry out an absolute orientation and all the elements associated with this process are usually set to their zero positions and remain undisturbed. The base length setting is set to as large a value as possible, or some convenient value, and is left unchanged as far as possible.

The method makes use of the positions of the two perspective centres and these are used as additional tie points in the connection of the models. It will be realised that the directions of the vertical or near-vertical lines such as O_iC_i and $O_{i+1}C_{i+1}$ in Fig 9.4 do in fact provide a sensitive way of relating the Φ and Ω tilts of one model with the next. In this method, therefore, the co-ordinates of the two perspective centres of the instrument must

198

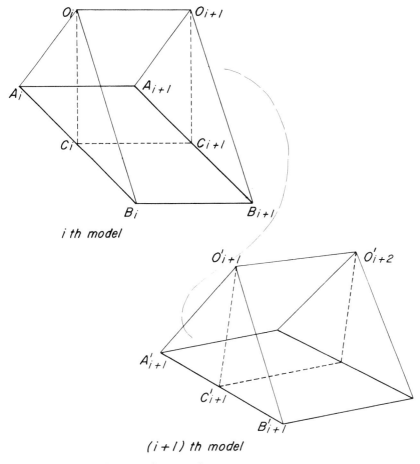

i th model

(i + 1) th model

Figure 9.4. *Aerial Triangulation Using Independent Models*

be determined. This will be done at least twice: before the commencement of the triangulation and on completion. During this time the co-ordinates will be assumed to have an accepted set of values with a low standard deviation. (In some techniques these co-ordinates are established each time a model is formed.)

The mathematical joining together of two adjacent models consists of three parts that can be carried out provided we know the co-ordinates of at least three points (such as O, A, B) common to both models. The three parts of the process are as follows.

(a) A bulk shift. The co-ordinates of the second model (based on O'_{i+1}) are transformed so that O'_{i+1} coincides with O_{i+1}, (see Fig 9.5(a)).

(b) Scaling of the second model, to bring the second model to the same scale as the first. The length of any common line may be used. In practice, the line AB is often used but the mean of two or more could also be used to establish the best value for the scale factor (e.g. use lines OA, OB, BA and perhaps OC).

(c) Rotation of the model, to bring the second model into correct orientation with the first. With a minimum of three common points (such as O, A, B) the operation can be thought of as making a plane O'A'B' coincide with a plane OAB as illustrated in part (b) of Fig 9.5.

Mathematically we can describe the above processes in the following way. Let any point n

199

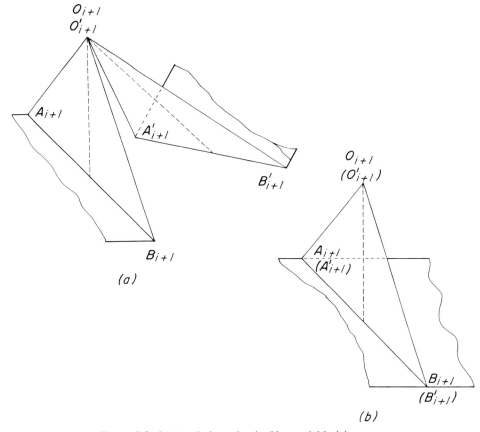

Figure 9.5. *Joining Independently Observed Models*

have co-ordinates in the ith model represented by $(X_n Y_n Z_n)_i$. In the next model of the strip this point has, before transformation, the observed co-ordinates $(X'_n Y'_n Z'_n)_{i+1}$. In order to transform these co-ordinates into the system of the preceding model, the following transformation must be carried out

$$\begin{pmatrix} X_n \\ Y_n \\ Z_n \end{pmatrix}_i = \lambda_{i+1} R_{i+1} \begin{pmatrix} X'_n \\ Y'_n \\ Z'_n \end{pmatrix}_{i+1} + \begin{pmatrix} \Delta X_o \\ \Delta Y_o \\ \Delta Z_o \end{pmatrix}_{i+1} \tag{9.1}$$

where λ is the scale factor, R is the rotation matrix and the Δ terms are the parameters of translation for the $(i+1)$th model. As the matrix R contains three independent unknowns (see Appendix A) there are, in all, seven unknown parameters associated with the above equation. To solve for these we therefore use the observed co-ordinates of at least three points common to both models.

A determination for the value of the scale factor can be obtained from two common points such as A and B

$$\lambda_{i+1} = \left[\frac{(X_B - X_A)^2 + (Y_B - Y_A)^2 + (Z_B - Z_A)^2}{(X'_B - X'_A)^2 + (Y'_B - Y'_A)^2 + (Z'_B - Z'_A)^2} \right]^{1/2} \tag{9.2}$$

More lines can be used to provide a better mean value, if this is thought necessary.

The nature of the elements of the rotation matrix depend on the form of the orthogonal matrix selected (see equations (A7) and (A14)), those leading to linear equations of the parameters being the easiest to handle. As a basis for the explanation given here, the Rodrigues–Cayley forms will therefore be used initially, for all the equations can be derived from simple algebraic manipulation. These equations have the forms given below. Using Rodrigues:

$$R_R = \frac{1}{\Delta} \begin{pmatrix} \Delta' & -c & b \\ c & \Delta' & -a \\ -b & a & \Delta' \end{pmatrix} + \frac{1}{2\Delta} \begin{pmatrix} a \\ b \\ c \end{pmatrix} (abc) \qquad (9.3)$$

where $\qquad\qquad \Delta = 1 + \frac{1}{4}(a^2 + b^2 + c^2)$

and $\qquad\qquad \Delta' = 1 - \frac{1}{4}(a^2 + b^2 + c^2)$

Using Cayley:

$$R_C = (I - S)(I + S)^{-1} \qquad (9.4)$$

where

$$S = \frac{1}{2} \begin{pmatrix} 0 & c & -b \\ -c & 0 & a \\ b & -a & 0 \end{pmatrix}$$

and I = unit matrix.

Since the parameters of the two equations are the same (one can be derived from the other – see section 6, appendix A) the calculation is carried out in the following way. Firstly, by using equation (9.4), the co-ordinates of the common tie points can be used to derive a set of linear equations involving the three parameters (a, b, c). Secondly, having evaluated these, their values can then be used to calculate the values of the elements of equation (9.3). We should note, however, that the co-ordinates to be used in the evaluation of R are derived from the observed values by taking the common perspective centre as the origin in both cases. The three parameters (a, b, c) approximate to the three rotations $(\omega, \varphi, \kappa)$ about the axes (X, Y, Z) only when they are small in magnitude. Their values are related in the following way

$$a = 2l \tan \theta/2$$
$$b = 2m \tan \theta/2$$
$$c = 2n \tan \theta/2$$

where θ is a rotation about a line whose direction cosines are (l, m, n).
Considering now the evaluation of R we have, from equation (9.1),

$$\begin{pmatrix} X_n \\ Y_n \\ Z_n \end{pmatrix}_i = R_{i+1} \begin{pmatrix} X'_n \\ Y'_n \\ Z'_n \end{pmatrix}_{i+1}$$

where n is any tie point with co-ordinates in both models now expressed in terms of a common origin (the common perspective centre) and brought to the same scale.

Using the form of R_C quoted in equation (9.4)

$$(I+S) \begin{pmatrix} X_n \\ Y_n \\ Z_n \end{pmatrix}_i = (I-S) \begin{pmatrix} X'_n \\ Y'_n \\ Z'_n \end{pmatrix}_{i+1}$$

i.e.

$$\begin{pmatrix} 1 & c/2 & -b/2 \\ -c/2 & 1 & a/2 \\ b/2 & -a/2 & 1 \end{pmatrix} \begin{pmatrix} X_n \\ Y_n \\ Z_n \end{pmatrix}_i = \begin{pmatrix} 1 & -c/2 & b/2 \\ c/2 & 1 & -a/2 \\ -b/2 & a/2 & 1 \end{pmatrix} \begin{pmatrix} X'_n \\ Y'_n \\ Z'_n \end{pmatrix}_{i+1}$$

On multiplying out, we find

$$\begin{pmatrix} X'_n - X_n \\ Y'_n - Y_n \\ Z'_n - Z_n \end{pmatrix} = \begin{pmatrix} \Delta X_n \\ \Delta Y_n \\ \Delta Z_n \end{pmatrix} = \begin{pmatrix} 0 & -\overline{Z}_n & \overline{Y}_n \\ \overline{Z}_n & 0 & -\overline{X}_n \\ -\overline{Y}_n & \overline{X}_n & 0 \end{pmatrix} \begin{pmatrix} a \\ b \\ c \end{pmatrix} \tag{9.5}$$

where

$$\overline{X}_n = \frac{X_n + X'_n}{2} \qquad \overline{Y}_n = \frac{Y_n + Y'_n}{2} \qquad \overline{Z}_n = \frac{Z_n + Z'_n}{2}$$

We have therefore three simple equations from which the parameters can be calculated. The matrix is singular and so, as we would expect, at least two model points are needed to solve these equations. If more than the minimum number of points are used (say O, A, B and C) then a least squares solution will give the most satisfactory set of values.

We are now in a position to carry out the transformation for all other points observed in the $(i+1)$th model. The parameters for scaling and rotation have been evaluated and it only remains to note that the bulk shift parameters are given by the strip co-ordinates of the common $(i+1)$th perspective centre.

The formation of strips from independent models by the Schut method is well documented[9.1,9.2] and is an advancement on the above in the sense that the derivation is rather more elegant and the computational procedures are simplified somewhat further. The method is now widely used and the Fortran programmes for its application are readily available. The main characteristics of the method are as follows.

The rotation matrix has the form

$$R = (dI - S)(dI + S)^{-1}$$

and

$$R = \frac{1}{d^2 + a^2 + b^2 + c^2} \begin{matrix} d^2 + a^2 - b^2 - c^2 & 2ab - 2cd & 2ac + 2bd \\ 2ab + 2cd & d^2 - a^2 + b^2 - c^2 & 2bc - 2ad \\ 2ac - 2bd & 2bc + 2ad & d^2 - a^2 - b^2 + c^2 \end{matrix} \tag{9.6}$$

The four parameters (a, b, c, d) are evaluated from linear equations of the form given below, each common point n yielding one set of equations:

(i) $\qquad a\Delta X_n + b\Delta Y_n + c\Delta Z_n \qquad = 0$

(ii) $\qquad\qquad -b\overline{Z}_n + c\overline{Y}_n + d\Delta X_n = 0$

(iii) $\qquad a\overline{Z}_n \qquad - c\overline{X}_n + d\Delta Y_n = 0$

(iv) $\qquad -a\overline{Y}_n + b\overline{X}_n \qquad + d\Delta Z_n = 0$

$$\tag{9.7}$$

202

As before, the co-ordinates used in the calculation of the parameters of rotation are those using the common perspective centre as origin. From this centre, the common vectors are reduced to unit length. A least square solution is used to determine the values of the parameters but the resulting normal equations will allow only their ratios to be found. In this situation one of the parameters can be made equal to unity. Usually d is chosen for this purpose and in this case the parameters a and b are functions of the φ and ω tilts respectively.

9.8 The accumulation of errors in strip triangulation

9.8.1 THE EFFECT OF EARTH CURVATURE

This does not give rise to an error in the strict sense of the word but it does produce systematic discrepancies between the height values of the analogue model and the true surface. It can be corrected for separately or, as is sometimes done in practice, taken care of when adjusting for the effects of the real errors arising from various sources. The discrepancy arises because the datum surface of the analogue instrument is in a plane (usually horizontal) surface whereas the ground datum surface is spherical in shape. The consequences of this are illustrated in Fig 9.6 and are most apparent in practice when carrying out a strip of aerial triangulation on a long bar Multiplex instrument.

We note that when two projectors using vertical photography taken from the same elevation are in absolute orientation the photograph planes are not coplanar but subject

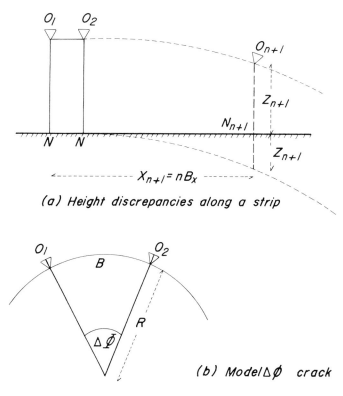

(a) Height discrepancies along a strip

(b) Model $\Delta\varnothing$ crack

Figure 9.6. Earth Curvature Effect

to a convergence in the form of a small relative φ tilt. The magnitude of this can be seen readily from an inspection of Fig 9.6(b), for we have

$$\Delta\varphi = \frac{B}{R}$$

$$= \frac{B_m}{R_m}\text{rad}$$

where B is the length of the air base and R is the radius of curvature of the earth; B_m and R_m are the same quantities expressed at model scale. In the successive connection of models, the air stations would therefore generate the curve $O_1, O_2, \ldots, O_{n+1}$ even although these were the result of an error-free process. The curve represents the correct path of the air stations in space and is the manifestation of the earth's curvature at model scale. We see, therefore, that after n models the discrepancy is such that

$$\Delta Z_{n+1} = -\frac{X_n^2}{2R_m} = -\frac{(nB_m)^2}{2R_m}$$

if we assume the first photograph was correctly orientated in the strip model.

For example, if we consider a strip of wide angle photographs ($f = 152$ mm) taken at 5000 m with the standard overlapping such that the air base is 92 mm in length at photo scale, we have for each model a convergence $\Delta\Phi$ given by

$$\Delta\Phi = \frac{92}{152} \times \frac{5000}{6400 \times 10^3} \text{ rad}$$

i.e. $\qquad\qquad\qquad \Delta\Phi = 97''$

After 10 models the discrepancy in height (again at photo scale) is given by

$$\Delta Z = -\frac{(10 \times 92)^2}{2 \times 6400 \times 10^3} \times \frac{5000}{152} \text{ mm}$$

i.e. $\qquad\qquad\qquad \Delta Z = -2.17$ mm

9.8.2 THE EFFECT OF ERRORS OF OBSERVATION

When forming a strip of aerial triangulation small random errors must be introduced into the process at each stage as a result of the observational errors that must inevitably occur. In analogue triangulation the relative orientation process is of paramount importance and the quality of the strip will depend largely on the ability of the observer to produce orientations with the smallest possible residual errors in y-parallax. Also, in the connecting and scaling procedures the observer must identify points common to different photographs and models with a high degree of skill. The overall effects of these observational errors can be conveniently regarded as a gradual accumulation of errors in the positions of the air stations (perspective centres) and the tilts of the photographs (projectors) as we progress along the strip. However, the nature of the propagation of errors should be such that we can regard the model as the smallest unit that need be considered. The various errors will therefore introduce errors in the location of each model ($\Delta X, \Delta Y, \Delta Z$), the scale of the model ($\Delta\lambda$) and the orientation of the model ($\Delta K, \Delta\Phi, \Delta\Omega$). It is the deformation of the strip model produced by the accumulation of these errors that we wish to investigate. Of course there will also be small residual errors in the orientation of each model and these will give rise to local model deformations but these have been discussed in chapter 7 and need not be considered again at this point.

In addition to the observational errors of a random nature there will be some other sources of error that will produce similar effects. For example in using inclinometers, the ability of these to record and reintroduce projector tilts is a source of further errors. The projectors themselves cannot carry out perfect projective transformations, and so these will contribute certain random and some systematic effects. This is particularly the case when the projectors are used in the base-in and then base-out configurations. Again, the photographs contain errors in position due to factors such as lens distortion and film shrinkage. However, whatever the cause, we can conveniently describe the effects of all errors as contributing in some way to the seven possible model errors noted above. This is of course only one of a number of possible approaches to the problem.

Since the seven model errors are made up of a number of random errors, it is reasonable to assume that they themselves are random errors with a low degree of correlation. Hence we can write as typical examples of their summations

$$\left.\begin{array}{l} \sum_{i}^{n} \Delta X_i = \Delta X_1 + \Delta X_2 + \ldots + \ldots \Delta X_n \\[2ex] \sum_{i}^{n} \Delta \lambda_i = \Delta \lambda_1 + \Delta \lambda_2 + \ldots \Delta \lambda_i + \ldots \Delta \lambda_n \\[2ex] \sum_{i}^{n} \Delta K_i = \Delta K_1 + \Delta K_2 + \ldots \Delta K_i + \ldots \Delta K_n, \text{ etc.} \end{array}\right\} \tag{9.8}$$

The resulting errors in the (i)th model are therefore single summations of the individual random errors. The accumulation of error in such summations is slow but they do in fact exhibit some quasi-systematic effects in cases where n is large. The arc sine law of statistics is the general expression of this property — see Refs 9.3, 9.4 and 9.5. In aerial triangulation, however, the value of n is comparatively small and likely to fall within the range of 5 to 25 in most cases. The mean square error of the sum is larger than the mean square error of the individual errors but not considerably so and can be estimated quite readily. A linear algebraic expression to denote the propagation is therefore quite satisfactory for practical purposes. An expression such as

$$\Delta \lambda_X = a_0 + a_1 X$$

might therefore be employed to describe the error $\Delta \lambda_X$ in the model scale, at a distance X along the strip. Such an adjustment is attempted for example when faced with an observed error in scale in the last model of a long bar Multiplex triangulation. In general, however, such a procedure is not possible because the control data provided does not lend itself to the direct recognition of such model errors. This is true for control data derived from ground surveys where the usual practice is to provide the ground co-ordinates of identified points on the photographs. It is not necessarily true for control information arising from the use of auxiliary instruments but this aspect is discussed at a later stage in chapter 12.

If we therefore consider control only in the form of photo control points then we are in the situation where we can observe the effects of the model errors on the strip model co-ordinates by comparing the strip co-ordinates obtained from triangulation with their known true values. In such circumstances it is most important to realise that these errors in model co-ordinates do not necessarily propagate along the strip in a simple linear fashion. They will do so only if they are the subject of a single summation of the type already noted. They may, however, be the subject of a more complex summation process, often referred to as a double summation but here described as a biased summation. This takes the following form:

$$\sum_{1}^{n} \Delta r = n \Delta r_1 + (n-1) \Delta r_2 + \ldots (n-i) \Delta r_{i+1} + \ldots 1 . \Delta r_n \tag{9.9}$$

where Δr_i is a series of small random errors as before. In this case the error terms from one model to the next have a strong correlation and the mean square error of the sum after n

models can be considerable. It cannot, therefore be the subject of a linear algebraic function. However, it is possible to demonstrate empirically that provided n is less than about 15 then a quadratic expression is an adequate description of the nature of the error propagation. For higher values of n, the number of cases where a third or higher order expression would be more satisfactory increases significantly. In photogrammetry the longest strips are usually to be found in small scale work and here 20–20 models would be considered as long. The employment of longer strips can lead to difficulties in adjustment.

As an example of a biased summation consider the effect of an accumulation of small scale errors $(\Delta\lambda_i)$ on the X co-ordinate after n models each of equal base length B. Let the first error $\Delta\lambda_1$ produce an error δX_1 in the nth model and so on. Hence we have, after n models,

$$\delta X_1 = nB \cdot \Delta\lambda_n$$

$$\delta X_2 = (n-1)B \cdot \Delta\lambda_2$$

$$\delta X_3 = (n-2)B \cdot \Delta\lambda_3$$

$$\delta X_i = (n-i+1)B \cdot \Delta\lambda_i$$
$$\vdots \qquad \vdots$$
$$\delta X_n = 1 \cdot B \cdot \Delta\lambda_n$$

$$\sum^n \delta X_i = B[n\Delta\lambda_1 + (n-1)\Delta\lambda_2 + \ldots 2 \cdot \Delta\lambda_{n-1} + \Delta\lambda_n$$

i.e.
$$\sum_1^n \delta X_i = B\sum_1^n (n-i+1)\Delta\lambda_i \qquad (9.10)$$

Because of the nature of the terms within the bracket, an algebraic expression can only be derived if we make some assumptions as to the nature of the errors. If, for example, we stipulate they are all of equal magnitude and of the same sign (i.e. not random at all) then, noting that $X = nB$, we have

$$\Sigma\delta X_i = B \cdot \Delta\lambda\Sigma n$$

$$= \frac{n(n+1)}{2} \cdot B\Delta\lambda$$

$$= \frac{X^2}{2B} \cdot \Delta\lambda + \frac{X}{2} \cdot \Delta\lambda \qquad (9.11)$$

In the case of both the errors in the Y and Z co-ordinate we have single summations for

$$\Sigma\delta Y_i = Y[\Delta\lambda_1 + \Delta\lambda_2 + \ldots \Delta\lambda_n]$$

and
$$\Sigma\delta Z_i = Z[\Delta\lambda_1 + \Delta\lambda_2 + \ldots \Delta\lambda_n]$$

the scale error in the nth model being a single summation of the previous errors. Again, if we take equal errors then

$$\Delta Y = \Sigma\delta Y_i$$

$$= Yn\Delta\lambda$$

$$= \frac{XY}{B} \cdot \Delta\lambda \qquad (9.12)$$

and
$$\Delta Z = \Sigma \delta Z_i$$

$$= Zn\Delta\lambda$$

$$= \frac{XZ}{B} \cdot \Delta\lambda \tag{9.13}$$

From equation (9.10) above we now see that in the case of a biased summation the contribution of each error to the total error after n models is directly related to the position of the error in the strip; the first error having the greatest effect, the next error a slightly smaller effect and so on. It is for this reason the term 'biased summation' has been used to describe this effect. There are four characteristics of this summation that are important for the understanding of the propagation of errors in aerial triangulation:

(a)   Small individual errors can produce much larger effects than one might have expected.

(b)   The accumulation of errors shows a strong systematic trend.

(c)   To produce a point of inflexion a run of errors all of the same sign is necessary, the number required to do this increasing progressively along the strip.

(d)   The signs and magnitudes of the first few errors are likely to determine the nature of the error curve.

In Fig 9.7 ten small random errors have been subject to both single and biased summation processes, and the corresponding error curves have been drawn to illustrate these points. In order to be able to devise equations that will adequately describe the effects of the seven model errors on the model co-ordinates, we must therefore examine each error and its effect on each co-ordinate in order to decide on the form of the equation we will use.

9.8.3 CORRECTION EQUATIONS FOR USE IN STRIP TRIANGULATION

It will be readily appreciated that the errors in position $(\Delta X_i, \Delta Y_i, \Delta Z_i)$ introduced at each model join are the subject of a single summation, hence the error after n models is such that

$$\Sigma\Delta X = \Delta X_1 + \Delta X_2 + \ldots \Delta X_i + \ldots \Delta X_n$$

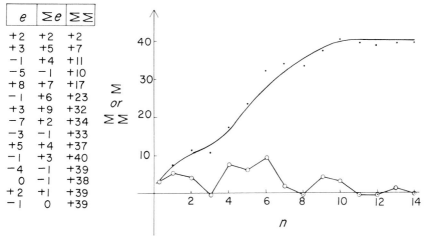

e	Σe	$\Sigma\Sigma$
+2	+2	+2
+3	+5	+7
−1	+4	+11
−5	−1	+10
+8	+7	+17
−1	+6	+23
+3	+9	+32
−7	+2	+34
−3	−1	+33
+5	+4	+37
−1	+3	+40
−4	−1	+39
0	−1	+38
+2	+1	+39
−1	0	+39

Figure 9.7. Single and Double Summation of Random Errors

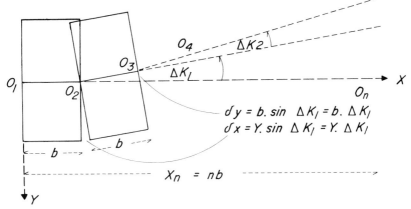

Figure 9.8. Effect of Kappa Error on Strip Model

And a suitable linear expression for photogrammetric purposes would be

$$\Sigma \Delta X = \Delta X_1 + aX \tag{9.14}$$

Similar equations would apply for the Y, Z co-ordinates.

Errors in scale affect all co-ordinates but this is a biased summation effect in the case of the X co-ordinate and a single summation in the case of the other two co-ordinates (see equations (9.10) et seq).

The effects of the model rotations ΔK_i can easily be established from an inspection of Fig 9.8 and we note in this case that, for small rotations:

(i)   the error in X is a single summation effect

(ii)   the error in Y is a biased summation effect

(iii) there is no error introduced into the Z co-ordinate.

For we have

$$\left.\begin{aligned}
\Delta X &= Y \sum_1^n \Delta K_i \\
\text{and} \quad \Delta Y_n &= \sum_1^n \delta Y_i \\
&= B \sum_1^n [(n-i+1)\Delta K_i]
\end{aligned}\right\} \tag{9.15}$$

In the case of the $\Delta \Phi_i$ errors, they are of a similar nature to those of ΔK and from an inspection of Fig 9.9 we see, for small angles:

(i)   the error in X is a single summation effect

(ii)   there is no error introduced into the Y values

(iii) there is a biased summation effect on Z values

For we have

$$\left.\begin{aligned}
\Delta X &= Z \sum_1^n \Delta \Phi_i \\
\text{and} \quad \Delta Z &= \sum_1^n \delta Z = B \sum_1^n [(n-i+1)\Delta \Phi_i]
\end{aligned}\right\} \tag{9.16}$$

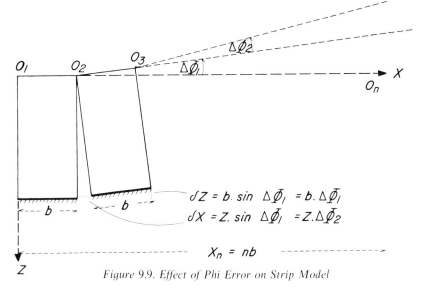

$$dZ = b.\sin\,\Delta\phi_i\ = b.\Delta\phi_i$$
$$dX = Z.\sin\,\Delta\phi_i\ = Z.\Delta\phi_2$$

$$X_n = nb$$

Figure 9.9. Effect of Phi Error on Strip Model

In the case of the $\Delta\Omega_i$ errors in model tilt, from an inspection of Fig 9.10 we see that, for small angles,

(i)   there is no error introduced into the X values

(ii)   there is a single summation effect on the Y co-ordinate values,

(iii) errors in the Z co-ordinate are due to a single summation of errors

For we have

$$\left.\begin{aligned}
\Delta Y &= Z\sum_{1}^{n}\Delta\Omega \\[2mm]
\Delta Z &= Y\sum_{1}^{n}\Delta\Omega
\end{aligned}\right\} \tag{9.17}$$

and

$$\Delta Z = Y.\sin\,\Delta\Omega\ = Y.\Delta\Omega$$
$$\Delta Y = Z.\sin\,\Delta\Omega = Z.\Delta\Omega$$

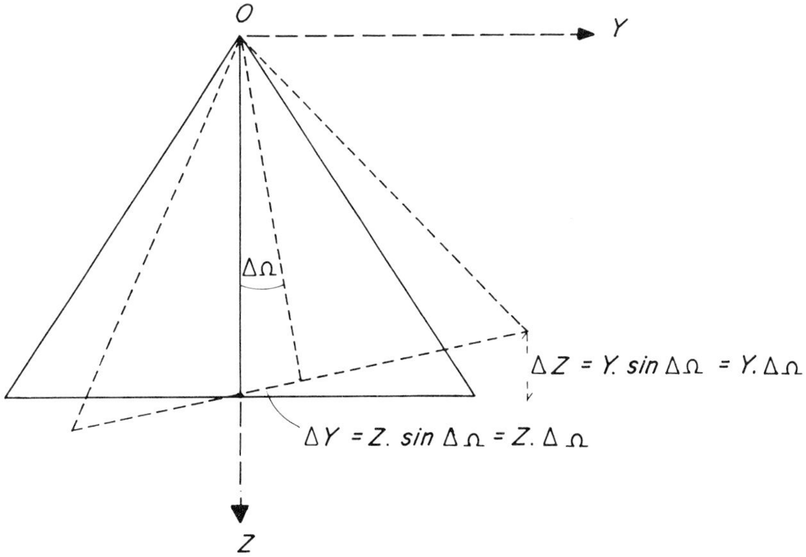

Figure 9.10. Effect of Omega Error on Strip Model

209

If we now combine the above effects we can write a composite equation of error for each of the three co-ordinates, as follows:

$$\left.\begin{aligned}
\Delta X &= \Sigma\Delta X_i + Y\Sigma\Delta K_i + Z\Sigma\Delta\Phi_i + B\Sigma[(n-i+1)\Delta\lambda_i] \\
\Delta Y &= \Sigma\Delta Y_i + B\Sigma[(n-i+1)\Delta K_i] + Z\Sigma\Omega_i + Y\Sigma\Delta\lambda_i \\
\Delta Z &= \Sigma\Delta Z_i + B\Sigma[(n-i+1)\Delta\Phi_i] + Y\Sigma\Delta\Omega_i + Z\Sigma\Delta\lambda_i
\end{aligned}\right\} \quad (9.18)$$

If in accordance with the comments made earlier we decide that all single summations can, for our purposes, be adequately described by a *linear* equation and that all biased summations can be replaced with, let us say, a *quadratic* expression in X then we have

$$\Sigma\Delta r_i = \rho X$$

and

$$B\Sigma[(n-i+1)\Delta r_i = (q_1 X + q_2 X^2)$$

Substituting expressions of these forms into equations (9.18) and also allowing for possible errors (ΔX_0, ΔY_0, ΔZ_0, ΔK_0, $\Delta\Phi_0$, $\Delta\Omega_0$, $\Delta\lambda_0$) in the first model we have

$$\left.\begin{aligned}
\Delta X &= (\Delta X_0 + a_1 X) + Y(\Delta K_0 + a_2 X) + Z(\Delta\Phi_0 + a_3 X) + [(\Delta\lambda_0 + a_4)X + a_5 X^2] \\
\Delta Y &= (\Delta Y_0 + b_1 X) + [(\Delta K_0 + b_2)X + b_3 X^2] + Z(\Delta\Omega_0 + b_4 X) + Y(\Delta\lambda_0 + b_5 X) \\
\Delta Z &= (\Delta Z_0 + c_1 X) + [(\Delta\Phi_0 + c_2)X + c_3 X^2] + Y(\Delta\Omega_0 + c_4 X) + Z(\Delta\lambda_0 + c_5 X)
\end{aligned}\right\} \quad (9.19)$$

Rearranging the terms a little we finally arrive at expressions of the following form

$$\left.\begin{aligned}
\Delta X &= A_0 + A_1 X + A_2 Y + A_3 Z + A_4 XY + A_5 XZ + A_6 X^2 \\
\Delta Y &= B_0 + B_1 X + B_2 Y + B_3 Z + B_4 XY + B_5 XZ + B_6 X^2 \\
\Delta Z &= C_0 + C_1 X + C_2 Y + C_3 Z + C_4 XY + C_5 XZ + C_6 X^2
\end{aligned}\right\} \quad (9.20)$$

As they stand, equations (9.20) require a minimum of seven control points in order to solve for the 21 unknowns present. However, we can reduce the amount of control required if we introduce a number of constraints. For example, if we assume that the changes in ground elevation are insignificant then we no longer need to take into account a variable Z. In this way the number of variables is reduced to five. This assumption is valid except in unusual terrain circumstances, and so equations of a familiar form are often used in practice to describe the propagation of co-ordinate error in strips, i.e.

$$\left.\begin{aligned}
\Delta X &= A_0 + A_1 X + A_2 Y + A_3 XY + A_4 X^2 \\
\Delta Y &= B_0 + B_1 X + B_2 Y + B_3 XY + B_4 X^2 \\
\Delta Z &= C_0 + C_1 X + C_2 Y + C_3 XY + C_4 X^2
\end{aligned}\right\} \quad (9.21)$$

In the above equations the coefficients are not of course identical with those of equation (9.20). Also, although some correlation between coefficients does exist and is indicated by the above analysis this has been ignored in these sets of equations. In recognition of the existence of some degree of correlation between coefficients the planimetric equations may be modified slightly to make them conformal. By doing this the planimetric shape of the surface will not be distorted by the adjustment process. To ensure this, the two Cauchy–Riemann conditions will need to be fulfilled:

$$\frac{\delta(\Delta X)}{\delta X} = \frac{\delta(\Delta Y)}{\delta Y}$$

$$\frac{\delta(\Delta X)}{\delta Y} = -\frac{\delta(\Delta Y)}{\delta X}$$

Examination will show that modified equations of the form given below will satisfy the two conditions quoted:

$$\left.\begin{array}{l} \Delta X = A_0 + A_1 X - B_1 Y + A_2(X^2 - Y^2) - 2B_2 X Y \\ \Delta Y = B_0 + B_1 X + A_1 Y + B_2(X^2 - Y^2) + 2A_2 X Y \end{array}\right\} \quad (9.22)$$

It will be noted that this process has required the introduction of terms involving Y^2 that were not suggested by our elementary attempts at error analysis. In some ways, therefore, the formulation of these polynomials is best regarded as an empirical error curve or surface fitting exercise. The number of terms that can be employed in practice is more often than not limited by economic factors; every extra coefficient introduced requiring additional control for its evaluation. An appreciation of the nature of the propagation of errors therefore centres around the shape of the curve produced by the biased or double summation effects. If this is such that over the range of n values used there is no point of inflexion in the general shape of the curve then it is likely that a second degree polynomial will provide an acceptable approximation. In practice, this seems to be the situation for many of the curves generated by strip aerial triangulation but it is always possible for the curve to exhibit at least one change in direction within the length of strips normally used. In this case a third degree curve would be much more likely to provide a better fit. Experience has indicated that there is little merit in employing equations of any higher degree. In the situation where more than sufficient control is available for the evaluation of the coefficients, it is therefore better in general to use a least squares technique to find the best values of a limited set of coefficients rather than introduce more coefficients involving higher order terms.

Some examples of the polynomial expressions that have been suggested and used in the past are given below.

Vermier

$$\Delta X = a_0 + a_1 X - b_1 Y + c_1 Z + a_2 X^2 - 2b_2 X Y + 2c_2 X Z$$
$$\Delta Y = b_0 + b_1 X + a_1 Y - d_1 Z + b_2 X^2 + 2a_2 X Y - 2d_2 X Z$$
$$\Delta Z = c_0 - c_1 X + d_1 Y + a_1 Z - c_2 X^2 + 2d_2 X Y + 2a_2 X Z$$

Schut

$$\Delta X = a_0 + a_1 X - b_1 Y + a_2(X^2 - Y^2) - 2b_2 X Y$$
$$\Delta Y = b_0 + b_1 X + a_1 Y + b_2(X^2 - Y^2) + 2a_2 X Y$$

Bervoet

$$\Delta X = aX^3 + bX^2 + cX + d - Y(3\rho X^2 + 2qX + r)$$
$$\Delta Y = \rho X^3 + qX^2 + rX + t - Y(3aX^2 + 2bX + c)$$

Lebanov

$$\Delta X = a_0 + a_1 X + a_2 Y + a_3 X^2 + a_4 X Y + a_5 X^2$$
$$\Delta Y = b_0 + b_1 X + b_2 Y + b_3 X^2 + b_4 X Y + b_5 Y^2$$
$$\Delta Z = c_0 + c_1 X + c_2 Y + c_3 X^2 + c_4 X Y + c_5 Y^2$$

9.8.4 GRAPHICAL METHODS OF ADJUSTMENT

Before the introduction of the modern computer with its large core the simultaneous solution of large numbers of equations was not a practical proposition. Photogrammetric adjustments therefore relied heavily on a number of graphical techniques. In cases where resources are limited or where the area to be mapped is of limited extent, these methods can still be usefully employed. They are also an instructive topic of study, for they can be an aid to the understanding of the mathematical techniques now being undertaken by the computer. Quite often in photogrammetric adjustments we find the Z co-ordinate being treated separately from the XY planimetric co-ordinates, and there are a number of practical reasons why this should be so. However, in graphical work it is well-nigh impossible to do otherwise and so we quite naturally deal with them separately.

9.8.5 PLANIMETRIC ADJUSTMENT USING STEREOTEMPLATES

In this technique each model, or very small group of models, is considered as the adjustment unit that is represented by a single stereotemplate. The stereotemplates are assembled to form a block of models in much the same way as ordinary slotted templates are assembled to form a block of photographs (see section 4.5). Within the block each model, defined by four or six model points, can shift position, rotate and change scale but it cannot change shape, i.e. the transformation allowed is a conformal one. When a block of templates is assembled over a number of fixed points, an overall adjustment of models is thereby carried out that provides the best overall fit to the control, within the mechanical and graphical limits of the process.

The templates are constructed when carrying out strip triangulation on the analogue plotting instrument. When each model has been brought into absolute orientation as well as possible, the positions of the four wing points are plotted along with any control points. (Sometimes, for convenience, the centre-line points are also recorded but this is not essential.) The scale of plotting should be map scale, or a little larger if this is practicable. From the plot, two slotted templates are produced each using opposite corner points as radiation centres, as shown in parts (b) and (c) of Fig 9.11. When the two halves are put together, one on top of the other, over four studs we have the stereotemplate. The centres of the four studs define the four wing positions. Using the common stud positions the templates are assembled on a laydown board in the usual way. One control point in each corner of the block is considered a reasonable minimum distribution of control for a square block of, say, 10 models long by 5 strips deep. For larger blocks practice has shown that extra control points around the periphery are more effective than points located in the centre of the block. It should be emphasised that the block size suggested above is merely a guide, and whether or not four points will provide an adequate frame of control depends on a number of factors. For example, for the mapping of rather featureless areas of 1:50 000 scale four corner points in a block 15 × 8 has proved to be quite adequate. On the other hand, for mapping of more important areas at larger scales a control point about every four or five models around the periphery of the block would be more satisfactory.[11.9]

Concerning the nature of the adjustment carried out by the process, it is true to say that

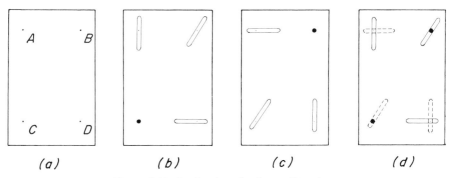

(a)   (b)   (c)   (d)

Figure 9.11. Production of a Stereo-Template

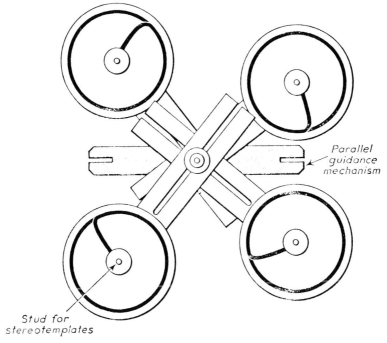

Parallel
guidance
mechanism

Stud for
stereotemplates

Figure 9.12. Spring Loaded Connector of Jerie Analogue Device

for well-made templates and a minimum of frictional forces each template is adjusted in position, azimuth and scale by roughly equal amounts. The adjustments of the four parameters $(\Delta X, \Delta Y, \Delta\lambda, \Delta K)$ are therefore of a linear nature, as discussed in the first part of section 9.8.2. If in the course of triangulation the absolute orientations remain reasonably correct, then the models themselves will contain only small residual errors of the type discussed earlier. However, if appreciable tilt errors do build up then the models will be the subject of an affine distortion that cannot be accurately compensated for by template adjustments. The main possibility here is the buildup of an appreciable error $(\Delta\Phi)$ in the Φ element of orientation towards the end of a long strip. In this case, the scale along the strip axis would be $S\cos\Delta\Phi$ when the scale value is S in the Y direction. However, this is not likely to be a source of appreciable error unless the slope of the model reaches the order of $2°$, an unlikely event in normal practice.

9.8.6 THE ITC–JERIE ANALOGUE COMPUTER FOR PLANIMETRIC ADJUSTMENTS
This semigraphical method was developed by Jerie in the mid-1950s and is a considerable improvement on normal stereotemplates from the point of view of accuracy. While stereotemplates are quite suitable for medium and small-scale work, the Jerie device provided an accuracy suitable for large-scale machine plotting. It does, however, use an iterative technique to achieve this accuracy and in consequence is rather time consuming in use. The large computer can now carry out the same type of adjustment very much more swiftly. The equipment, however, still has something to offer since it does provide a readily understood mechanical model of this adjustment technique. A detailed description of the equipment is given in Ref 0.4, the main features of which might be summed up as follows. Stereotemplates are made for each model as before but instead of the common wing points being joined directly they are connected by means of the four spring-loaded studs of a multiplet unit (see Fig 9.12). The displacement of a stud from its zero position represents a discrepancy in the co-ordinates of a tie point provided by each template. Because of the

213

manner of construction, these discrepancies can be displayed at a much larger scale than the original stereotemplate scale. After each laydown the changes in scale, azimuth and position introduced into each template are measured and a new set of model co-ordinates calculated. From these, the shift in each tie point position is introduced into the multiplet stud position. This can be done at increasingly large scales as the transformed model co-ordinates approach their 'correct' values. After a few iterations the recorded template movements become much reduced and represent insignificant changes in model parameters. At this stage the mean of the four sets of transformed tie point co-ordinates are accepted as the adjusted values. Because the studs are spring loaded, at every laydown the whole unit takes up a minimum energy position. That is to say, the sum of the squares of the displacements of the multiplet studs is a minimum. At the final laydown with the accepted modified co-ordinates of all tie points the discrepancies are small, even when introduced at a large scale, hence the technique provides an adjustment with a high degree of precision.

This apparatus also includes a mechanical method of adjusting height values for strip points. This is mentioned in more detail in the section concerned with graphical height adjustments.

9.8.7 THE POLYNOMIAL METHOD OF ADJUSTING PLANIMETRY

If we compare the plan co-ordinates of individual model points with their correct values we can of course carry out an adjustment using a polynomial expression of the type mentioned in section 9.8.3. For single, isolated strips such an adjustment could be carried out graphically and would be identical to that described in the next section in connection with the graphical adjustment of the Z co-ordinate. Separate graphs would need to be drawn for the X and Y co-ordinates. However, if more than one strip is involved then the stereotemplate technqiue is to be preferred.

A polynomial adjustment for a block can be carried out analytically and this method is described in chapter 11, where we consider the formation of blocks.

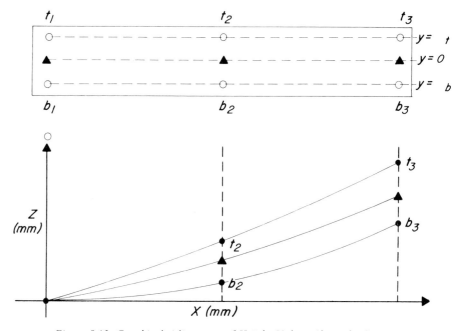

Figure 9.13. Graphical Adjustment of Height Valves Along the Strip

9.8.8 THE GRAPHICAL ADJUSTMENT OF HEIGHT VALUES

In this case we compare differences in height over the model surface with true differences in height. These discrepancies will not usually be a linear function of the X co-ordinate. It is not easy in practice to construct an error (or correction) surface although in fact this has been done in the case of the ITC–Jerie height adjustment device. The simple graphical methods therefore define the error surface by considering at least two longitudinal sections each of constant Y value. Using one control point as a datum point, the discrepancies between the model differences in height and those of the ground are plotted with the X co-ordinate as abscissa. This may be done using either true scale or model scale but in both cases the strip model scale is required to be known; Fig 9.13 illustrates the situation. With six known points, the top and bottom curves can be drawn up at some convenient scale if we assume a second order curve of some form for the propagation of error, for example

$$\Delta Z_t = (C_0 + C_2 Y_t) + (C_1 + C_3 Y_t)X + (C_4 + C_5 Y_t)X^2$$

and

$$\Delta Z_b = (C_0 + C_2 Y_b) + (C_1 + C_3 Y_b)X + (C_4 + C_5 Y_b)X^2$$

In these two equations we have merely regrouped some of the terms and introduced the YX^2 term to the equations (9.21) above. As there are only five unknowns contained in these equations, then six suitably distributed points would be sufficient to provide for their evaluation. Clearly six points distributed as shown is a most convenient arrangement for a simple graphical solution. Further, because we have assumed a linear propagation in the Y direction, these two curves are sufficient to define the form of the surface. In practice, this assumption has proved to be satisfactory for most work. We note that if a further three control points are provided along the centre line then the top and bottom curves should be found to be symmetrically disposed about the centre line curve. Such a pattern has been much used in practice for it provides some form of check on the accuracy of the adjustment. As to the length of strips and the number of models between control points these factors, as always, are influenced to some extent by the nature of the mapping requirement and the scale of mapping. For 1 : 50 000 mapping with 50 ft contours, strips of between 15–20 models in length have been used a great deal in the past. For larger scales and more accurate work a maximum distance of no more than five models between height control points is sometimes quoted. For longer strips than those quoted above the addition of extra control points should be considered.

Sometimes the main difficulty when providing control by normal ground survey is the initial one of bringing control into an area. Having established this, the extra survey work involved in providing extra control points is minimal. In such a case one would therefore consider providing full height control on models rather than say a minimum of two points. For example, absolute orientation of the first model of a strip can be of some advantage for the magnitude of the errors in subsequent models is then at a minimum. Also, if one has an idea of the order of magnitude these errors might be, the detection of gross errors and blunders is much facilitated. Additional control on any model has the advantage that it provides an immediate check on the quality of the control. An error in an identification or a co-ordinate value can be isolated and rejected before it is incorporated into the adjustment process.

In the formation and adjustment of blocks of aerial triangulation the top and bottom points would be arranged to fall in the common overlap of strips. When observing these strips it is good practice to observe one strip from left to right and then the next strip from right to left. Each tie point in the common overlap will therefore be provided with two adjusted height values. If the values lie within certain limits then a simple mean can be accepted as the best value available. Such a mean would be considered acceptable for example if the spread was less than one quarter of the mapping contour interval.

215

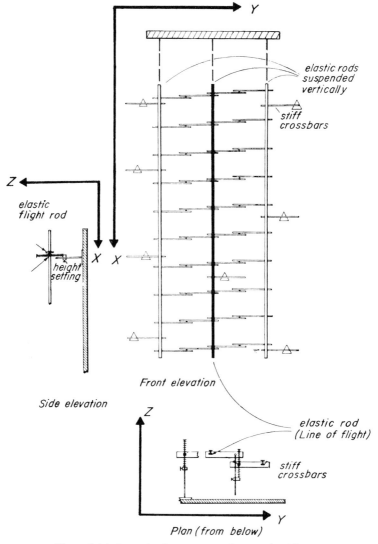

Figure 9.14. *Jerie Analogue Device for Height Adjustment*

The results of strip adjustments for height can be treated rather better than the above with the aid of the height adjustment apparatus of the ITC–Jerie analogue computer. Here the individual models are defined mechanically by selecting wing point positions on stiff crossbars fixed to a central flexible spline representing the centre line of the strip. These splines are hung vertically and each strip model is displaced with respect to the adjacent strips so that the common wing points of the two strips can be connected with spring-loaded connectors, as shown in Fig (9.14). (See Ref 0.4 from which this diagram was taken.) Control points can also be introduced along the strips, again using spring-loaded connectors. After assembly the apparatus is vibrated a little to ensure a state of mechanical equilibrium is reached. Again a minimum energy, that is a least square condition, is achieved. Some care is needed in the selection of the strength of springs and the cross section and mechanical properties of the spline material. In this way a correlation can be maintained between the various deformations that take place and strength of the photogrammetric observations. The height discrepancies are read off at the connectors by noting the displacements that have taken place at the various junctions.

216

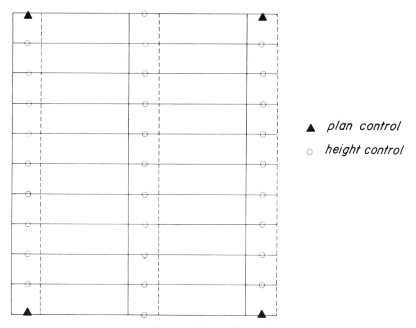

Figure 9.15. Distribution of Control for Block Adjustment

Control requirements for graphical adjustments

For a block of limited size, say of the order of 8–16 models long and 4–8 strips deep the pattern of ground control required should be no less than that shown in Fig 9.15.

Four plan points will suffice, although doubling up within the four models will increase the strength of the block and facilitate the analogue triangulation and adjustment. The establishment of an accurate model scale at the commencement of the work is particularly useful. Four corner points, well sited, will of course help to control any adjacent blocks that might also be observed.

The pattern of height control repeats itself from strip to strip and occurs in three narrow bands through the block. Again the edge strips would also be used to control any blocks on either side. The formation of a block from the strips does not really allow any reduction in the height control requirement despite the fact that there are numerous tie points connecting one strip to the next. In theory it would be possible to introduce height control into only every other strip but this would not in general be regarded as good practice.

When the observations are the subject of a mathematical adjustment some reduction in control is possible, as we shall see in chapter 11, where we consider the analytical formation of blocks. The results of triangulation by independent models of course requires such a treatment but any analogue method that provides a digital output composed of strip model co-ordinates can be most satisfactorily adjusted in this same way.

10: Analytical photogrammetry

10.1 Introduction

In the first chapter it was emphasised that in photogrammetry the camera was essentially a device for recording a set of directions in space and that the concept of central perspective projection was a most convenient model to employ when discussing its function. In analogue instruments the directions required are always made manifest, either by using the rays of light passing from a projector lens or by defining them in some way with the aid of mechanical devices. It is true to say that the optical projector resembles in a number of ways the actual camera while the mechanical projector is more closely allied to the mathematical model. Many of the basic differences between the two approaches arise from this fact. A third possible way of defining the required directions is a mathematical one using vectors. In analytical photogrammetry we therefore define the model point as the intersection of two or more vectors. The single model, the strip model and the block of models can all be defined in terms of sets of condition equations. The single model, for example, requires only five equations to describe the condition for relative orientation, but a block of models to be formed and adjusted to a given set of control points might well involve the simultaneous solution of thousands of equations. It is for this reason that the practical applications of the technique had to wait until the large digital computer became available.

In this section we first of all examine the methods used to formulate models and strips mathematically before we examine the comparators that are required to provide the plane photo co-ordinates. In studying analytical photogrammetry for the first time it should be remembered that all the basic operations of photogrammetry have already been described in terms of analogue models and that these are always available to us, as an aid, when we are seeking to understand some particular analytical technique.

10.2 Inner orientation and the definition of the vector quantities

The plane photograph co-ordinates provided by direct measurement are of course in two-dimensional space. In order to introduce the third dimension we need the data provided by camera calibration. These data locate the position(s) of the inner nodal point of the lens in relation to a focal plane defined mechanically within the body of the camera. If, for a while, we put off any consideration of the effects of lens distortion then the nodal point does in fact become a fixed point and so the nodal point and the focal plane can be equated to a perspective centre and projection plane respectively. In this situation, it is most useful to make the perspective centre (the node) the origin of the plane (photograph) co-ordinates, as illustrated in Fig 10.1.

The foot of the perpendicular from the perspective centre to the plane gives us the position of the principal point (p) whose co-ordinates are therefore, $(0, 0, +f)$. All points in the photo plane have a constant z value and so the co-ordinates of any photo point, a, are given by (x_a, y_a, f). From the diagram we see that the line Oa defines the direction of the vector **OA** and so all points such as, a, will define the directions of the bundle of rays centred on O. Now, because this point has been chosen as an origin, a direction such as Oa with

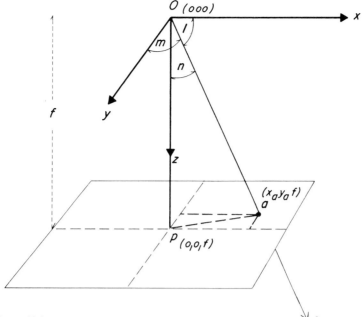

Figure 10.1. *Perspective Centre as an Origin of Co-ordinates*

respect to O and the chosen set of axes is immediately related to the point co-ordinates. The direction cosines of the line Oa being given by,

$$\cos l = \frac{x_a}{\sqrt{(x_a^2 + y_a^2 + f^2)}} = \frac{x_a}{Oa}$$

$$\cos m = \frac{y_a}{Oa} \quad \text{and} \quad \cos n = \frac{f}{Oa}$$

We can express this fact in the form of a column matrix

$$\mathbf{Oa} = \begin{pmatrix} x_a \\ y_a \\ f \end{pmatrix}$$

If we rewrite this equation in a more general form involving all points (k) on the photograph associated with a perspective centre (O_i), we have an expression equivalent to the condition of inner orientation

$$\mathbf{O_i k} = \begin{pmatrix} x_k \\ y_k \\ f \end{pmatrix}_i$$

where co-ordinates $(x_k y_k)$ are with respect to the origin O_i.

In practice, it might not be immediately convenient to use p as an origin of co-ordinates on the photograph. An arbitrary set of axes may be used. For example the fiducial marks provide a convenient system; in such a case we have

$$\mathbf{O_i k} = \begin{pmatrix} x_k' - x_\rho \\ y_k' - y_\rho \\ f \end{pmatrix}_i \qquad (10.1)$$

where all photo co-ordinates are expressed in any plane rectangular system.

219

10.3 Space resection: collinearity equations

In this problem the co-ordinates of a number of points are known in both the ground and photograph co-ordinate systems. Provided at least three such sets of co-ordinates are known, the location of the air station (i.e. the perspective centre) and the tilt of the photograph (i.e. the orientation of the photo axial system) can be determined with respect to the ground system. Or, if it is convenient, vice versa. Analytical photogrammetry is largely concerned with the transformation of sets of points from one system of axes to another. Provided the nature of the transformations are clearly understood, the operations tend to be more tedious than intellectually demanding. The space resection problem is a good example of this process, and Fig 10.2(a) and (b) illustrates this. In part (a) of the figure the two separate axial systems (X, Y, H) and (x, y, z) are depicted.

In the photograph system the point O is the origin; in the ground system it has co-ordinates $(X_0 Y_0 H_0)$. Point A on the ground has co-ordinates $(X_A Y_A H_A)$ and its photo point on the tilted photograph has co-ordinates $(x_a y_a)$ in the photograph system. In the part (b) of Fig 10.2 we show a set of transformations that will readily provide a solution to the problem and we note the following:

(i)   Both sets of axes are now parallel to one another and coincide at O.

(ii)   To lie in the plane of a vertical photograph at O the photopoint a'' of any point A must be collinear with points O and A but it must also fall in the plane $z'' = f$.

If we now examine the relationship between these sets of transformed co-ordinates we have

$$\left.\begin{array}{l} \dfrac{x_o''}{f} = \dfrac{\Delta X_A}{\Delta Z_A} \\[2mm] \quad = \dfrac{X_A - X_0}{Z_A - Z_0} \\[4mm] \dfrac{y_a''}{f} = \dfrac{\Delta Y_A}{\Delta Z_A} \\[2mm] \quad = \dfrac{Y_A - Y_0}{Z_A - Z_0} \end{array}\right\} \tag{10.2}$$

The rotation of the photo axes results in the transformation

$$\begin{pmatrix} x_a' \\ y_a' \\ z_a' \end{pmatrix} = R \begin{pmatrix} x_a \\ y_a \\ f \end{pmatrix}$$

where R is an orthogonal matrix of rotation that at this moment we will represent by the array of direction cosines (see appendix A) as follows,

$$\begin{pmatrix} x_a' \\ y_a' \\ z_a' \end{pmatrix} = \begin{pmatrix} a_{11} & a_{12} & a_{13} \\ a_{21} & a_{22} & a_{23} \\ a_{31} & a_{32} & a_{33} \end{pmatrix} \begin{pmatrix} x_a \\ y_a \\ f \end{pmatrix} \tag{10.3}$$

The points, however, still lie in the inclined plane and so in order to produce the required vertical photo co-ordinates we must apply a scale factor to each vector, such that

$$\lambda_a = \frac{\text{O}a''}{\text{O}a}$$

i.e.

$$\lambda_a = \frac{f}{z_a'}$$

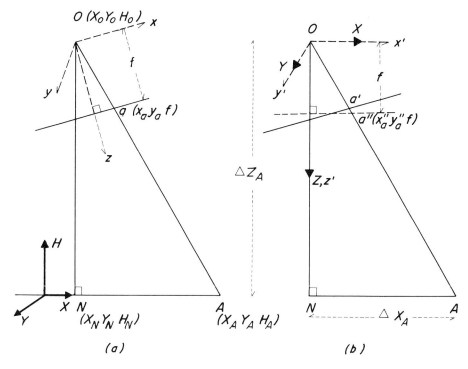

Figure 10.2. Rotation of a System of Axes

Now, from equation (10.3) we have

$$z'_a = a_{31}x + a_{32}y + a_{33}f$$

Hence combining this with equations (10.2) and (10.3) yields

$$\frac{x''_a}{f} = \frac{\lambda_a x'_a}{f}$$

$$= \frac{\lambda_a}{f}(a_{11}x + a_{12}y + a_{13}f)$$

i.e.

$$\frac{x''_a}{f} = \frac{a_{11}x + a_{12}y + a_{13}f}{a_{31}x + a_{32}y + a_{33}f}$$

$$= \frac{\Delta X_A}{\Delta Z_A}$$

And

$$\frac{y''_a}{f} = \frac{a_{21}x + a_{22}y + a_{23}f}{a_{31}x + a_{32}y + a_{33}f}$$

$$= \frac{\Delta Y_A}{\Delta Z_A}$$

(10.4)

If we consider the case of a horizontal ground surface the ΔZ has a constant value and the two equations of (10.4) take on the form of a projective transformation from one plane to another.

Now, if we rotate the ground system into that of the photo system then we must carry

221

out similar rotations and scale changes on the ground co-ordinates instead and these transformations will produce two equations similar to those above:

$$\left.\begin{array}{l} \dfrac{x_a}{f} = \dfrac{b_{11}\Delta X_A + b_{12}\Delta Y_B + b_{13}\Delta Z_A}{b_{31}\Delta X_A + b_{32}\Delta Y_B + b_{33}\Delta Z_A} \\[3mm] \dfrac{y_a}{f} = \dfrac{b_{21}\Delta X_A + b_{22}\Delta Y_A + b_{23}\Delta Z_A}{b_{31}\Delta X_A + b_{32}\Delta Y_A + b_{33}\Delta Z_A} \end{array}\right\} \tag{10.5}$$

and

Equations (10.4) and (10.5) are often referred to as the collinearity equations. Although there are nine parameters in the direction cosine matrix these elements are not independent and in fact there are just three independent unknowns here, directly related to the tilt of the photo plane with respect to the ground datum plane. The other three unknowns are (X_0, Y_0, Z_0), the co-ordinates of the air station. Hence given a minimum of three points with known co-ordinates in each system it is possible to solve for these six unknowns.

If we also treat the parameters of calibration (f, x_p, y_p) as unknowns then the total number of unknowns becomes nine (i.e. three elements of rotation, (X_0, Y_0, Z_0) and (x_p, y_p, f).

10.3.1 THE LINEARISATION OF EQUATIONS

A difficulty arises in the solution of these equations due to the fact that they are not linear in the unknowns. This means that the usual procedures for calculation cannot be adopted until a linearisation process of some form has been carried out. To do this we can make use of the Taylor expansion

$$f(x) = f(a) + (x-a)f'(a) + \frac{(x-a)^2}{L^2}f''(a) + \dots \tag{10.6}$$

If $(x-a)$ is small, then

$$f(x) \simeq f(a) + (x-a)f'(a)$$

And if we regard a as the estimated value of the unknown x then we would wish to evaluate the quantity

$$\Delta x = (x-a)$$

as being the correction to be applied to our estimated value.

If the function involves more than one variable the same process can be applied using partial derivatives.

If
$$F(p, q, r, s \dots) = A$$

then
$$A \approx F(p_0, q_0, r_0 \dots) + \frac{\partial F}{\partial p} \cdot \Delta p_0 + \frac{\partial F}{\partial q} \cdot \Delta q_0 + \frac{\partial F}{\partial r} \cdot \Delta r_0 + \dots \tag{10.7}$$

where p_0, q_0, r_0, etc. are the estimated values of the unknowns we wish to evaluate, and Δp_0, Δq_0, Δr_0, etc. are the corrections to be applied to these estimates and for which we solve a set of equations of the form of equation (10.7). In this way, better estimates of the unknowns are obtained:

$$p_1 = p_0 + \Delta p_0$$
$$q_1 = q_0 + \Delta q_0$$
$$r_1 = r_0 + \Delta r_0 \text{ etc.}$$

These values are now substituted in equations (10.7) and the process of solution gone through once again to find further (smaller) corrections ($\Delta p_i, \Delta q_i, \Delta r_i$, etc.) to our latest estimated values. By this process of iteration the values of the unknowns can be found. As the equations involving Δp, Δq, Δr, etc. are linear the standard procedures for their solution can be applied.

10.3.2 THE SOLUTION TO THE SPACE RESECTION EQUATIONS

We now apply the technique of the previous section to the equations (10.5). In this case we can write our function as

$$F = x_i[b_{31}\Delta X_i + b_{32}\Delta Y_i + b_{33}\Delta Z_i] - f[b_{11}\Delta X_i + b_{12}\Delta Y_i + b_{13}\Delta Z_i] = v_x$$

The coefficients b_{11}, b_{12} etc. will now need to be replaced by the equivalent expressions involving the three parameters of rotation and these we can obtain from equations (A.8) of appendix A. The residual v_x has been introduced because in the overdetermined situation there will be no set of values of the unknowns that will give a zero identity for all points. A similar equation would be written for v_y, involving the y co-ordinate.

In order to quantify the terms of the function we must carry out the process of partial differentiation and at the same time substitute estimated values for the six variables. Let these initial estimates be as follows:

$$\Delta\kappa_0 = \Delta\varphi_0 = \Delta\omega_0 = 0$$

$$(\Delta X_i)_0 = X_i - X_N \quad : \quad (\Delta Y_i)_0 = Y_i - Y_N \quad : \quad (\Delta Z_i)_0 = Z_i - Z_N$$

We will then have an expression of the following form:

$$v_x = t_1 + t_2\Delta\kappa + t_3\Delta\varphi + t_4\Delta\omega + t_5\Delta X + t_6\Delta Y + t_7\Delta Z \tag{10.8}$$

where

$$t_1 = F_0$$

$$= x_i(\Delta Z_i)_0 - f(\Delta X_i)_0$$

$$t_2\Delta\kappa_0 = \frac{\partial F}{\partial\kappa}\cdot\Delta\kappa_0$$

$$= [x_i(0) - f(\Delta Y_i)_0]\Delta\kappa_0$$

$$= -f(\Delta Y_i)_0\Delta\kappa_0$$

$$t_3\Delta\varphi_0 = \frac{\partial F}{\partial\varphi}\cdot\Delta\varphi_0$$

$$= [x_i(\Delta X_i)_0 + f(\Delta Z_i)_0]\Delta\varphi_0$$

$$= [x_i(\Delta X_i)_0 + f(\Delta Z_i)_0]\Delta\varphi_0$$

$$t_4\Delta\omega_0 = \frac{\partial F}{\partial\omega}\cdot\Delta\omega_0$$

$$= [x_i - (\Delta Y_i)_0 - f(0)]\Delta\omega_0$$

$$= -x_i(\Delta Y_i)_0\cdot\Delta\omega_0$$

$$t_5\Delta X_0 = \frac{\partial F}{\partial(\Delta X)}\cdot\Delta X_0$$

$$= [x_i(0) - f(1)]\Delta X_0$$

$$= -f\Delta X_0$$

223

$$t_6 \Delta Y_0 = \frac{\partial F}{\partial (\Delta Y)} \cdot \Delta Y_0$$

$$= [x_i(0) - f(0)] \Delta Y_0$$

$$= 0 \cdot \Delta Y_0$$

$$t_7 \Delta Z_0 = \frac{\partial F}{\partial (\Delta Z)} \cdot \Delta Z_0$$

$$= [x_i(1) - f(0)] \Delta Z_0$$

$$= x_i \Delta Z_0$$

Hence we have from equation (10.8)

$$v_x = [x_i(\Delta Z_i)_0 - f(\Delta X_i)_0] - f(\Delta Y_i)_0 \Delta \kappa_0 +$$

$$[x_i(\Delta X_i)_0 + f(\Delta Z_i)_0] \Delta \varphi_0 - x_i(\Delta Y_i)_0 \Delta \omega_0 - f \cdot \Delta X_0 + x_i \Delta Z_0$$

A similar expression can be obtained for v_y for every point. Given three or more points the values $\Delta \kappa_0 \ldots \Delta Z_0$ can be determined in the usual way from the set of six or more equations. Having found these initial values new values can be calculated for the terms $t_1 \ldots t_7$ and once again a solution obtained for the variables $\Delta \kappa_i \ldots \Delta Z_i$. The process is thus repeated until no further convergence in values is obtained.

10.3.3 COLLINEARITY EQUATIONS IN RELATIVE ORIENTATION

It will be appreciated that the collinearity equations are of a fundamental nature since they can be used directly in many of the processes of photogrammetry. The formation of models will be introduced in this section while the formation of strips and blocks using these equations is discussed in chapter 11.

If we take the first camera station as a local origin and assume that the first photograph is vertical, then we can form a set of equations that will describe the one projector relative orientation process. In this we note that the number of unknowns associated with the unit model are the orientations of the second camera station and its co-ordinates $(Y_0, Z_0)_2$. (The X co-ordinate is not required.) In addition, for every model point we introduce there will be the three unknown model co-ordinates $(XYZ)_i$. As each point generates two observation equations we see that a minimum of five points must be used, for we have:

Number of points $= 5$

Number of equations generated $= 5 \times 2 \times 2 = 20$

Number of model unknowns $= 5$

Number of point unknowns $= 5 \times 3$

∴ Total number of unknowns $= 20$

The number of equations involved is therefore quite large, although they are of a simple nature. As before, the equations need to be linearised in order to facilitate their solution.

10.4 The coplanarity condition

There are a number of variations by which the condition for the relative orientation of two photographs can be expressed in mathematical terms. For example we can write equations that express the following properties:

(i)   The y-parallax at five or more points is zero,

(ii)   Five pairs of vectors intersect in model space.

(iii)   Five pairs of vectors are coplanar.

We will discuss here in some detail the third condition and this is most conveniently treated using concepts taken from vector algebra.

10.5 The vector dot and cross products

Figure 10.3 shows a triangular figure composed of the three vector quantities $\mathbf{a}$, $\mathbf{b}$, and $\mathbf{c}$.

By definition the dot product of any two vectors is a scalar quantity whose magnitude is given by

$$\mathbf{a} \cdot \mathbf{b} = ab \cos \theta$$

$$= \mathbf{b} \cdot \mathbf{a}$$

If we let $\mathbf{a} = a_1 \mathbf{x} + a_2 \mathbf{y} + a_3 \mathbf{z}$ etc., where $\mathbf{x}$, $\mathbf{y}$, $\mathbf{z}$ are unit vectors in the direction of the three axes, then the dot product could be written as below and expanded accordingly

$$(a_1\mathbf{x} + a_2\mathbf{y} + a_3\mathbf{z}) \cdot (b_1\mathbf{x} + b_2\mathbf{y} + b_3\mathbf{z}) = a_1b_1 + a_2b_2 + a_3b_3$$

By definition, the cross product is a vector quantity whose direction is at right angles to the two vectors (it should be noted that these do not necessarily lie in one plane) and is given by,

$$\mathbf{a} \times \mathbf{b} = ab \cdot \sin \theta \cdot \mathbf{n}$$

Hence

$$(a_1\mathbf{x} + a_2\mathbf{y} + a_3\mathbf{z}) \times (b_1\mathbf{x} + b_2\mathbf{y} + b_3\mathbf{z}) = (a_1b_2)\mathbf{z} - (a_1b_3)\mathbf{y} - (a_2b_1)\mathbf{z}$$
$$+ (a_2b_3)\mathbf{x} + (a_3b_1)\mathbf{y} - (a_3b_2)\mathbf{x}$$

Therefore

$$\mathbf{a} \times \mathbf{b} = \begin{vmatrix} \mathbf{x} & \mathbf{y} & \mathbf{z} \\ a_1 & a_2 & a_3 \\ b_1 & b_2 & b_3 \end{vmatrix}$$

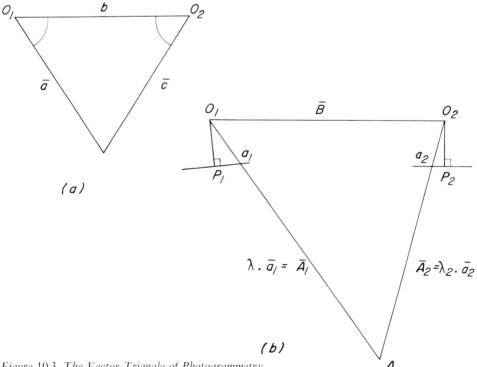

(a)

$\lambda \cdot \bar{a}_l = \bar{A}_l$

$\bar{A}_2 = \lambda_2 \cdot \bar{a}_2$

(b)

Figure 10.3. The Vector Triangle of Photogrammetry

225

Finally, by definition the triple scalar product is a scalar quantity given by the product of three vectors that are subjected to both a cross and a dot product. We note that if the operation is to have any meaning the cross multiplication must be carried out first. Hence

$$\mathbf{a} \cdot \mathbf{b} \times \mathbf{c} = \mathbf{a} \cdot (bc \sin \alpha)\mathbf{n}$$

Hence we have

$$(a_1\mathbf{x} + a_2\mathbf{y} + a_3\mathbf{z}) \cdot \begin{vmatrix} \mathbf{x} & \mathbf{y} & \mathbf{z} \\ b_1 & b_2 & b_3 \\ c_1 & c_2 & c_3 \end{vmatrix}$$

$$= a_1(b_2c_3 - b_3c_2) - a_2(b_1c_3 - b_3c_1) + a_3(b_1c_2 - b_2c_1)$$

i.e.

$$\mathbf{a} \cdot \mathbf{b} \times \mathbf{c} = \begin{vmatrix} a_1 & a_2 & a_3 \\ b_1 & b_2 & b_3 \\ c_1 & c_2 & c_3 \end{vmatrix}$$

It can be shown that the position of the dot and cross is immaterial, for

$$\mathbf{a} \cdot \mathbf{b} \times \mathbf{c} = \mathbf{a} \times \mathbf{b} \cdot \mathbf{c} = [\mathbf{abc}]$$

also

$$[\mathbf{abc}] = -[\mathbf{cba}]$$

where the terms in the brackets indicate their cyclic order.

10.5.1 THE TRIPLE SCALAR PRODUCT AND THE CONDITION FOR RELATIVE ORIENTATION

If we consider the cross product of vectors $\mathbf{A}$, and $\mathbf{B}$ of Fig 10.3(b) which represents the photogrammetric situation, we have

$$\mathbf{A}_1 \times \mathbf{B} = (A_1 B \sin \theta)\mathbf{n}$$

Where $\mathbf{n}$ is a unit vector normal to the plane defined by these two vectors. If the third vector $\mathbf{A}_2$ lies in this same plane then the unit vector is also normal to it and hence their dot product must be zero. This therefore gives us the condition for the coplanarity of the three vectors, i.e.

$$[\mathbf{A}_1\mathbf{A}_2\mathbf{B}] = 0$$

$$= \mathbf{A}_1 \cdot \mathbf{A}_2 \times \mathbf{B}$$

$$= \mathbf{A}_1 \times \mathbf{A}_2 \cdot \mathbf{B} \text{ etc.}$$

$$\begin{vmatrix} A_{1x} & A_{1y} & A_{1z} \\ A_{2x} & A_{2y} & A_{2z} \\ B_X & B_Y & B_Z \end{vmatrix} = 0 \tag{10.9}$$

where $\mathbf{A}_{1x}$ is the component of $\mathbf{A}_1$ in the x direction, etc.

Consider now the condition for a one-projector relative orientation. In such a case we can regard the first photograph as vertical, the base components and the elements of rotation of the second set of photo co-ordinates have therefore to be evaluated in accordance with the condition set out in equation (10.9).

Now

$$\begin{pmatrix} a'_{2x} \\ a'_{2y} \\ a'_{2z} \end{pmatrix} = R \begin{pmatrix} a_{2x} \\ a_{2y} \\ f \end{pmatrix}$$

where R is the orthogonal matrix of rotation, the elements of which are required to be determined. Hence the required condition can be written out as

$$\begin{vmatrix} a_{1x} & a_{1y} & f \\ a'_{2x} & a'_{2y} & a'_{2z} \\ B_x & B_y & B_z \end{vmatrix} = 0 \qquad (10.10)$$

The magnitude of Bx is not relevant to this problem and may be written in as unity. There are therefore five unknowns contained in the equation (10.10) and these are κ_2, φ_2, ω_2, B_Y/B_X, B_Z/B_X.

As in the previous case of collinearity the equation is not linear in the unknowns. There are, however, a minimum of five equations this time but each involves the five variables. To solve such a set of equations it is necessary to carry out a linearisation process, and as before this can be done by evaluating the corrections to estimated values from a function of the form

$$F_0 + \frac{\partial F}{\partial \kappa} \cdot \Delta\kappa + \frac{\partial F}{\partial \varphi} \cdot \Delta\varphi + \frac{\partial F}{\partial \omega} \cdot \Delta\omega + \frac{\partial F}{\partial B_Y} \cdot \Delta B_Y + \frac{\partial F}{\partial B_Z} \cdot \Delta B_Z = v \qquad (10.11)$$

The required values are found by a process of iteration using modified values for the estimated parameters after each solution.

As an alternative technique we could also linearise our equation by using the approximate orthogonal matrix of equation (A.9), for this has the linear terms required. The values of the unknowns can, in this way, be evaluated directly but with some degree of approximation depending on the amount of rotation required to satisfy the condition equation. In this method the values so obtained are then used in a truly orthogonal matrix (such as equation (A.13) or (A.14)) to produce a set of transformed photo co-ordinates for use in the condition equation (10.10) at the next iteration. The approximate condition equation derived below by this technique is identical to the one obtained by partial differentiation.

From the condition equation (10.10) we have,

$$(a_{1y}a'_{2z} - fa'_{2y}) - \frac{B_Y}{B_X}(a_{1x}a'_{2z} - fa'_{2x}) + \frac{B_Z}{B_X}(a_{1x}a'_{2y} - a_{1y}a'_{2x}) = 0$$

We also have

$$\begin{pmatrix} a'_{2x} \\ a'_{2y} \\ a'_{2z} \end{pmatrix} \simeq \begin{pmatrix} 1 & \Delta\kappa & -\Delta\varphi \\ -\Delta\kappa & 1 & \Delta\omega \\ \Delta\varphi & -\Delta\omega & 1 \end{pmatrix} \begin{pmatrix} a_{2x} \\ a_{2y} \\ f \end{pmatrix}$$

Hence

$$a'_{2x} = a_{2x} + a_{2y}\Delta\kappa - f\Delta\varphi$$
$$a'_{2y} = -a_{2x}\Delta\kappa + a_{2y} + f\Delta\omega$$
$$a'_{2z} = a_{2x}\Delta\varphi - a_{2y}\Delta\omega + f$$

Substituting these values in the condition equation we get,

$$0 = a_{1y}(a_{2x}\Delta\varphi - a_{2y}\Delta\omega + f) - f(-a_{2x}\Delta\kappa + a_{2y} + f\Delta\omega)$$

$$- \frac{B_Y}{B_X} \cdot a_{1x}(a_{2x}\Delta\varphi - a_{2y}\Delta\omega + f) + \frac{B_Y}{B_X}f(a_{2x} + a_{2y}\Delta\kappa - f\Delta\varphi)$$

$$+ \frac{B_Z}{B_X}a_{1x}(-a_{2x}\Delta\kappa + a_{2y} + f\Delta\omega) - \frac{B_Z}{B_X}a_{1y}(a_{2x} + a_{2y}\Delta\kappa - f\Delta\varphi)$$

If we now ignore terms of the second order of magnitude such as $B_Y/B_X \cdot \Delta\varphi$ etc. the equation simplifies to a linear form suitable for solution by least squares that will make the sum of the squares of the residuals (Σv^2) a minimum.

Hence

$$v = (a_{1y}f - fa_{2y}) + (fa_{2x})\Delta\kappa$$
$$- (a_{1y}a_{2y} + ff)\Delta\omega + (a_{1y}a_{2x})\Delta\varphi$$
$$+ (fa_{2x} - a_{1x}f)\frac{B_Y}{B_X} + (a_{1x}a_{2y} - a_{1y}a_{2x})\frac{B_Z}{B_X} \quad (10.12)$$

Given five or more sets of photo co-ordinates, values for the five unknowns can be determined. (We note that model co-ordinates do not appear in this treatment as they would in a collinearity method.) These values for the elements of rotation are then used in a truly orthogonal matrix to produce a set of transformed co-ordinates for the right-hand photograph.

10.5.2 THE VECTOR TRIANGLE AND THE CONDITION FOR RELATIVE ORIENTATION
A simpler vector equation can be used for obtaining the condition equation for relative orientation if we write the condition as one for the minimum misclosure of a vector triangle:

$$\mathbf{O_1O_2} + \mathbf{O_2A} - \mathbf{O_1A} = 0$$

In order to proceed we have to introduce the scale factors λ_1 and λ_2, hence

$$\mathbf{B} + \lambda_2\mathbf{a_2} - \lambda_1\mathbf{a_1} = 0$$

If we again take a one-projector orientation as an example then we can write

$$\lambda_1 \begin{pmatrix} x_{1a} \\ y_{1a} \\ f \end{pmatrix} - \lambda_2 \begin{pmatrix} x'_{2a} \\ y'_{2a} \\ z'_{2a} \end{pmatrix} = \begin{pmatrix} B_X \\ B_Y \\ B_Z \end{pmatrix} \quad (10.13)$$

From the three equations of (10.13) we can obtain expressions for the scale factors, i.e.

$$\left. \begin{aligned} \lambda_1 &= \frac{z'_{2a}B_X - x'_{2a}B_Z}{x_{1a}z'_{2a} - x'_{2a}f} \\ \lambda_2 &= \frac{fB_X - x_{1a}B_Z}{x_{1a}z'_{2a} - x_{2a}f} \end{aligned} \right\} \quad (10.14)$$

If we now substitute for λ_1 and λ_2 in equation (10.13) we again obtain the required condition equation and we then proceed as before to obtain values for the right-hand side co-ordinates using first an approximate orthogonal matrix and then a correct one. More details of the method can be found in Ref 10.1.

10.6 Scale factors and scale restraint conditions
The above analyses do not give a complete relative orientation, for as we have seen scale factors have not been introduced at any stage. In order to obtain model co-ordinates values are required for the factors λ_1 and λ_2. If we assume that all vectors will in fact intersect in space (i.e. we ignore for a moment the idea of any residual y-parallaxes in the model) then the two factors can be obtained directly from equations (10.14). If, however, we consider that the two vectors can be slightly skew then the above treatment provides scale factors that give to the terminal point of each vector the same X and Z co-ordinates; any misclosure being solely in the Y-direction. Starting from either O_1 or O_2 two values for

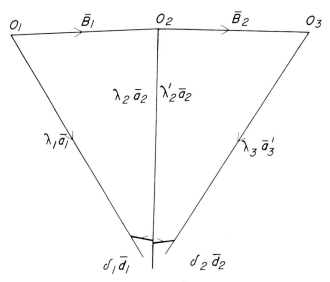

Figure 10.4. *Scale Restraint Condition for Adjacent Models*

Y will be found and their mean value is therefore taken as the best value. This is analogous to the operation carried out on a plotting instrument by an operator when faced with some visible residual Y-parallax.

If we now consider the misclosure of the vector triangle in a little more detail, from an examination of Fig 10.4, we see that in general the misclosure would be $\delta_1 \mathbf{d}_1$ where $\mathbf{d}_1$ is a unit vector whose direction is perpendicular to both $\mathbf{a}_1$ and $\mathbf{a}_2$.

By definition

$$\delta_1 \mathbf{d}_1 = \lambda_1 \mathbf{a}_1 \times \lambda_2 \mathbf{a}_2'$$

Also

$$\lambda_1 \mathbf{a}_1 - \lambda_2 \mathbf{a}_2' - \delta_1 \mathbf{d}_1 = \mathbf{B}_1 \tag{10.15}$$

That is to say

$$\lambda_1 a_{1x} - \lambda_2 a_{2x}' - \delta_1 d_{1x} = (B_X)_1$$

$$\lambda_1 a_{1y} - \lambda_2 a_{2y}' - \delta_1 d_{1y} = (B_Y)_1$$

$$\lambda_1 a_{1z} - \lambda_2 a_{2z}' - \delta_1 d_{1z} = (B_Z)_1$$

In this set of three equations we have the three unknown scale factors. By Crammers rule we have therefore, for example,

$$\lambda_2 = \frac{\begin{vmatrix} a_{1x} & (B_X)_1 & d_{1x} \\ a_{1y} & (B_Y)_1 & d_{1y} \\ a_{1z} & (B_Z)_1 & d_{1z} \end{vmatrix}}{\begin{vmatrix} a_{1x} & a_{2x}' & d_{1x} \\ a_{1y} & a_{2y}' & d_{1y} \\ a_{1z} & a_{2z}' & d_{1z} \end{vmatrix}} = \frac{\mathbf{a}_1 \cdot \mathbf{B}_1 \times \mathbf{d}_1}{\mathbf{a}_1 \cdot \mathbf{a}_2 \times \mathbf{d}_1} \tag{10.16}$$

If, as discussed above, $\mathbf{d}$ lies purely in the Y-direction then from equation (10.15) we have

$$\delta_1 d_{1y} = \lambda_1 a_{1y} - \lambda_2 a_{2y}' - (B_Y) \tag{10.17}$$

From equation (10.16) we see that

$$\lambda_2 = \frac{a_{1x}(B_Z)_1 - a_{1z}(B_X)_1}{a_{1x}a'_{2z} - a_{1z}a'_{2x}}$$

which of course is the same result as equation (10.14).

Referring once again to Fig 10.4 and considering the second model based on the common vector, here we have a second misclosure $\delta_2 \mathbf{d}_2$ and we can write as before

$$\delta_2 \mathbf{d}_2 = \lambda'_2 \mathbf{a}'_2 \times \lambda_3 \mathbf{a}'_3$$

Also, using the triple scalar products we have

$$\lambda'_2 = \frac{\mathbf{a}'_3 \cdot \mathbf{B}_2 \times \mathbf{d}_2}{\mathbf{a}'_2 \cdot \mathbf{a}'_3 \times \mathbf{d}_2} \tag{10.18}$$

In order to carry the correct scale forward from model 1 to model 2 we therefore have the scale condition

$$\lambda_2 = \lambda'_2$$

the values of which are given by equations (10.16) and (10.18).

10.7 The measurement of photo co-ordinates

For the purposes of analytical photogrammetry we need to measure the two sets of photo co-ordinates, and for this purpose some form of comparator is used. If the measurements are carried out separately on each photograph then a monocomparator is used. If the measurements are taken simultaneously with the aid of stereoscopic viewing then the instrument is termed a stereocomparator.

We consider first the mono-instrument and with this identified photo points can be co-ordinated with the aid of a viewing microscope. In the simplest designs the instrument would take the form of a travelling microscope moving along a single accurately calibrated linear scale; the second co-ordinate being obtained by rotating the photograph with respect to the linear scale through exactly 90°. Much more commonly, two precise lead screws working at right angles to one another would be used to carry the viewing head to any point on the picture format. Accurate lead screws with a pitch of 1 mm and measuring devices capable of estimating point positions to a 1/1000th part of one revolution can provide a 1 μm precision in the readout. The relative movements between measuring head and photograph can be allocated in a number of ways depending on the design of the instrument. In order to cover the whole format of a standard size photograph and take in the sets of fiducial marks, a travel of about 240 mm is required in both directions.

As an alternative to making purely mechanical measurements, optical scales viewed through low powered microscopes can be used. In such cases the coarse readings of the main scale can be subdivided by using a microscope fitted with a mechanical vernier reading device or perhaps a rotating parallel plate micrometer. In more recent years, measurement techniques using some form of electro-optical readout have been more widely employed. For example, the Moiré fringes generated by two fine gratings held at a small inclination one to the other can be used by employing a suitable fringe counting technique. Linear and circular binary encoded scales have also been used in some instruments. One advantage of these systems is the reduction in wear because the method of transportation and fine movement no longer form part of the measurement process. In order to reduce greatly the length of the measuring scales necessary a precise glass graticule or reseau can be incorporated in the instrument. By their use, measurement is restricted to distances from the nearest reference point or line. Such a technique is used in the Zeiss (Oberkochen) PSK Stereocomparator for example. As an alternative method the reseau may be part of the photo imagery if the taking camera was of a type using a glass plate marked with a reseau pattern to define the focal plane. This technique was used in the Cambridge stereocomparator of 1937 and in the Hilger–Watts instrument more recently.

In the absence of any particular reference points the four fiducial marks situated in the four corners of the format are used to define a system of rectangular axes. One point is selected as an origin and the best mean fitting set of orthogonal axes is calculated for these.

With a monocomparator the points to be co-ordinated must be selected beforehand. These points will fall on other overlapping photographs and it is therefore essential that the identical points should be defined on all these photographs with the very minimum of error. Sometimes the points will take the form of points defined with adequate precision by the lines or shapes of natural features. In this category we could at this point include premarked points that have been photographed as part of the natural landscape. (The considerable advantage of having such points available in the super-overlapping parts of the photography cannot be overemphasised when considering this aspect of the measurement process.) If suitable points are not available in the required areas then artificial points must be introduced, using one of the point marking devices that can produce small well-defined marks on the photograph negatives or diapositives. Two such pieces of equipment are the Zeiss Snap Marker and the Wild PUG. The Wild instrument consists of a viewing stereoscope fitted with a floating mark system. The two stages on which the photographs are placed can each be moved in unison in the $\mathbf{XY}$ directions and relatively by the small amounts $\mathbf{\Delta X}$ and $\mathbf{\Delta Y}$. Any part of the format can therefore be examined and any X- and Y-parallax eliminated at that point. In this way a floating mark can be placed on the ground surface in any desired area. When required two small drills can be brought into operation that will produce marks on the two photographs in the positions apparently occupied by the two halves of the floating mark. The size of mark produced can be varied between 40 and 250 μm. The Zeiss Snap Marker produces small indentations in the emulsion surface of about 100 μm diameter, using the same principle in a less elaborate form.

In the case of the stereocomparator, two photographs are viewed stereoscopically and the two sets of photo co-ordinates are measured at the same time when the floating marks are placed over the point and cleared of Y-parallax. When using artificial points, only one photograph need be marked in order to locate a point. When observing to such points (either in analytical or analogue instruments) the floating mark is put on to the surface in the immediate vicinity of the point and all Y-parallax removed. The photographs, or measuring marks, are then moved together until the marks lie exactly over the point.

The output from the comparator can take a number of forms. The photo co-ordinates of a point may be expressed in terms of the reseau system of each photograph or an axial system based on the fiducial marks appearing on each photograph. In some simpler instruments the sets of co-ordinates of a point are expressed in terms of one system of co-ordinates together with the differences ΔX, ΔY.

10.8 Analytical plotters

In the analytical plotter photo co-ordinates are measured by some form of stereocomparator. These are then fed directly into a minicomputer which solves the basic photogrammetric equations and provides a readout of model or ground co-ordinates. If required, these can be used to drive a digital plotting table which can provide a graphical output at a wide variety of scales. This type of instrument has been under active development for the past 18 years or more, and much of the development work is associated with the name of U. V. Heleva.[10.2, 10.3] One of the earliest instruments of this type to be developed was the OMI AP/C3 and this has now been followed (in 1977) by the AP/C4 instrument. However, for many years instruments of this category were not produced in any quantity due mainly to their relatively high cost and the constraints imposed by the computers available at the time. With the very rapid development of computer technology the computers have become more simple to programme, swifter in computing, more reliable and much cheaper. Such factors have now resulted in a situation where the analytical plotter can now compete even on cost terms with the larger analogue instruments. Many of the major manufacturers are now including such instruments in their range. As the larger plotters have in recent years been increasingly more concerned with a digital output rather than a graphical one, this would seem to be a logical step in the development of photogrammetric instrumentation. The analytical instrument can have the precision of the first order analogue machine and also display an increasing versatility as software technology expands.

One of the basic operating programmes of the analytical plotter is the loop programme that in essence replaces the space rods of the analogue instrument. When the instrument is correctly orientated, movements of the hand and foot wheels indicate a change in the model point position and its co-ordinates. The resulting impulses sent to the computer are transformed into plate co-ordinates for both the left-hand and right-hand photographs. These in turn result in impulses being transmitted to the left- and right-hand photo carriages that move the photographs by the correct amounts to ensure the continued absence of x- and y-parallaxes. This whole cycle is carried out in a matter of a few milliseconds.

Before the model is produced the processes of interior, relative and absolute orientation need to be carried out by analytical methods analogous to those of the mechanical plotter. To carry out the process of interior orientation, the mark is set accurately over the four fiducial marks in turn. These four sets of photo co-ordinates therefore define the position of the principal point and the orientation of the photograph in the co-ordinate system of that photo carriage. The focal length and a radial lens distortion function entered into the computer then completes the process. For the process of relative orientation the six standard points, and others if necessary, are used. The instrument usually moves automatically to the standard points and the operator removes the y-parallax at each point in turn. From these data the five parameters of relative orientation are computed using either the coplanarity or collinearity equations. The residual parallaxes at the points used are displayed for inspection by the operator before the next stage is carried out. For absolute orientation, the co-ordinates of the control points (plan, height or both) are fed into the instrument. After an accurate pointing at each control point in the model, the computer will provide the corrected parameters of orientation based on a least squares computation if more than the minimum amount of control has been made available.

Aerial triangulation can also be carried out since the orientation elements of each photograph are stored in the computer. By using the classical method of procedure with a 'base-in and base-out' observation routine, the scale can be transferred from one model to the next by observing common points. It is claimed that the formation of strips in this way is largely error free, for the method allows numerical checks at all stages. The programmes for block formation and adjustment can therefore be somewhat simpler and these are at present under investigation.

In order to give some idea of the characteristics of these instruments, a few salient details concerning just a few instruments are given below. Details of a large selection of analytical plotters are to be found in an edition of *Photogrammetric Engineering* devoted to the analytical plotter.[10.4]

The OMI AP/C4 has a precision rather better than a first-order plotting instrument and is capable of mathematical solutions for near-vertical photography for any focal length with a format up to 230×230 mm. Corrections can be introduced for film shrinkage, lens distortion, earth curvature and any other known systematic error. The tasks it can carry out include the following.

(i)   Conventional plotting of high precision over a very wide range of model to plot ratios. The model can be scanned using either the hand wheels or the free-hand 'joy stick'.

(ii)   Digital plotting in time or distance mode.

(iii) Digital terrain modelling in any predetermined array and terrain sampling at any density or alignment.

(iv)   Cross sections for road construction.

(v)   Strip triangulation in the ground co-ordinate system.

The instrument uses a 15 μm diameter floating mark. It incorporates a PDP 11/03 minicomputer and CRT display. A wide range of peripheral equipment is available.

Another instrument of a similar specification is the Zeiss (Oberkochen) C1000 Planicomp. This instrument uses a Hewlett–Packard HP 21 MX computer. A feature of the instrument is its calibration and functional checkout system. Grid test measurements over 24 points give a precision equivalent to $0.004°/_{oo}$ Z with wide angle photography. A comprehensive description of this instrument has been produced by Dr-Ing D. Hobbie.[10.5]

Most analytical instruments at the present time are in the first order category but an interesting exception is the Zeiss (Oberkochen) G-2 Stereocord. This instrument is based on the Zeiss approximate plotter known as the Stereotop. Linear encoders measure the co-ordinates of the left-hand photograph with a standard deviation of 20 μm. The photographs are base lined in the usual way. Measurements of x-parallax are recorded with a standard deviation of 10 μm; these measurements of X, Y and P_x being carried out by the Direc-1 unit. The output of this unit can be interfaced with a calculator and in this instrument a Hewlett–Packard 9810 or 9830 calculator is used. The transformation equations can be based on the simple parallax equation or on equations such as (6.17) that give a good approximate solution provided the photo tilts are not too large. A rigorous solution is possible for special purposes. The instrument has been designed with agricultural, forestry, geological and similar applications in mind. For photogrammetric purposes if the computer is fitted with an XY plotter then point-to-point model plotting can be carried out as suggested in section 4.2.2. The instrument can also be used as a comparator.[10.6]

11: The formation and adjustment of blocks

11.1 Introduction
In chapter 9 the concept of the formation of blocks from strips or models was introduced when we discussed the various analogue techniques that had been evolved for this purpose. In general, it is true to say that these methods were more successful in the formation of planimetric blocks than in the simultaneous adjustment of height values. Stereo templates can produce results of an acceptable accuracy for the purposes of medium and small-scale mapping. The Jerie analogue device is capable of even better results but it is very time consuming to use. With the advent of the large digital computer in the 1960s, these analogue techniques were almost universally replaced by analytical methods. In some cases the nature of the adjustments carried out are the equivalent of their analogue predecessors. The computer merely carries out the same operation in a more precise and very much swifter manner. The only disadvantage would seem to be in the detection and elimination of errors, for the nature of the graphical techniques is such that errors over a certain magnitude readily draw attention to themselves in a visual way. In analytical solutions it is therefore vital that routines are introduced that detect and eliminate gross errors as early as possible at each stage of the computational process.

11.2 The methods used in the formation of blocks
A homogeneous block of co-ordinates can be produced in a number of ways:

(a) From strips of photography that we have used to form a strip model adjusted onto control with the aid of polynomial expressions; the end product being a homogeneous set of X or Y or Z co-ordinates of each tie point of the block. For example, the Jerie analogue device for height adjustment attempts this for the Z co-ordinate.

(b) From individual models, or small groups of models (sections), that have been brought into approximate absolute orientation; a linear adjustment of their planimetric co-ordinates can be carried out. For example, stereo templates of one form or another attempt such an adjustment. This together with (a) for Z values will provide a three-dimensional adjustment.

(c) From individual models or sections a three-dimensional non-linear spatial transformation can be carried out. A mechanical method for such an adjustment is most difficult to realise.

(d) From individual photographs, a three-dimensional transformation of each bundle of rays can be carried out using a technique that is sometimes referred to as the fully analytical method of block adjustment. As an example of this technique we could say that in the process of long bar Multiplex aerial triangulation, the one-projector orientation is a process of adjusting the bundle of rays of that projector until these intersect the rays from other bundles.

Whenever part of the analogue process is replaced by an analytical one, a set of equations must be generated that provides an adequate mathematical description of the mechanical process that has been superseded. In section 9.7 for example we saw how the mechanical process of joining one model to the next was replaced by spatial transformations of the form given by equation (9.1). For each of the procedures (a) to (d) mentioned here we will therefore need to devise suitable sets of equations by means of which strips, or models, or photographs can be joined together to produce a homogeneous set of co-ordinates for all points appearing on the block of photographs. This formulation of block co-ordinates must be the subject of errors in kind similar to those experienced by the analogue processes. Uncertainties in the observation of the photo or model co-ordinates will give rise to errors of a random nature. Factors such as film shrinkage, atmospheric refraction, earth curvature, discrepancies in camera calibration data, imperfections in analogue reprojection and so on, can produce errors of a more systematic nature in the block co-ordinates. In all situations, therefore, a distribution of control points (or control data in some other form) is required in order to restrict the buildup of errors in block co-ordinates. As we shall see, a little later in this chapter, it is often possible to carry out the main block formation and adjustment processes simultaneously.

Figure 11.1, based on a diagram that appears in Ref 11.1, shows the various combinations of analogue and analytical processes that can be used to form and adjust a block of co-ordinates. The diagram illustrates the different stages at which mathematical procedures need to be introduced in connection with the four techniques listed above. In the following sections of this chapter we will examine the nature of these four main analytical approaches. It will be appreciated from an examination of the diagram that there are a number of other possibilities. A paper by Thompson[11.24] gives an excellent review of the various methods that can be used for the calculation of aerial triangulation.

11.3 Block adjustment using polynomial expressions

As a first example we will take the case of the simultaneous adjustment of a set of height values derived from the observation of a series of strips of aerial triangulation. The nature of the analytical adjustment carried out is similar to that attempted by the Jerie analogue device (see section 9.8.8).

Notation: i denotes the point number (which is therefore unique)
j denotes the strip number

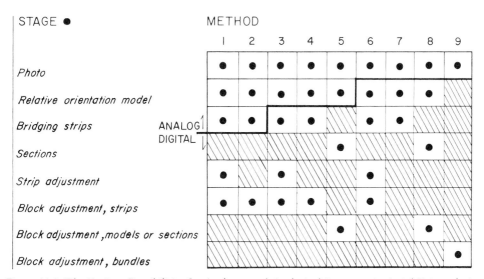

Figure 11.1. The Various Possibilities for Analogue and Analytical Processes in Aerial Triangulation

235

We will take as our example a second-order polynomial expression of a type commonly used, especially for shorter strip lengths,

$$\Delta z_{ij} = c_0 + x_{ij}c_1 + x_{ij}^2 c_2 + y_{ij}c_3 + x_{ij}y_{ij}c_4$$

$$\Delta z_{ij} = z_{ij} - Z_i$$

$$= \text{(observed strip value)} - \text{(true value at strip scale)}$$

It should be noted that although the phrase commonly used to describe this process is one of 'polynomial adjustment', the equations are of course linear expressions of the unknowns ($c_0, c_1 \ldots c_4$) and have been written out in the manner above to emphasise this fact. If they were not linear, a direct least squares treatment would not be possible.

As an example of this form of adjustment we will take a small block of the form illustrated in Fig 11.2. Over the area formed by the two strips are the five control points (1, 2, 3, 4 and 5) and the six tie points (6, 7, 8, 9, 10 and 11). From observations carried out with the analogue plotting instrument the strip co-ordinates (x, y, z) are known for each point at the strip scale and in the strip frame of reference. The true height values of the limited number of control points are also known. When making use of these height data it is necessary to work in terms of height differences using the same datum point in the strip and ground systems. The equations can be framed in terms of either ground or strip model co-ordinates but it is often more convenient to work in terms of the latter.

There are therefore two forms of observation equation depending on whether the point is a control point of known height or a tie point, the adjusted height of which is required.

The total number of unknowns is made up therefore of the unknown heights of tie points (Z_i) and the unknown values of the coefficients ($c_0 \ldots c_4$)$_j$. From an examination of Fig 11.2 we see the following.

Every control point and every tie point gives rise to one observation equation each time it appears in a strip, hence we have:

Control points 1, 2, 4, 5 give one equation each   = 4
Control point 3 gives two equations   = 2
Tie points 6–11 give two equations each   = 12
∴ Total number of equations   = 18

Each strip has five unknown coefficients   = 2×5 unknowns
Each tie point gives rise to one unknown   = 1×6 unknowns
∴ Total number of unknowns   = 16

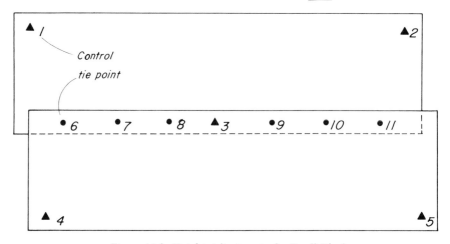

Figure 11.2. Height Adjustment of a Small Block

Hence we have a slightly overdetermined system. We note that the minimum number of control points is 5 and that the minimum number of tie points for this case is therefore 4.

The following are typical examples of equations.

For control point 1 we have

$$\Delta z_{11} = (1 \ \ x_1 \ \ x_1^2 \ \ y_1 \ \ x_1 y_1)_1 (c_0 \ \ c_1 \ \ c_2 \ \ c_3 \ \ c_4)_1^T$$

where
$$\Delta z_{11} = z_{11} - Z_1$$

For control point 3 we have two equations:

$$\Delta z_{31} = (1 \ \ x_3 \ \ x_3^2 \ \ y_3 \ \ x_3 y_3)_1 (c_0 \ \ c_1 \ \ c_2 \ \ c_3 \ \ c_4)_1^T$$

and
$$\Delta z_{32} = (1 \ \ x_3 \ \ x_3^2 \ \ y_3 \ \ x_3 y_3)_2 (c_0 \ \ c_1 \ \ c_2 \ \ c_3 \ \ c_4)_2^T$$

For a typical tie point 6 we would have

$$z_{61} = (1 \ \ x_6 \ \ x_6^2 \ \ y_6 \ \ x_6 y_6)_1 (c_0 \ \ c_1 \ \ c_2 \ \ c_3 \ \ c_4)_1^T + Z_6$$

$$z_{62} = (1 \ \ x_6 \ \ x_6^2 \ \ y_6 \ \ x_6 y_6)_2 (c_0 \ \ c_1 \ \ c_2 \ \ c_3 \ \ c_4)_2^T + Z_6$$

In an abbreviated form we can write these equations in the form

$$A_{11} P_1 = \Delta z_{11}$$
$$A_{31} P_1 = \Delta z_{31}$$
$$A_{32} P_2 = \Delta z_{32}$$
$$A_{61} P_1 + Z_6 = z_{61}$$
$$A_{62} P_2 + Z_6 = z_{62}$$

For the general case, a control point equation takes the form

$$A_{ij} \cdot P_j = L_{ij}$$

(coefficient matrix)(column matrix of unknowns)=(column matrix of constants)

For the general case of a tie point we have

$$A_{ij} \cdot P_j + Q_i = L_{ij}$$

Figure 11.3 shows the complete list of observation equations. Note the order in which the unknowns have been listed in the column matrix. From an examination of the diagram we see that

$$A \cdot P = L$$

$$(18 \times 16) \cdot (16 \times 1) = (18 \times 1)$$

Assuming no correlation and all observations of equal weight, we can form the normal equations and solve them in the usual way:

$$A^T \cdot A \cdot P = A^T \cdot L$$

$$(16 \times 18)(18 \times 16)(16 \times 1) = (16 \times 18)(18 \times 1)$$

i.e.
$$N \cdot P = A^T \cdot L$$

$$(16 \times 16)(16 \times 1) = (16 \times 1)$$

i.e.
$$P = N^{-1} \cdot A^T \cdot L$$

Planimetric values

In a similar way to the above we can adjust the planimetric values using polynomial

237

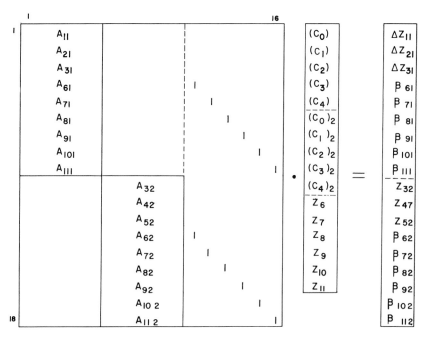

Figure 11.3. *Table of Observation Equations for Polynomial Height Adjustment*

expressions. Again we will select a second-order polynomial and for this case we will use transformation expressions of the following type,

$$X = a_0 + (x)a_1 + (x^2)a_2 - (y)b_1 - (2xy)b_2$$

and

$$Y = b_0 + (x)b_1 + (x^2)b_2 + (y)a_1 + (2xy)a_2$$

These are similar to the Vermier equations quoted at the end of section 9.8.3. In this pair of equations we have six unknowns for each strip and we assume the normal lateral overlap between strips. For our purpose the equations are better expressed in matrix form, as below:

$$\begin{pmatrix} X_i \\ Y_i \end{pmatrix} = \begin{pmatrix} 1 & x_i & x_i^2 & 0 & -y_i & -2x_iy_i \\ 0 & y_i & 2x_iy_i & 1 & x_i & x_i^2 \end{pmatrix} \begin{pmatrix} a_0 \\ a_1 \\ a_2 \\ b_0 \\ b_1 \\ b_2 \end{pmatrix}$$

As before, we note the following points.
Each control point generates two equations each time it appears in a strip. Hence points in lateral overlaps provide four equations. The equations are of the form

$$A_{ij} \cdot P_j = L_i$$

Each tie point will yield four equations of the form

$$A_{ij} \cdot P_j - Q_i = 0$$

In Fig 11.4 we show three strips of photography with a distribution of five control points and five tie points. From an examination of the figure we note the following,

238

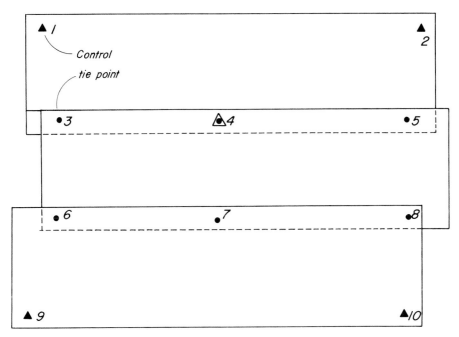

Figure 11.4. *Planimetric Adjustment of a Small Block*

Number of equations $= 32$
Number of unknowns $= (3 \times 6) + (5 \times 2)$
$\qquad\qquad\qquad = 28$

A few typical equations are given below.
For the control point 1 we have

$$\begin{pmatrix} 1 & x_1 & x_1^2 & 0 & -y_1 & -2x_1y_1 \\ 0 & y_1 & 2x_iy_1 & 1 & x_1 & x_1{}^2 \end{pmatrix}_1 \begin{pmatrix} a_0 \\ a_1 \\ a_2 \\ b_0 \\ b_1 \\ b_2 \end{pmatrix}_1 = \begin{pmatrix} X_1 \\ Y_1 \end{pmatrix}$$

For the tie point 7 in strip 3 we would have

$$\begin{pmatrix} 1 & x_7 & x_7^2 & 0 & -y_7 & -2x_7y_7 \\ 0 & y_7 & 2x_7y_7 & 1 & x_7 & x_7^2 \end{pmatrix}_3 \begin{pmatrix} a_0 \\ a_1 \\ a_2 \\ b_0 \\ b_1 \\ b_2 \end{pmatrix}_3 - \begin{pmatrix} X_7 \\ Y_7 \end{pmatrix} = 0$$

The full set of observation equations is written out in tabular form in Fig 11.5, and again we have

$$A \cdot P = L$$

$$(32 \times 28) \cdot (28 \times 1) = (32 \times 1)$$

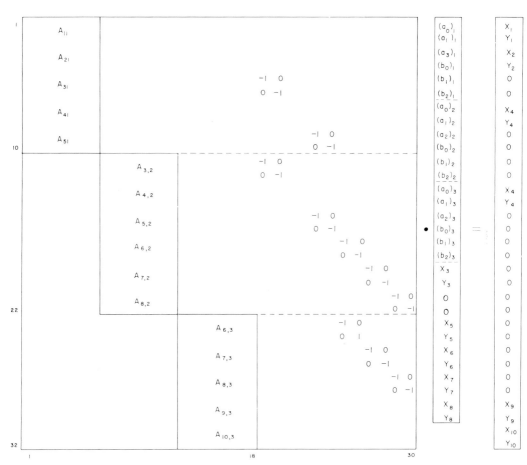

Figure 11.5. *Table of Observation Equations for Polynomial Planimetric Adjustment*

And again we have the formation of the normal equations, assuming unit weight matrix:

$$A^T \cdot A \cdot P = A^T \cdot L$$

i.e.

$$N \cdot P = A^T \cdot L$$

Hence

$$P = N^{-1} \cdot A^T \cdot L$$

Papers by Schut[11.2,11.3] describe the use of polynomial methods of adjustment developed by the National Research Council of Canada.

11.4 Block adjustment using models as the basic unit

In this case the basic unit is the model or a section made up of a small number of models (usually two or four). Within this unit the errors are considered negligible and so each unit can only be adjusted as an entity. The scale, position and orientation of the unit may be changed but not its shape. For the purposes of adjustment a model or section is defined by the co-ordinates of its four corner points in the case of a planimetric adjustment and by six points (i.e. the four corner points and the two perspective centres) in the case of a spatial adjustment. The model co-ordinates will usually have been obtained from independent model observations, as described in section 9.7, but it would be possible to derive them also from stereocomparator observations and the application of either the collinearity

240

equations or the coplanarity condition, as described in sections 10.3.4 and 10.4.3 respectively. Such possibilities are shown diagrammatically in Fig 11.1. In all cases the models obtained from each strip of photography can then be joined together using an analytical technique based on the spatial transformation equation (9.1). In this way all models of the strip are brought into the co-ordinate system of the first (or any selected) model of that strip. In order to obtain a first set of block co-ordinates the co-ordinates of the second strip can be transformed approximately into those of the first using a minimum of two common points in their lateral overlap; one at each end of the strip. In this way, all models of all strips can be brought approximately into a single frame of reference. Some organisations prefer to use more than the minimum of two common tie points and carry out a second order rather than a linear transformation of the strip co-ordinates. Such a process provides a closer agreement between the two sets of co-ordinates obtained for common tie points. Any gross errors in co-ordinates can therefore more readily be identified and eliminated at this stage. Differences in the strip co-ordinates of common points should be small and build up in a regular way between the points used for the transformation.

Having obtained a set of block co-ordinates for each model that are approximately homogeneous a least squares adjustment of each model or section can now be carried out as an alternative to the strip procedure outlined in section 11.3. In this case, however, all systematic deformations should as far as possible be eliminated before proceeding with any least squares adjustment. The effects of any lens distortion, atmospheric refraction, Earth curvature, etc. should therefore be taken into account at this stage whenever possible and where applicable. As we shall see, because the basic unit is now much smaller a block adjustment using models or sections generates many more observation and normal equations than the equivalent adjustment by strips. The method does also provide a much more flexible approach and is particularly valuable when the control available is well in excess of the minimum requirement. Both methods are capable of giving excellent results and the various characteristics of the two methods are discussed by Schut.[11.4]

In the adjustment of models there are two methods of approach. The first and less complex method involves the separate adjustments of planimetry and height values. By

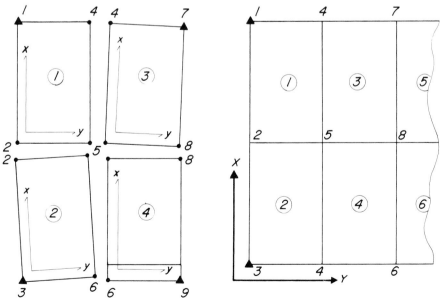

Figure 11.6. Planimetric Transformation of Models

241

using this technique direct solutions for the model transformation parameters are possible. As an alternative, three-dimensional transformations can be used to give a simultaneous adjustment of all three co-ordinates. Because such equations involve non-linear orthogonal matrices the observation equations so produced cannot be the subject of a least squares treatment until some form of linearisation has been carried out. An iterative form of solution is therefore necessary.

If we now consider the first of the methods referred to above then we can see from an examination of Fig 11.6 that initially each individual model has its own local origin and set of tie point co-ordinates. The adjustment brings them together and provides the best overall set of values for the tie points. For the planimetry we use the following linear and conformal transformation equations:

$$X = ax + by + c$$

$$Y = ay - bx + d$$

where (X, Y) are the required block co-ordinates and (x, y) are the model or section point co-ordinates. That is, we have four adjustments composed of a scale change, rotation and shift of origin:

$$a = \lambda \cos \theta \qquad b = \lambda \sin \theta$$

$$c = \Delta X_0 \qquad d = \Delta Y_0$$

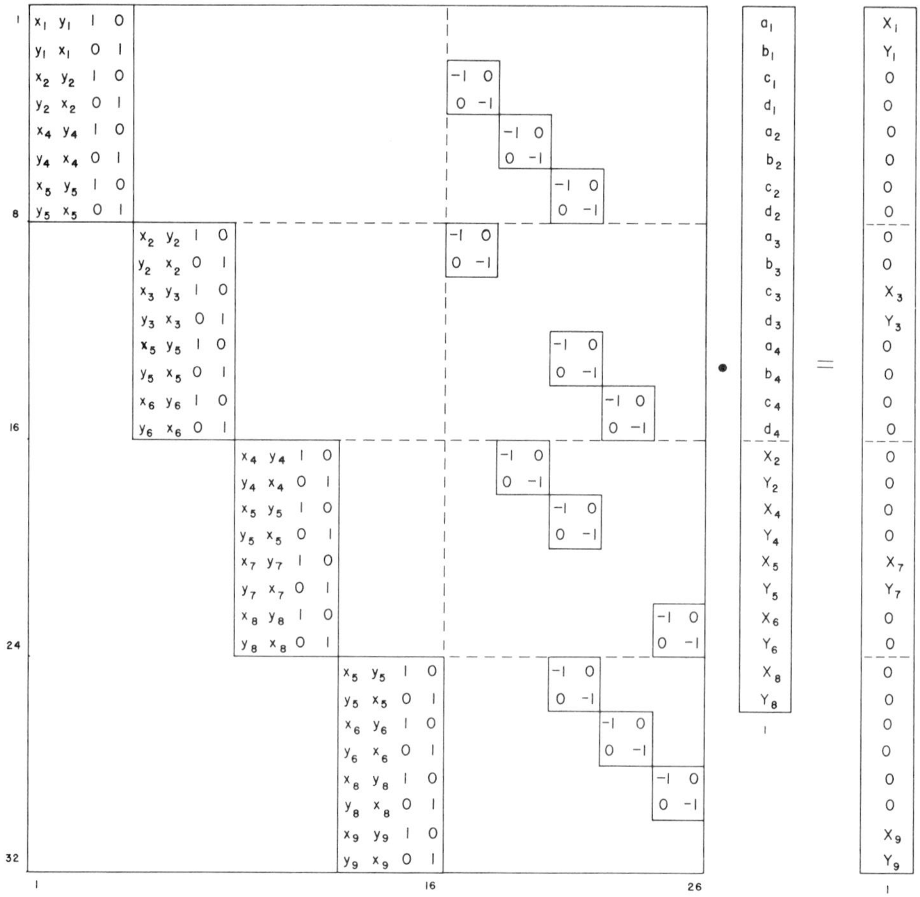

```
 1 | x₁ y₁  1  0                                                          |  a₁ |     | X₁
   | y₁ x₁  0  1                                                          |  b₁ |     | Y₁
   | x₂ y₂  1  0                     -1  0                                |  c₁ |     |  0
   | y₂ x₂  0  1                      0 -1                                |  d₁ |     |  0
   | x₄ y₄  1  0                           -1  0                          |  a₂ |     |  0
   | y₄ x₄  0  1                            0 -1                          |  b₂ |     |  0
   | x₅ y₅  1  0                                 -1  0                    |  c₂ |     |  0
 8 | y₅ x₅  0  1                                  0 -1                    |  d₂ |     |  0
   |        x₂ y₂  1  0              -1  0                                |  a₃ |     |  0
   |        y₂ x₂  0  1               0 -1                                |  b₃ |     |  0
   |        x₃ y₃  1  0                                                   |  c₃ |  •  | X₃   =
   |        y₃ x₃  0  1                                                   |  d₃ |     | Y₃
   |        x₅ y₅  1  0                           -1  0                   |  a₄ |     |  0
   |        y₅ x₅  0  1                            0 -1                   |  b₄ |     |  0
   |        x₆ y₆  1  0                                 -1  0             |  c₄ |     |  0
16 |        y₆ x₆  0  1                                  0 -1             |  d₄ |     |  0
   |              x₄ y₄  1  0        -1  0                                |  X₂ |     |  0
   |              y₄ x₄  0  1         0 -1                                |  Y₂ |     |  0
   |              x₅ y₅  1  0                     -1  0                   |  X₄ |     |  0
   |              y₅ x₅  0  1                      0 -1                   |  Y₄ |     |  0
   |              x₇ y₇  1  0                                             |  X₅ |     | X₇
   |              y₇ x₇  0  1                                             |  Y₅ |     | Y₇
   |              x₈ y₈  1  0                           -1  0             |  X₆ |     |  0
24 |              y₈ x₈  0  1                            0 -1             |  Y₆ |     |  0
   |                    x₅ y₅  1  0              -1  0                    |  X₈ |     |  0
   |                    y₅ x₅  0  1               0 -1                    |  Y₈ |     |  0
   |                    x₆ y₆  1  0                    -1  0              |   1 |     |  0
   |                    y₆ x₆  0  1                     0 -1              |     |     |  0
   |                    x₈ y₈  1  0                          -1  0        |     |     |  0
   |                    y₈ x₈  0  1                           0 -1        |     |     |  0
   |                    x₉ y₉  1  0                                       |     |     | X₉
32 |                    y₉ x₉  0  1                                       |     |     | Y₉
   |                    16                        26                     |
```

Figure 11.7. *Anblock. Adjustment of Planimetry. Formation of Observation Equations*

242

(Figure A.2 and equation (A.3) of appendix A illustrate the nature of the rotation.) Hence the observation equations for the block formation are

$$\begin{pmatrix} x_i & y_i & 1 & 0 \\ y_i & -x_i & 0 & 1 \end{pmatrix}_j \begin{pmatrix} a \\ b \\ c \\ d \end{pmatrix}_j = \begin{pmatrix} X_i \\ Y_i \end{pmatrix}$$

Examining Fig 11.6 we see that each model has four unknowns while each tie point produces two unknowns. Every point, control or tie, produces two equations each time it appears in a model. We have therefore

$$\begin{aligned} \text{Number of equations} \quad &= 4 \times 2 \times 4 \\ &= 32 \\ \text{Number of unknowns} \quad &= (4 \times 4) + (5 \times 2) \\ &= 26 \end{aligned}$$

Again the equations will take on one or other of two standard forms.

For control points, $\qquad\qquad\qquad A_{ij} \cdot P_j = L_i$

For tie points $\qquad\qquad\qquad A_{ij} \cdot P_j - Q_i = 0$

The nature of the observation equations is shown in Fig 11.7.

The formation of the normal equations and their solution follows the usual form and is illustrated in Figs 11.8 and 11.9.

The paper by Van den Hout[11.5] describes the Anblock method of adjustment, which is based on the approach described above. Another similar method is described by Amer.[11.9] The relaxation method of computation used in this is similar in effect to the form of adjustment carried out by the Jerie analogue device. The Amer method of adjustment has been used successfully for many years by the Ordnance Survey of Great Britain.

11.4.1 THE USE OF SPATIAL TRANSFORMATION EQUATIONS
In this second method of adjustment by models or sections we make use of the three-dimensional observation equation

$$\lambda_j R_j \begin{pmatrix} x_i \\ y_i \\ z_i \end{pmatrix}_j + \begin{pmatrix} X_0 \\ Y_0 \\ Z_0 \end{pmatrix} = \begin{pmatrix} X_i \\ Y_i \\ Z_i \end{pmatrix} \qquad (11.1)$$

where λ_j is the scale factor of the jth model or section, R_j is the 3×3 rotation matrix of the section, X_0, Y_0, Z_0 are the three shifts of the section origin, X, Y, Z are the required block co-ordinates of point (x, y, z).

Because this equation is not linear we must carry out a linearisation process similar to that mentioned in section 10.3.2. Figure 11.10 illustrates the nature of the connections required.

If we let

$$F = \lambda_j R_j \begin{pmatrix} x_i \\ y_i \\ z_i \end{pmatrix} + \begin{pmatrix} X_C \\ Y_0 \\ Z_0 \end{pmatrix}$$

244

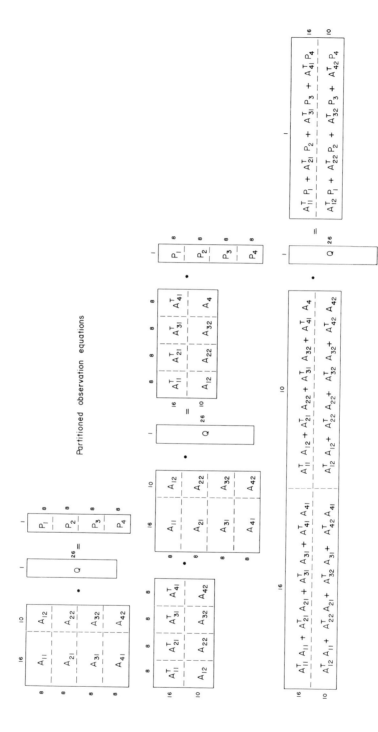

Figure 11.8. Anblock. Formation of Normal Equations. Partitioned Matrices

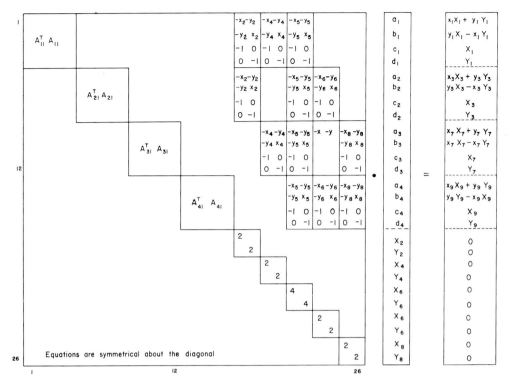

Figure 11.9. *Anblock Normal Equations*

then we can write, as before, an expression for F in the form of a Taylor expansion involving only the first differential coefficients:

$$F = F_0 + \left(\frac{\partial F}{\partial \lambda}\right)\Delta\lambda + \left(\frac{\partial F}{\partial \kappa}\right)\Delta\kappa + \left(\frac{\partial F}{\partial \varphi}\right)\Delta\varphi + \left(\frac{\partial F}{\partial \omega}\right)\Delta\omega$$

$$+ \left(\frac{\partial F}{\partial X_0}\right)\Delta X_0 + \left(\frac{\partial F}{\partial Y_0}\right)\Delta Y_0 + \left(\frac{\partial F}{\partial Z_0}\right)\Delta Z_0 \qquad (11.2)$$

In order to evaluate the differential terms involving the scale factor and the elements of rotation we can conveniently make use of the linear approximate orthogonal matrix

$$\lambda_j R_j = \begin{pmatrix} \lambda & \kappa & -\varphi \\ -\kappa & \lambda & \omega \\ \varphi & -\omega & \lambda \end{pmatrix}_j$$

As our initial estimate of the unknowns, we will make the scale factor unity and all other unknowns equal to zero. This is reasonable, for we remember that each section has been brought into approximately the correct absolute orientation when forming the crude block of co-ordinates. For this case we therefore find that

$$F_0 = \begin{pmatrix} X_i \\ Y_i \\ Z_i \end{pmatrix}_0 = \begin{pmatrix} x_i \\ y_i \\ z_i \end{pmatrix}$$

245

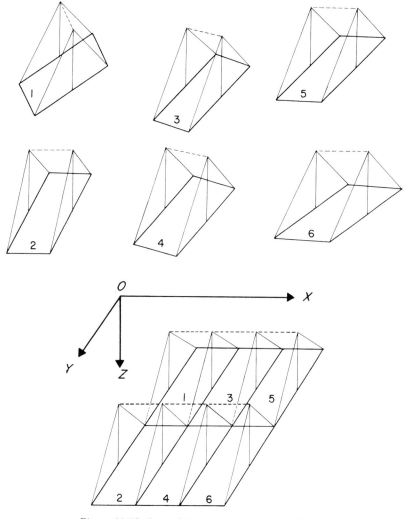

Figure 11.10. *Spatial Transformation of Models*

And the values of the seven differential coefficients are as follows,

$$\left(\frac{\partial F}{\partial \lambda}\right)_0 = \begin{pmatrix} 1 & 0 & 0 \\ 0 & 1 & 0 \\ 0 & 0 & 1 \end{pmatrix} \begin{pmatrix} x_i \\ y_i \\ z_i \end{pmatrix}$$

$$\left(\frac{\partial F}{\partial \kappa}\right)_0 = \begin{pmatrix} 0 & 1 & 0 \\ -1 & 0 & 0 \\ 0 & 0 & 0 \end{pmatrix} \begin{pmatrix} x_i \\ y_i \\ z_i \end{pmatrix}$$

$$\left(\frac{\partial F}{\partial \varphi}\right)_0 = \begin{pmatrix} 0 & 0 & -1 \\ 0 & 0 & 0 \\ 1 & 0 & 0 \end{pmatrix} \begin{pmatrix} x_i \\ y_i \\ z_i \end{pmatrix}$$

$$\left(\frac{\partial F}{\partial \omega}\right)_0 = \begin{pmatrix} 0 & 0 & 0 \\ 0 & 0 & 1 \\ 0 & -1 & 0 \end{pmatrix} \begin{pmatrix} x_i \\ y_i \\ z_i \end{pmatrix}$$

$$\left(\frac{\partial F}{\partial X_0}\right)_0 = \begin{pmatrix} 1 \\ 0 \\ 0 \end{pmatrix} \quad , \quad \left(\frac{\partial F}{\partial Y_0}\right)_0 = \begin{pmatrix} 0 \\ 1 \\ 0 \end{pmatrix} \quad , \quad \left(\frac{\partial F}{\partial Z_0}\right)_0 = \begin{pmatrix} 0 \\ 0 \\ 1 \end{pmatrix}$$

Substituting in equation (11.2) and rearranging the unknowns into a more convenient form, we have

$$F = \begin{pmatrix} X_i \\ Y_i \\ Z_i \end{pmatrix} = \begin{pmatrix} x_i \\ y_i \\ z_i \end{pmatrix} + \begin{pmatrix} x_i & y_i & -z_i & 0 & 1 & 0 & 0 \\ y_i & -x_i & 0 & z_i & 0 & 1 & 0 \\ z_i & 0 & x_i & -y_i & 0 & 0 & 1 \end{pmatrix} \begin{pmatrix} \Delta\lambda \\ \Delta\kappa \\ \Delta\varphi \\ \Delta\omega \\ \Delta X_0 \\ \Delta Y_0 \\ \Delta Z_{0\ 0j} \end{pmatrix} \quad (11.3)$$

Using the notation previously employed, the observation equations take the form given below.

For control points $\qquad\qquad\qquad A_{ij} \cdot P_j = L_i - (F_o)$

For tie points $\qquad\qquad\qquad\qquad A_{ij} \cdot P_j - Q_i = -(F_o)$

The equations (11.3) are therefore the required observation equations involving the seven unknowns for the first iteration involving zero estimates of the rotations and shifts. In the usual overdetermined situation we would form the normal equations and solve these in the usual way. However, we note that in this case the two perspective centres of each model are now included as model tie points. It will be realised that the co-ordinates of these points are not observed in the same manner as ordinary model points. The question arises therefore as to whether these points should be given different weights. Although in this introductory treatment of the subject we will not introduce different weights, it will be appreciated that this could be done. In such a case all model points would be assigned a certain weight value, all air stations a different weight value and control point weight value in accordance with the strength of its determination in the field.

Having solved the normal equations of the first iteration we have the first set of corrections to our initial estimates. Hence we have

$$(\lambda)_1 = (\lambda)_0 + \Delta\lambda_0 = 1 + \Delta\lambda_0$$

$$(\kappa)_1 = 0 + \Delta\kappa_0 \qquad (\varphi)_1 = 0 + \Delta\varphi_0 \qquad (\omega)_1 = 0 + \Delta\omega_0$$

$$(X_0)_1 = 0 + \Delta X_0 \qquad (Y_0)_1 = 0 + \Delta Y_0 \qquad (Z_0)_1 = 0 + \Delta Z_0$$

Using these values we now calculate a better estimate of the function (F_o), and to do this we should use a correct orthogonal matrix:

$$(F_o)_1 = (\lambda_j)_1 (R_j)_1 \begin{pmatrix} x_i \\ y_i \\ z_i \end{pmatrix} + \begin{pmatrix} X_0 \\ Y_0 \\ Z_o \end{pmatrix}_1$$

i.e.

$$(F_o)_1 = \begin{pmatrix} x_i \\ y_i \\ z_i \end{pmatrix}_1$$

247

These new values for $(F_0)_1$ and $(x_i y_i z_i)_1$ can now be used in equations (11.3) for the second iteration. The differential coefficients are the same as before; calculated for unit scale factor and zero rotations. With these modifications the normal equations can be formed and solved and once again used to determine the second and further corrections to the estimated values of the unknowns;

$$(\lambda)_n = (\lambda)_{n-1} + (\Delta\lambda)_{n-1}$$

$$(\kappa)_n = (\kappa)_{n-1} + (\Delta\kappa)_{n-1} \ldots \text{etc.}$$

The final accepted values are obtained when no further significant convergence of values is produced by additional iterations. The number of iterations required will vary with the distribution of the control points available.

A programme developed on the above method of computation has been produced by Ackermann and his associates at Stuttgart University and has the designation PAT-M-7.[11.6, 11.7] This programme, involving the use of a very large computer, is highly flexible and can handle large number of models. The SPACE-M programme, developed by Blais, is a similar programme using rather a different computational approach.[11.8]

11.4.2 THE SEPARATE SIMULTANEOUS ADJUSTMENT OF PLAN AND HEIGHT CO-ORDINATES

A considerable saving in computer requirements can be achieved by separating out the planimetric and height adjustments. In this way a direct solution for the unknowns is possible. This can be an advantage where, due to a poor distribution of control, the iterative process may not converge satisfactorily. The equations (11.3) are rearranged as follows:

$$\begin{pmatrix} x_i & y_i & 1 & 0 \\ y_i & -x_i & 0 & 1 \end{pmatrix} \begin{pmatrix} \lambda \\ \kappa \\ X_0 \\ Y_0 \end{pmatrix} = \begin{pmatrix} X_i \\ Y_i \end{pmatrix} + \begin{pmatrix} z_i & 0 \\ 0 & -z_i \end{pmatrix} \begin{pmatrix} \varphi \\ \omega \end{pmatrix} \tag{11.4}$$

and

$$(x_i \quad -y_i \quad 1) \begin{pmatrix} \varphi \\ \omega \\ Z_0 \end{pmatrix} = Z_i - z_i \lambda \tag{11.5}$$

We will call the second terms of the right-hand sides of equations (11.4) and (11.5) the cross factor terms C and C' respectively. In equation (11.4) when C has a zero value we have the Anblock equations of section 11.4. This set of equations is therefore set out and solved in the usual way to give the first approximate values of the unknowns λ, κ, X_0, Y_0. This value obtained for λ is then used to calculate the factor C' of equations (11.5). These are then solved for the unknowns φ, ω, and Z_0 and the values so obtained used to calculate the value of the factor C of equation (11.4). The process of solution is continued until the values of the unknowns stabilise. Figure 11.11 illustrates this process in the form of a flow chart.

A programme using the above technique has been developed by Ackermann and his colleagues at Stuttgart University and has been designated PAT-M-43.[11.10]

11.5 Block formation and adjustment using bundles

The collinearity conditions investigated in section 10.3.1 are simple in the sense that each equation involves only the one variable, but the use of these conditions involves the processing of many equations. However, it is also one of the most flexible techniques that can be employed.

248

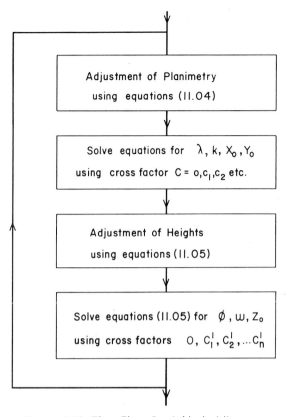

Figure 11.11. *Flow Chart for Anblock Adjustment*

Each point i on photograph j gives rise to two collinearity equations:

$$\frac{x_{ij}}{f} = \frac{\mathbf{X}_i}{\mathbf{Z}_i} \tag{11.6}$$

$$\frac{y_{ij}}{f} = \frac{\mathbf{Y}_i}{\mathbf{Z}_i} \tag{11.7}$$

where

$$\left.\begin{array}{l} \mathbf{X}_i = a_{11}[X_i - X_0] + a_{12}[Y_i - Y_0] + a_{13}[Z_i - Z_0] \\ \mathbf{Y}_i = a_{21}[X_i - X_0] + a_{22}[Y_i - Y_0] + a_{23}[Z_i - Z_0] \\ \mathbf{Z}_i = a_{31}[X_i - X_0] + a_{32}[Y_i - Y_0] + a_{33}[Z_i - Z_0] \end{array}\right\} \tag{11.8}$$

$(x_i y_i)$ are photo co-ordinates and (f) is the principal distance, $(X_i Y_i Z_i)$ are the ground co-ordinates of the point, $(X_0 Y_0 Z_0)$ are the ground co-ordinates of the air station. $(\mathbf{XYZ})$ are ground co-ordinates in a system of axes parallel to those of the photograph system (see Fig 11.12).

Again, these equations are not linear and we therefore must carry out some approximating process that will provide a workable set of linear equations. As before, we will have to commence with an estimated set of values for the unknowns and then improve these systematically by the calculation of sets of corrections obtained from iterative solutions of the normal equations.

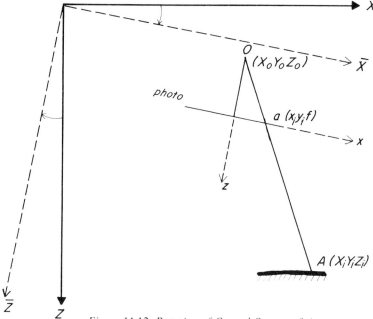

Figure 11.12. *Rotation of Ground System of Axes*

Let the first set of values of unknowns be as follows: $(\kappa)_0 = (\varphi)_0 = (\omega)_0 = 0$ are the parameters of rotation; $(X_0)_0$, $(Y_0)_0$, $(Z_0)_0$ are the first estimated co-ordinates of the air station; $(X_i)_0$, $(Y_i)_0$, $(Z_i)_0$ are the first estimated co-ordinates of the ground point.

We will take only equation (11.6) as our example of the method of procedure and we say

$$F_x = \frac{x_{ij}}{f} = (F_0)_0 + \left[\frac{\partial F}{\partial \kappa}\right]_0 \Delta\kappa + \left[\frac{\partial F}{\partial \varphi}\right]_0 \Delta\varphi + \left[\frac{\partial F}{\partial \omega}\right]_0 \Delta\omega$$

$$+ \left[\frac{\partial F}{\partial X_0}\right]_0 \Delta X_0 + \left[\frac{\partial F}{\partial Y_0}\right]_0 \Delta Y_0 + \left[\frac{\partial F}{\partial Z_0}\right]_0 \Delta Z_0$$

$$+ \left[\frac{\partial F}{\partial X_i}\right]_0 \Delta X_i + \left[\frac{\partial F}{\partial Y_i}\right]_0 \Delta Y_i + \left[\frac{\partial F}{\partial Z_i}\right]_0 \Delta Z_i \qquad (11.9)$$

With zero values for the unknown rotations and using the approximate orthogonal matrix to evaluate the differential coefficients, we can calculate these coefficients in the following way.

We have

$$\begin{pmatrix} \mathbf{X}_i \\ \mathbf{Y}_i \\ \mathbf{Z}_i \end{pmatrix} = R \begin{pmatrix} X_i - X_0 \\ Y_i - Y_0 \\ Z_i - Z_0 \end{pmatrix}$$

hence

$$\begin{pmatrix} \mathbf{X}_i \\ \mathbf{Y}_i \\ \mathbf{Z}_i \end{pmatrix} \simeq \begin{pmatrix} 1 & -\kappa & \varphi \\ \kappa & 1 & -\omega \\ -\varphi & \omega & 1 \end{pmatrix} \begin{pmatrix} X_i - X_0 \\ Y_i - Y_0 \\ Z_i - Z_0 \end{pmatrix}$$

250

Now from equation (11.6)

$$F_x = \frac{\mathbf{X}_i}{\mathbf{Z}_i}$$

and hence

$$(F_x)_0 = \left(\frac{\mathbf{X}_i}{\mathbf{Z}_i}\right)_0$$

Therefore using the approximate orthogonal matrix and zero values $\kappa_0 = \varphi_0 = \omega_0 = 0$, we have

$$(F_x)_0 = \left(\frac{X_i - X_0}{Z_i - Z_0}\right)$$

Also

$$\frac{\partial F_x}{\partial \kappa} = \frac{\partial}{\partial \kappa}\left(\frac{\mathbf{X}_i}{\mathbf{Z}_i}\right)$$

$$= \frac{\mathbf{Z}_i(\partial \mathbf{X}_i/\partial \kappa) - \mathbf{X}_i(\partial \mathbf{Z}_i/\partial \kappa_0)}{\mathbf{Z}_i^2}$$

Hence, if we carry out a partial differentiation for all nine variables and then introduce the initial zero estimates, we obtain the following differential coefficient values:

$$\left[\frac{\partial(F_x)}{\partial \kappa}\right]_0 = -\frac{Y_i - Y_0}{Z_i - Z_0} \quad : \quad \left[\frac{\partial(F_x)}{\partial \varphi}\right]_0 = 1 + \frac{(X_i - X_0)^2}{(Z_i - Z_0)^2} :$$

$$\left[\frac{\partial(F_x)}{\partial \omega}\right]_0 = -\frac{(X_i - X_0)(Y_i - Y_0)}{(Z_i - Z_0)^2}$$

$$\left[\frac{\partial(F_x)}{\partial X_0}\right]_0 = -\frac{1}{(Z_i - Z_0)} \quad : \quad \left[\frac{\partial(F_x)}{\partial Y_0}\right]_0 = 0 \quad : \quad \left[\frac{\partial(F_x)}{\partial Z_0}\right]_0 = \frac{(X_i - X_0)}{(Z_i - Z_0)^2}$$

$$\left[\frac{\partial(F_x)}{\partial X_i}\right]_0 = \frac{1}{(Z_i - Z_0)} \quad : \quad \left[\frac{\partial(F_x)}{\partial Y_i}\right]_0 = 0 \quad : \quad \left[\frac{\partial(F_x)}{\partial Z_i}\right]_0 = -\frac{(X_i - X_0)}{(Z_i - Z_0)^2}$$

The first estimated values for the ground co-ordinates of air stations and ground points can be obtained from a rough assembly of the photographs into a block. For this purpose, the photographs are assumed to be verticals and have a constant average photo scale. Using these assumptions, a set of approximately homogeneous ground co-ordinates can readily be calculated for all air stations and ground points. Some organisations find it worthwhile to carry out this assembly with a little more rigour, for by doing so the co-ordinates obtained are sufficiently accurate to be of use in the detection of gross errors in the main computer programme.

The coefficients of the linearised form of equation (11.7) are obtained in a similar manner to the above. Hence we have two observation equations for the nine unknowns generated for each point on the photograph, as follows:

$$F_x = \left[\frac{x_{ij}}{f}\right]_0$$

$$= \left[\frac{(X_i - X_0)}{(Z_i - Z_0)}\right]_0 - \left[\frac{(Y_i - Y_0)}{(Z_i - Z_0)}\right]_0 \Delta\kappa + \left[1 + \frac{(X_i - X_0)^2}{(Z_i - Z_0)^2}\right]_0 \Delta\varphi$$

$$- \left[\frac{(X_i - X_0)(Y_i - Y_0)}{(Z_i - Z_0)^2}\right]_0 \Delta\omega - \left[\frac{1}{(Z_i - Z_0)}\right]_0 \Delta X_0 + \left[\frac{(X_i - X_0)}{(Z_i - Z_0)^2}\right]_0 \Delta Z_0$$

$$+ \left[\frac{1}{(Z_i - Z_0)}\right]_0 \Delta X_i - \left[\frac{(X_i - X_0)}{(Z_i - Z_0)^2}\right]_0 \Delta Z_i \qquad (11.10)$$

$$F_y = \left[\frac{y_{ij}}{f}\right]_0$$

$$= \left[\frac{(Y_i - Y_0)}{(Z_i - Z_0)}\right]_0 + \left[\frac{(X_i - X_0)}{(Z_i - Z_0)}\right]_0 \Delta\kappa + \left[\frac{(X_i - X_0)(Y_i - Y_0)}{(Z_i - Z_0)^2}\right]_0 \Delta\varphi$$

$$- \left[1 + \frac{(Y_i - Y_0)^2}{(Z_i - Z_0)^2}\right]_0 \Delta\omega - \left[\frac{1}{(Z_i - Z_0)}\right]_0 \Delta Y_0 + \left[\frac{(Y_i - Y_0)}{(Z_i - Z_0)^2}\right]_0 \Delta Z_0$$

$$+ \left[\frac{1}{(Z_i - Z_0)}\right]_0 \Delta Y_i - \left[\frac{(Y_i - Y_0)}{(Z_i - Z_0)^2}\right]_0 \Delta Z_i \qquad (11.11)$$

If we rewrite these two equations using the letters $(a, b, c \ldots i)$ and $(a', b', c' \ldots i')$ to denote the 18 elements of the coefficient matrix, we obtain observation equations of a familiar form:

$$\begin{pmatrix} a & b & c & d & 0 & f & g & 0 & i \\ a' & b' & c' & 0 & e' & f' & 0 & h' & i' \end{pmatrix} \begin{pmatrix} \Delta\kappa \\ \Delta\varphi \\ \Delta\omega \\ \Delta X_0 \\ \Delta Y_0 \\ \Delta Z_0 \\ \hline \Delta X_i \\ \Delta Y_i \\ \Delta Z_i \end{pmatrix}_j = \begin{matrix} \left[\dfrac{x_{ij}}{f}\right] - \left[\dfrac{X_i - X_0}{Z_i - Z_0}\right]_0 \\[2em] \left[\dfrac{y_{ij}}{f}\right] - \left[\dfrac{Y_i - Y_0}{Z_i - Z_0}\right]_0 \end{matrix} \qquad (11.12)$$

Each photo point will therefore produce two equations for each photo on which it appears. As before, the nature of the equation will change slightly, depending on whether the point is a control point of known co-ordinates $(X_i Y_i Z_i)$ or a tie point. Each photograph of the block will introduce six unknown parameters of resection, while the ground point associated with each photo point will also introduce three further unknown co-ordinates unless it is a control point. In such a case the elements g, i and h', i' of the coefficient matrix will be entered as zero values and the known values introduced on the right-hand side into the column matrix of constants.

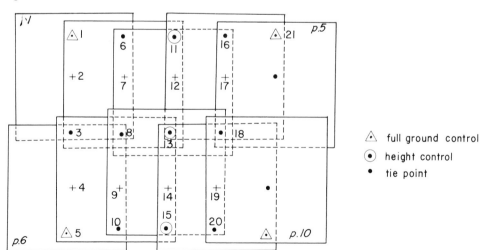

Figure 11.13. Bundle Method of Adjustment for a Small Block

In Fig 11.13 we show a simple block of 10 photographs. For the sake of simplicity the super overlaps of the two strips coincide nicely so that the number of tie points in the lateral overlaps is a minimum. The points can be numbered in either direction but it should be noted that in fact the method adopted here using the shorter direction will produce a narrower bandwidth for the coefficient matrix. (A narrower bandwidth is less demanding of computer resources and is therefore an important point of consideration when large numbers of equations are involved and the capacity of the computer must be used with maximum efficiency.)

From an examination of the diagram we note the following:

$$
\begin{aligned}
\text{The number of unknowns} &= 10 \times 6 \text{ (for the photographs)} \\
&+ 18 \times 3 \text{ (for the tie points)} \\
&+ 3 \times 2 \text{ (for the height points)} \\
\hline
&120 \quad \text{total number of unknowns}
\end{aligned}
$$

$$
\begin{aligned}
\text{The number of equations} &= 2 \times 15 \times 3 \text{ (for points on three photos)} \\
&+ 2 \times 10 \times 2 \text{ (for points on two photos)} \\
\hline
&130 \quad \text{total number of equations}
\end{aligned}
$$

The structure of the observation equations for the block is shown diagrammatically in Fig 11.14. From these, the normal equations are obtained in the usual way. Rather than solve for all the unknowns directly, it is usually the custom to conserve on computer requirements and solve for the photo parameters only in the first instance. This can be done by partitioning the matrix and solving for a reduced set of normal equations as explained in section 11.6. The solution to the normal equations will give the first set of corrections to the estimated values of the unknowns. Using the corrected values for the rotation elements, each set of photo co-ordinates is now transformed using exact orthogonal matrices of the form R^T. In this way the left-hand sides of the equations (11.10) and (11.11) are suitably modified. Again, zero values for rotations can be assumed and so the coefficients of the right-hand sides of the above equations can also be recalculated using the revised estimates for the remaining unknowns to give revised values for the coefficients $(a, b, c \ldots i)$ and $(a', b', c' \ldots i')$. As before, a set of normal equations are generated and solved in the way described to provide further (smaller) corrections to the latest estimated values of the unknowns. The process is continued until iteration produces no significant corrections. The flow chart shown in Fig 11.15 shows the nature of the calculations.

Papers by Bauer and Muller[11.20] and Brown[11.21, 11.22] describe the bundle method and provide details of the capabilities of this method. A comparison between polynomial and bundle adjustments is contained in Ref 11.17.

11.6 The formation and solution of normal equations

From the photogrammetric point of view we might regard the identification and formation of a satisfactory set of observation equations as a suitable end point. After this stage, the subsequent steps involving the formation of normal equations and their solution by a least squares method are more general mathematical techniques. However, for the sake of completeness, a few comments on matters of this nature are included here.

The use of partitioned matrices is sometimes useful when handling very large matrices that tend to take the form of regular arrays. Such submatrices can be handled in the same way as single elements. The method of partitioning two matrices is restricted only by the

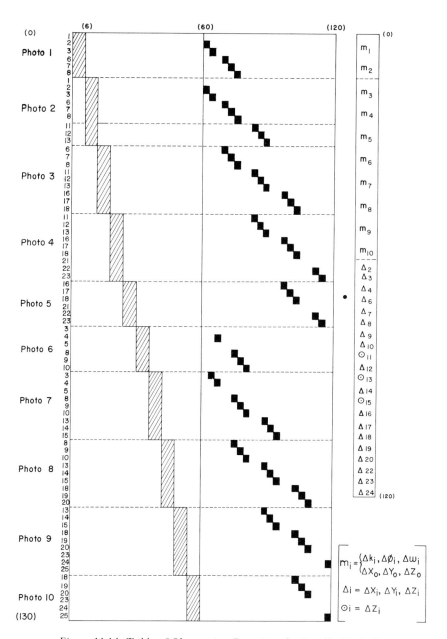

Figure 11.14. *Table of Observation Equations for Bundle Method*

requirements of compatibility for the processes of addition and multiplication. The following examples illustrate this.

(i)
$$(A) + (B) = \left(\frac{A_1 \mid A_2}{A_3 \mid A_4} \right) + \left(\frac{B_1 \mid B_2}{B_3 \mid B_4} \right)$$

$$= \left(\frac{A_1 + B_1 \mid A_2 + B_2}{A_3 + B_3 \mid A_4 + B_4} \right)$$

254

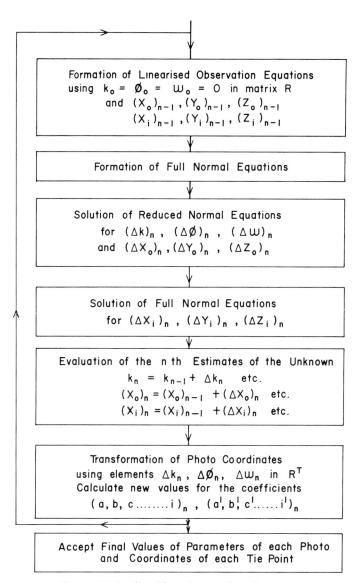

Figure 11.15. *Flow Chart for Bundle Adjustment*

(ii)

$$(A)(B)^{\mathrm{T}} = \left(\begin{array}{c|c} A_1 & A_2 \\ \hline A_3 & A_4 \end{array}\right)\left(\begin{array}{c|c} B_1^{\mathrm{T}} & B_3^{\mathrm{T}} \\ \hline B_2^{\mathrm{T}} & B_4^{\mathrm{T}} \end{array}\right)$$

$$= \left(\begin{array}{c|c} A_1 B_1^{\mathrm{T}} + A_2 B_2^{\mathrm{T}} & A_1 B_3^{\mathrm{T}} + A_2 B_4^{\mathrm{T}} \\ \hline A_3 B_1^{\mathrm{T}} + A_4 B_2^{\mathrm{T}} & A_3 B_3^{\mathrm{T}} + A_4 B_4^{\mathrm{T}} \end{array}\right)$$

To carry out these operations we must have

A_1 of order $m \times n$ with B_1^{T} of order $n \times m$

A_2 of order $m \times p$ with B_2^{T} of order $p \times m$

A_3 of order $q \times n$ with B_3^{T} of order $n \times q$

A_4 of order $q \times p$ with B_4^{T} of order $p \times q$

and so on.

255

A typical use of partitioning is found in the solving of some sets of normal equations. For example, in the case where the column matrix of unknowns is composed of transformation parameters and the unknown co-ordinates of the points as in Fig 11.9. The full solution of the equations is rather long so we can solve for a reduced set of equations involving only the transformation elements in the first instance.

$$\left(\begin{array}{c|c} A_{11} & A_{12} \\ \hline A_{12}{}^{T} & A_{22} \end{array}\right)\left(\begin{array}{c} P_1 \\ \hline P_2 \end{array}\right) = \left(\begin{array}{c} L \\ \hline 0 \end{array}\right)$$

i.e.
$$A_{11}P_1 + A_{12}P_2 = L$$

and
$$A_{12}^{T}P_1 + A_{22}P_2 = 0$$

If we multiply the second of these equations by A_{22}^{-1} we have

$$A_{22}^{-1}A_{12}^{T}P_1 + A_{22}^{-1}A_{22}P_2 = 0$$

i.e.
$$P_2 = -A_{22}^{-1}A_{12}^{T}P_1$$

We now substitute for P_2 in the first equation, to produce

$$A_{11}P_1 - A_{12}A_{22}^{-1}A_{12}^{T}P_1 = L$$

i.e.
$$(A_{11} - A_{12}A_{22}^{-1}A_{12}^{T})P_1 = L$$

which is a reduced set of equations involving only 16 unknowns.

As a further example of the use of partitioning, the observation equations of the Anblock adjustment of the small block illustrated in Fig 11.7 have been written out in this manner in Fig 11.8. The normal equations resulting from this matrix are shown in Fig 11.9. This process is of value when a large number of elements of the observation equations are zero. In this instance this is the case and so the equations of Fig 11.8 are readily determined.

In this treatment a knowledge of least squares has been assumed. A paper by Thompson[11.23] gives a sound theoretical introduction to this subject. The textbook by Mikhail[0.12] is recommended, for it contains examples taken from analytical photogrammetry and considers such items as reduced normal equations, partitioning of matrices and bonded matrices.

11.7 The control requirements for block formation

In this section we will discuss in quite general terms the distribution of control points required for the satisfactory formation of a block of homogeneous co-ordinates regardless of the method of block formation used. To a limited extent the various methods react in different ways to different densities and patterns of control.[11.4] At this stage, however, we will note merely that with a high density of control those methods employing the most observation equations are most likely to produce the lowest standard deviations. The term 'satisfactory' used above is of course a variable, for what can be regarded as an adequate standard deviation in the absolute residuals at one scale and for one purpose may not be considered so in other circumstances. For the sake of comparison, it is best to quote such deviations at photo scale. A considerable number of empirical tests have been carried out on this topic, and those quoted in the bibliography are typical but by no means exhaustive. Because of the nature of the propagation of errors it is necessary to discuss the planimetric and height co-ordinates separately.

11.7.1 PLANIMETRIC CONTROL

It has been well established that in the provision of planimetric control very little improvement is obtained by adding control to the interior of the block. A distribution of

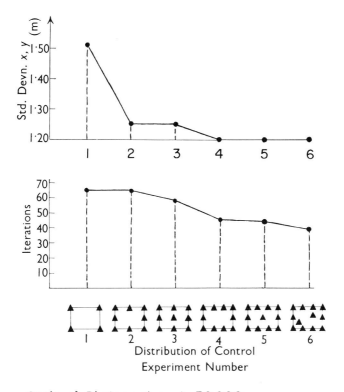

Scale of Photography: 1:30 000
Block size : 20 models in 7 strips

Figure 11.16. *Control Requirements for Planimetric Block Adjustment*

control around the perimeter of the block is therefore considered to be the optimum pattern of planimetric control. Beyond a certain density of control there is no significant improvement in accuracy. Figure 11.16 illustrates these points and is taken from Ref 11.9, which reports on experiments carried out on a block of 1:30 000 photographs composed of seven strips each of 20 models. The results quoted for the absolute residuals should be used for internal comparisons only. Results reported for the Oberschwaben test area of the European Organisation for Experimental Photogrammetric Research (OEEPE)[11.14] indicate quite wide variations in accuracy by different institutions using the same photogrammetric material.

For large-scale work, control points every five or six models (or sections) would seem to be sufficient for most purposes. For small-scale work of the order of 1:50 000 scale planimetric control every 15–20 models is not uncommon.[11.18]

11.7.2 HEIGHT CONTROL

Because of a less favourable error propagation the pattern of height control needed for a satisfactory block formation is greater than that required for planimetry. In this case height control within the block is necessary. This extra control is usually provided in bands of control running across the direction of the flight line – see Fig 9.15, for example. In this case the height control is in three bands as would be required for a second order polynomial adjustment. Four bands evenly spaced could, therefore be related to a third order polynomial form of adjustment. These will greatly reduce the magnitude of any systematic errors that can arise when using three bands only with a spacing of greater than about 6 models. Most authorities seem to consider a minimum density of height control

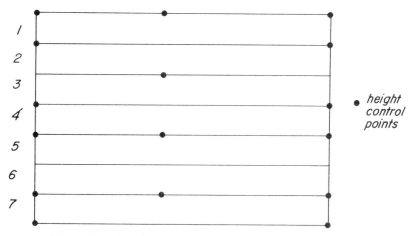

Figure 11.17. *Height Control Requirement for Block Adjustment*

for any interior band of at least one point in every other overlap, and with rather higher density than this in the two end bands. Figure 11.17 shows such a form of control. (This has been set out so that in fact each fourth strip can be the subject of an independent second degree polynomial adjustment.)

Discussions on the provision of height control are contained in Refs 11.13, 11.16. In the paper by Allam and others[12.10] a comprehensive set of tests on two large blocks of photography were carried out to investigate the influence of cross bands of vertical control. Very briefly the results of these tests indicated that a spacing of six models between bands was about the minimum required, the accuracy falling away in a linear manner with spacings of six to 20 models. With a control point in each lateral overlap along a band, a spacing of every eight models gave adequate results and seemed to be an optimum. However, an increase in spacing of between 12 to 16 models is possible, with no significant decrease in accuracy, provided an extra four to six control points were placed in the top and bottom strips.

11.7.3 THE USE OF TIE STRIPS
If we begin to think of bands of control for aerial triangulation it is natural to then consider the possible uses of additional cross or tie-strips of photography flown at regular intervals across the main flight lines. Certainly for height control purposes such strips can be of value. There are, however, some difficulties in introducing these into an analytical block adjustment. This topic is the subject of a comprehensive paper by Smith, Miles and Verall,[11.19] where the values of tie strips and the observation equations required for their incorporation are described.

12: Auxiliary instruments for providing control data

12.1 Introduction

In addition to the camera there are a variety of instruments that may also be used on the photographic sortie. These fall into two main categories, those concerned with the accurate navigation of the aircraft along the required flight lines and those concerned with the provision of data that will prove useful in the control of models in the strip or block formation.

In the first category of equipment we have such items as the Doppler type of navigation aid, the various radio-wave positioning systems such as Decca, Hiran, Shoran and the microwave systems such as Aerodist. These instruments are not discussed in this text but they are given a most comprehensive treatment in Ref 0.7.

In this chapter we will examine the main types of instrument that have been used as a source of control data for photogrammetric mapping purposes. The data we require are such that will enable us to solve more easily and more reliably the space resection problem; that is the determination of the space co-ordinates of the camera at the moment of exposure together with its orientation and tilt with respect to the vertical direction. Much of the information of this sort can be provided by inertial guidance systems but it would seem that these are both expensive and bulky and so at the present stage of development they are not in general use for photogrammetric purposes. The instruments in common use, the subject of this chapter, are optical, mechanical or electromagnetic devices.

12.2 The horizon camera

Part (a) of Fig 12.1 shows in diagrammatic form the nature of this instrument. It is comprised of four small-format cameras whose axes are perpendicular to the optical axis of the main camera. They are directed into the fore-and-aft and port and starboard directions with respect to the camera alignment. Each takes a photograph of part of the horizon trace at each exposure of the main camera. With the main axis near vertical, each horizon camera will therefore record the position of the horizon trace in a particular direction. In part (b) of Fig 12.1 we illustrate the elementary relationship between the factors involved:

$$hp = d$$
$$= f_h \tan \theta$$

(f_h being the focal length of the cameras)

Hence

$$\frac{\partial d}{\partial \theta} = f_h \sec^2 \theta$$
$$= f_h$$

on the camera axis where $\theta = 0$.

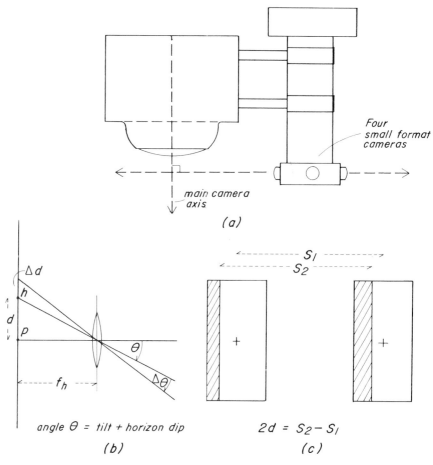

Figure 12.1. *Diagram of Horizon Camera*

If we assume a focal length value of say 50 mm and can measure the distance d with a precision of about 10 μm then we have

$$\Delta d = 10$$

$$= 50 \times 10^3 \Delta\theta \text{ rad on the axis}$$

i.e.
$$\Delta\theta \simeq 40''$$

In the instrument the situation is more favourable than the above because cameras in diametrically opposed directions are providing complementary information. We therefore measure the difference in the position of the horizon trace on one photograph with respect to the other and so we find a value for $2d$. A precision of the order quoted above is therefore equivalent to some 20″ of arc in the angular determination. In one form of measuring device, the two small photographs are viewed stereoscopically and a floating mark device very similar to a parallax bar is used to measure the distance $2d$ (see Fig 12.1(c)). We note that, in this method of measurement, errors due to refraction will be eliminated in a homogeneous atmosphere situation.

A further improvement is possible if we consider only the differences in the $2d$ measurements between successive exposures. In this way we are detecting the difference in tilt ($\Delta\varphi, \Delta\omega$) between adjacent exposure stations. We can start with a known orientation of the second photograph of a strip if we have sufficient ground control data to carry out an absolute orientation of the first model. The orientations of the third and subsequent air stations can therefore be found by a process of simple summation. Further, if at the end of the strip we can again establish the absolute orientation of the last two air stations then we have a measure of the angular misclosure for the φ and ω tilts at the end of the strip. A linear adjustment of values can then be applied. This technique was devised and used in conjunction with Kern PG2 plotting instruments for the production of 1:50 000 mapping in Nigeria by J. M. Zarzycki.[12.1] It is in effect an unusual and interesting form of aerial triangulation.

The horizon camera can only be used under favourable climatic conditions when the horizon trace is not obscured by clouds. The line formed by a cloud formation may be a tempting image to use but cloud formation and buildup is a dynamic process and it is therefore a dangerous datum for anything other than very short period observations. In the Nigerian work the horizon trace was in fact sometimes obscured by a thick belt of dust caused by the seasonal harmattan. Tests showed that the top line of this layer was in fact sufficiently stable for the purposes required.

Some of the earliest uses of horizon camera data were in the provision of control for optical rectifiers producing controlled photo mosaics, as described by Löfstrom.[12.2]

12.3 The statoscope and hypsometer

The statoscope commonly takes the form of a constant volume gas thermometer maintained at a constant temperature to within very close limits. Under such conditions. the gas in the container maintains a constant pressure value that can be used as a source of comparison with the external pressure that may fluctuate. Figure 12.2 shows the apparatus in diagrammatic form with a simple monometer to record the difference in pressure. In practice such a monometer is not very suitable for this purpose and some form of transducer responding to small pressure changes is to be preferred. From this an electrical output signal is required for recording purposes.

Before the aircraft reaches the working altitude the value of the instrument remains open. In this way the pressure within the vessel remains at that of the atmosphere. When

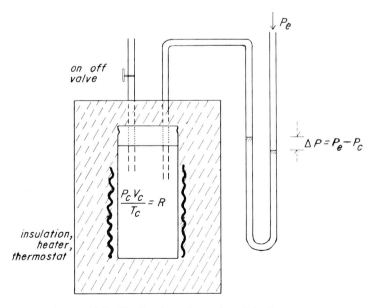

Figure 12.2. The Working Principles of the Statoscope

the required working altitude is reached the valve is then closed and in consequence the pressure within the vessel now remains at that pressure thereafter, provided conditions of constant temperature and volume can be maintained, i.e.

$$P_c V_c = R T_c$$

where R is the universal gas constant and T_c the absolute temperature.

Now, at sea level, 0.1 mb change in pressure corresponds to a change in elevation of about 1 m. If we take the air pressure at sea level as 1013 mb at 20°C then we have

$$\frac{P_1}{P_2} = \frac{T_1}{T_2}$$

i.e.

$$\frac{P_1}{P_1 - \Delta P} = \frac{T_1}{T_1 - \Delta T}$$

i.e.

$$1 + \frac{0.1}{1013} = 1 + \frac{\Delta T}{293}$$

i.e.

$$\Delta T = 0.03°C$$

which corresponds to 0.1 mb change in pressure. At altitude, a 0.1 mb change would correspond to a greater change in elevation than the 1 m at sea level. The calculation shows therefore that a temperature stability of the order of 100th of a degree centigrade is required if measurements with a precision of ± 0.1 mb are to be made. This stability is achieved in later instruments by the use of electrical heating elements, good insulation and a very sensitive thermistor to provide the thermostatic control.

The output signal from the manometer is fed to the pen of a roll-chart recorder. In this way a continuous record of the variations in aircraft height can be recorded with respect to the isobaric surface. A blip on the chart indicates the occurrence of each exposure of the survey camera. It will be appreciated that the isobaric surface may not be parallel to the sea level surface. However, tests over long lines have indicated that the use of a linear correction function is quite adequate.[12.3] We can therefore think of any correction as that due to a linear slope of the surface with respect to datum level. As with the horizon camera, if we have a sound absolute orientation for the first and last models of the strip then the height of the aircraft is known at both ends of the strip. The slope of the surface can therefore be determined. The corrected statoscope readings can then be used to calculate the changes in elevation between air stations with the aid of one of the formulae based on the Laplace equation relating changes in elevation to changes in atmospheric pressure. Tests have indicated that accuracies of the order of ± 1–2 m are possible with this instrument. The report by Ackermann[12.4] on the results of the OEEPE tests indicates the influence statoscope data can have on block adjustments.

Changes in barometric pressure and the velocity of the geostrophic wind are of course related. This in turn influences the heading of any aircraft flying in it, hence we have an equation sometimes referred to as the Henries equation:

$$\Delta H = \frac{2\omega}{g} V \sin \varphi \sin \delta \times 10^3$$

where ΔH is the isobaric gradient along the flight line, ω is the angular velocity of rotation of the Earth, g is the acceleration due to gravity, V is the ground speed of the aircraft, φ is the latitude of the flight path, δ is the observed angle of drift of the aircraft.

Hence from the observed angle of drift of the aircraft and a knowledge of its ground speed a value for the gradient can be computed. If the velocity is expressed in metres per second the gradient will be given in metres per kilometre.

Some instruments used for measuring variations in the heights of air stations with reference to the isobaric surface use the principle of the change of the boiling point of a liquid with change in pressure, i.e. the change in elevation within the atmosphere. Such instruments are sometimes referred to as hypsometers, although the word has a more general meaning than this. In this type of instrument the liquid is boiled using a heated electrical element. A highly sensitive thermistor in the vapour of the liquid measures the temperature of the vapour. Once again, very accurate temperature measurements are required to produce the precision needed. For example, at sea level, for water, a change in boiling point of 0.01 °C requires a 3 m change in elevation. In practice, a more appropriate liquid is used but in order to measure changes in elevation to ± 1 m it is necessary to record changes in boiling point temperature to a precision of about 0.01 °C.

It will be seen that this type of instrument can provide changes of elevation between air stations, i.e. values of the base inclination $\Delta\Phi$. When carrying out analogue triangulation, if the first model is orientated correctly the statoscope data can be used to introduce the absolute Φ component of each subsequent orientation. Using this technique the chance of a large buildup in φ tilt error is avoided. Although the statoscope information does not provide a levelling as accurate as that possible with the photogrammetric orientation process, the errors produced are not subject to a double summation effect such as that experienced in aerial triangulation. The usual error curve for the Z co-ordinate does not therefore develop.

12.4 The radar altimeter and airborne profile recorder

A narrow beam of microwaves in the form of a pulsed signal is emitted from the aircraft in a vertical direction parallel to the axis of the survey camera. A small portion of the energy arriving at the Earth's surface is reflected back along the incidence path and is detected by the instrument. The instrument measures the time taken for the pulse to travel the double path distance and so, knowing the velocity of electromagnetic waves in air, the distance can be computed. The precision of the instrument is such that a standard velocity can be assumed for average working conditions. The microwave signal used is in the 10 cm waveband and the dimensions of the antenna are such that the divergence of the beam is of the order of one degree only. This means, however, that the diameter of the circle of ground illuminated by the signal increases at the rate of about 17.5/1000 m. The return pulse therefore decreases in sharpness as the altitude increases. Further, if the terrain is not comparatively flat over the area of illumination then some sort of average reading will be obtained. If the axis of propagation is not truly vertical then the slant range will be measured. A cosine correction factor would be required to allow for this and this can be applied when the elements of orientation of the camera are finally evaluated. The measurements from the instrument are converted into electrical signals that deflect the pen of a continuous chart recorder. The statoscope/hypsometer data and the terrain profile trace are usually recorded simultaneously on the one chart. Tests have shown that an accuracy of about 2–3 m are possible with this equipment.

In order to define ground points on the centre line a 35 mm camera, operating at a higher frequency than the main camera, takes a series of small photographs; the axis of the camera being parallel to the direction of signal propagation. Each exposure produces a blip on the record trace that allows it to be identified on that trace. About eight exposures are taken along each base line. From a careful comparison of the 35 mm photograph and the main camera photograph the point of measurement is defined. Not all measurements prove to be suitable. Ideally, flat open ground or a water surface provides data of the most reliable nature.

The radar altimeter data alone provide ground clearance data along the centre line of flight. This information can therefore be used to control the scale of each model of a strip. Any buildup of scale error can therefore be avoided. However, the most useful information is obtained when the results are combined with those of the statoscope or hypsometer. In this case a profile of the ground surface can be obtained, as demonstrated in Fig 12.3.

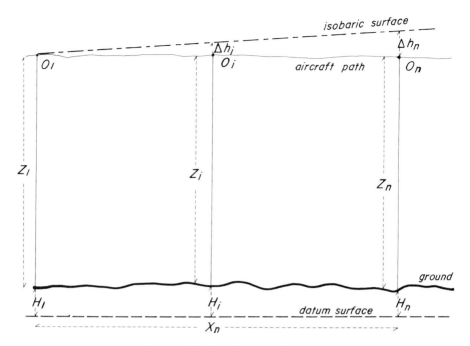

Figure 12.3. *The Use of APR Data*

At the first air station 1, the statoscope instrument is turned on and the measurement Z_1 is made with the radar altimeter to the ground point of known height. At the end of the line, at point n, the ground elevation H_n is known while the range Z_n is recorded together with Δh_n, the statoscope reading. From these measurements we see that the inclination of the barometric surface with respect to the datum is given by

$$\tan \beta = \frac{(\Delta h_n + Z_n + H_n) - (Z_1 + H_1)}{X_n}$$

At any point i of unknown ground height we have

$$H_i = (Z_1 + H_1 + X_i \tan \beta) - (\Delta h_i + Z_i)$$

In this way the height of all intermediate points are found.

Tests have shown that heights to within approximately ± 3 m can be determined. A later form of instrument employs a modulated laser beam in place of the microwaves. Such a beam can be much more highly collimated and accuracies of the order of ± 2 m have been reported.[12.5]

The profile recorder produces a series of heighted points along the base line. Although these points are relatively imprecise compared with photogrammetric determinations, they are direct determinations and do not suffer from any unfavourable error propagation. Hence when used in conjunction with aerial triangulation significant increases in precision can be obtained.[12.6]

The paper by Zarzycki[12.7] gives details of an application of auxiliary data for the provision of control and mapping of an inaccessible region of Guyana.

12.5 The use of auxiliary data in analytical work

For some time the data from auxiliary instruments was little used in analytical photogrammetry and this was probably due to the difficulty of incorporating them into the

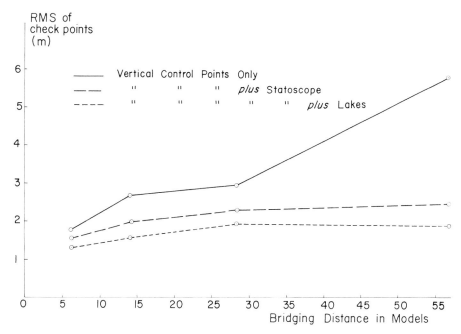

Figure 12.4. Effect of Statoscope Data on Adjustment

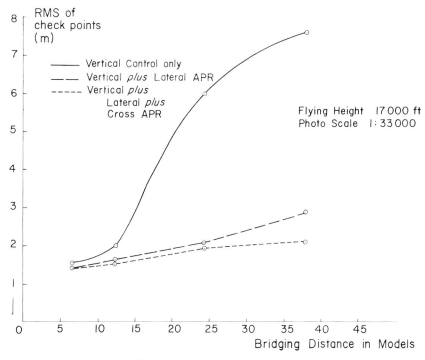

Figure 12.5. Effect of Cross Profiles and APR on Adjustment

block adjustment programmes available. Over the past decade or so we have seen a great development in such programmes together of course with developments in the computers themselves. In more recent times much more attention has therefore been given to the use of such data.

Miles and Smith (1969)[12.8] give methods for the use of data in a rigorous adjustment based on the collinearity equations.

Faig[12.9] gives details concerning the nature of the extra observation equations that should be incorporated into the PAT-M-43 program in order to take into account the data provided by lake shore, statoscope and APR observations. The results of the tests described indicate that with the aid of statoscope data the height control can be limited to a perimeter distribution similar to that required for the planimetry. Figure 12.4, taken from Faig's article, shows the nature of the results obtained with 1:33 000 scale photography. The tests concerned with APR data indicated that the accuracy decreased very slowly with the bridging distance between control when longitudinal APR was used. Lateral cross profiles obtained by using APR at a lower altitude further improved the situation as indicated by Fig 12.5.

In a comprehensive set of tests carried out by Allam and others[12.10] the above results were confirmed and the importance of cross profiles at spacings of every 10 models was indicated. In using APR heighted points along the strip a high density of points is not necessary. A point every one or two base lengths is sufficient.

13: Ground control

13.1 The use of natural features

In earlier chapters, the various distributions of ground control required for different photogrammetric processes have been described in turn. In this chapter we examine the nature of the points themselves and consider the problems they pose for the ground surveyor. On the ground the surveyor will select his photopoint, define it with a mark of very small dimensions and then proceed to find the co-ordinates of this mark in the ground frame of reference, perhaps to very high standards of accuracy. At some other time, a photogrammetric observer will observe a point on the photograph that is supposed to be the homologue of the point on the ground. He too will carry out a precise set of observations to establish the co-ordinates of this point in some local frame of reference. From this we realise that unless the two sets of measurements are concerned with exactly the same point then inaccuracies can develop in the mapping process. The identified photo point is therefore the important link between accurate ground survey and precise machine observing. Unfortunately, the selection of a satisfactory planimetric photopoint can range in difficulty from the straightforward to the impossible, depending mainly on the nature of the ground surface but also on the time lapse between photography and ground survey. Furthermore, in the difficult situations where a check of some sort on the accuracy of the identification would be reassuring, none is possible. Two or more doubtful identifications over a small area are really no improvement on one.

In order to clarify the problem somewhat, the following points can be made. First, a point (in the mathematical sense) can be defined only by the intersection of two or more lines or the estimation of a centre of area. Both on the ground and on the photograph the observer is therefore attempting to define points in this way from the patterns of detail present. In the built-up environment often enough there are patterns of high contrast that provide adequate lines or delineations of areas. Natural landscapes, however, rarely provide such detail. The problem of photo identification therefore centres round satisfying the three following criteria, within certain limits.

(i)   The detail pattern seen on the ground must be capable of defining a point to within a certain tolerance.

(ii)   The detail pattern on all photographs of the ground should be capable of defining a point to within the same tolerance as (i) expressed at photo scale. (It might for example be $\pm 0.01\%Z$, i.e. 15 μm for wide angle photography.)

(iii)   The positions defined by (i) and (ii) should be of the same point exactly.

Just a few examples will suffice to illustrate the significance of these three statements. In a rural area, the only 'lines' available may be the estimated centre lines of hedges or paths. These can be very much a matter of opinion. Their intersection can therefore lie within a large area of uncertainty.[13.1] In addition, the view from the air portrayed by the photograph is a very different aspect from that of the ground surveyor. The apparent centre lines of the photo detail may therefore be physically different from those estimated on the ground. Again, the photopoint may be the estimated centre of the crown of a small tree.

But the shape of this can change on the photographs depending on the position of the camera with respect to the tree. Also, with small-scale photography that tree on the ground may not be so small! If it is more than 2 m or so in height the ground surveyor's view is very different from that of the air photograph. It should also be borne in mind that the pattern recorded on the photograph is a result of the illumination present at that time and on the spectral response of the photographic emulsion to the wavelengths reflected. With unusual lighting effects or unusual emulsions the pattern of contrast recorded may be quite different from that experienced by the ground surveyor. For example with an infrared emulsion a patch of shallow clear water would show up as a dark image. It would not appear so on the ground – or on a black and white panchromatic photograph.

Difficulties over criterion (iii) are more often than not caused by the lapse of time between photography and ground survey. Hedge and footpath widths can change in a cyclic manner with the season but also in a more random way over a number of seasons. It is not always safe to assume that organic growth or decay is regular and systematic about some central line or stem.

In addition to the above considerations concerning the nature of the definition of the points, there is also a further consideration concerning pointing accuracy. On the ground there is no problem, the surveyor having decided on the location of the photopoint, puts in his ground mark and takes all his readings to that point. The photogrammetric operator is not in such a fortunate position; his measuring mark is a circular dot often apparently 40–50 μm in diameter at photo scale. To use this device with any precision the photopoint pattern must provide a suitable target for this. For example, if the mark is 50 μm in diameter then a photopoint at the centre of a light-coloured circle of about 70–90 μm diameter at photo scale would give a pointing accuracy of the highest possible precision. Note that this precision depends on the relationship between the two dimensions not on the dimensions themselves, see Fig 13.1(a) and (b).

An alternative to a circular mark would be a T-shape or cross mark. In this case again the widths of the limbs would be slightly in excess of the floating mark diameter in order to give a precise pointing. In such a case the point is well defined as the intersection of the centre lines of the two limbs. It should be noted that some patterns tend to introduce a systematic pointing error on behalf of the observer. This tendency has been shown in an exaggerated form in Fig 13.1(c). The incorrect alignments occur because of a tendency to divide the target into parts of equal area. References 13.2 and 13.3 discuss such problems in some detail.

13.2 The use of pre-marked points

In view of the uncertainties that can be introduced by the use of natural features as photo-points a strong case can sometimes be made for providing manmade marks just prior to the photographic sortie. These would be patterns of high contrast in the form of circles, crosses, Y-shapes or similar. Their size must be related to the diameter of the floating mark and the scale of the photography to be taken. However, to be of maximum use these points will need to appear in the lateral overlaps of strips and in the overlapping areas of adjacent models. This will necessitate a more rigid schedule for the photographic sortie. Flight lines will have to be laid down and adhered to, within close limits and the location of exposure stations will be similarly constrained. This will add to the cost of the photography. At larger scales and where there are existing maps on which flight lines can be planned, the extra cost may be of little consideration. At small scales over unmapped territory things may be very different and pre-marking may not be such a practical proposition. The availability of one of the navigational aids may be a factor in deciding on the feasibility or otherwise of carrying out a satisfactory pre-marking exercise.

With regard to the materials used for the construction of pre-marks, there are factors such as cost and durability to be taken into account. For a large area, marks will have to be

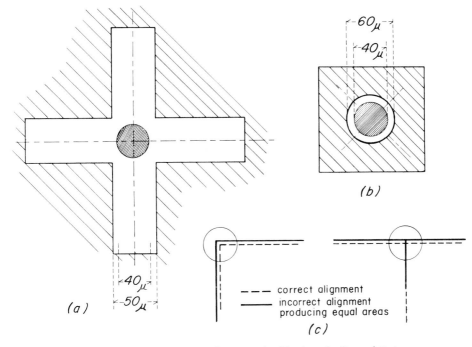

Figure 13.1. *Geometric Considerations for Planimetric Control Points*

set out well in advance of the programmed time of photography. The marks will have to withstand the weather for a limited time and also perhaps the attentions of the local inhabitants. The aim always will be to try to make use of local materials of low value whenever possible.

It should be noted that points may be pre-marked for reasons other than those of control. Sometimes features may be marked only because they are of particular importance, and their size and colour is such that unless marked in some way it would be difficult to identify them on the photographs. This is most likely to occur with large-scale surveys. For example, in the photogrammetric survey of a quayside area, the accurate location of water, electricity and fuel points may be important enough to warrant such special attention. In smaller scale work, triangulation stations may be marked although their positions are not required for photogrammetric control purposes.

13.3 Height control

Heighted points are of no use in photogrammetry because of the finite size of the floating mark, which at natural scale may cover a large area on the ground. Height information concerning an area, flat within a certain tolerance, is therefore the requirement. As a guide, if we think of the operation the operator is carrying out when he attempts to place the floating mark on the model surface, we might say that the area in question should be at least about ten times greater than the area of the floating mark at natural scale. (If the area is circular on the photograph the diameter should be about three times that of the floating mark.) To be a satisfactory photogrammetric point, however, mere flatness alone is not sufficient. In addition, the area must have texture, the stronger the better, and especially with firm delineations in the Y direction. (The reasons for this were discussed in section 10.6 when we considered the mechanism of the floating mark observations.) Over truly featureless areas the floating mark cannot be placed on the model surface with any

269

precision whatsoever. We can encounter such conditions with water features, sand, snow and so on. Pre-marking is occasionally carried out for the purposes of ensuring maximum precision in obtaining photogrammetric spot height determinations. This may be done, for example, when providing spot heights on concrete aircraft runways. While on this topic it should be recalled that this same problem of accurate pointing arises in the use of artificial point markers such as the Wild PUG and the Zeiss Snap Marker. When using these devices the accuracy of transfer of any artificial mark from one photograph to another is very dependent on the existence of good photo image texture. It is surprising just how well a skilled operator can carry out this operation over nearly featureless areas. It is also surprising just how poor these transfers become after some undefinable threshold has been crossed.

Finally, an unusual but apparently effective method of point transfer using very high powers of optical magnification is described by Eden.[13.2] In this method natural points are selected and used with the aid of optical magnification of × 40.

13.4 Requirements of accuracy

Wherever possible, the precision in the determination of the ground co-ordinates of a photo point should be superior to that of the photogrammetric process by a factor of about 4. In surveying we are concerned with determinations of differences in co-ordinates. If the precision of the ground surveying is such that the standard error is just half that of the photogrammetric measurements then, in the worst case, an error in the difference of co-ordinates would be detectable by the photogrammetric process. In order to place the control beyond question it is therefore desirable to have control to the precision stated above.

If we take the case of a first-order analogue plotting instrument with a quoted precision of, say, 16 μm then for the ground survey it is desirable to establish the ground control to an accuracy equivalent to 4 μm at photo scale. For example, with wide angle photography taken from a height of 500 m then a precision of ± 5 cm can be expected from the photogrammetric spot height measurements. Ground surveying giving spot heights to ± 1.25 cm would therefore be appropriate.

Overall, for the purposes of the present considerations, it is sufficiently correct to assume that the precision of the planimetric control will be of the same order of magnitude as the spot heighting. In the above example, therefore, the same figure of ± 1.25 cm can be used also for the planimetric co-ordinates. If we think in terms of values at photo scale we see that we are implying a precision of little under ± 4 μm, taking the most common case and using wide angle photography with a focal length of 152 mm. If we consider the situation with respect to photogrammetric work of the highest precision, using a stereo-comparator, then we find that pointing errors of the order of ± 1–2 μm are now being quoted. Coordinate determinations with an accuracy of 3–5 μm (at photo scale) are therefore possible. For such purposes ground control determinations should be done to single micron accuracy.

Appendix A: Orthogonal matrices

A.1 Introduction

In photogrammetric processes there are many occasions when rotations using orthogonal transformations are required. Sometimes we wish to rotate a set of vectors within a fixed frame of reference, as in a relative orientation procedure. At other times we need to rotate the frame of reference of a set of points (i.e. a model) in order to align these axes with some fixed directions, as for example in a model-to-model connecting process. Such transformations are most easily carried out using matrix methods and there are a number of orthogonal matrices that can be employed for this purpose. We therefore need to examine the main characteristics of the matrices most commonly employed in photogrammetric analysis.

A.2 The properties of orthogonal matrices

A linear transformation is orthogonal if the lengths of vectors remain unchanged in the process. If we represent such a transformation by the matrix R then the matrix and its elements have a number of special properties, which are as follows:

(i) The transpose is equal to the inverse,

i.e. $$R^T = R^{-1}$$

i.e. $$RR^T = RR^{-1} = I$$

(where I is unit matrix).

(ii) If

$$\begin{pmatrix} X' \\ Y' \\ Z' \end{pmatrix} = R \begin{pmatrix} X \\ Y \\ Z \end{pmatrix}$$

then $$(X^1)^2 + (Y^1)^2 + (Z^1)^2 = (X^2 + Y^2 + Z^2)$$

(iii) If

$$R = \begin{pmatrix} r_{11} & r_{12} & r_{13} \\ r_{21} & r_{22} & r_{23} \\ r_{31} & r_{32} & r_{33} \end{pmatrix} \tag{A.1}$$

then the following are well-known properties of the elements.

The sum of the squares of the elements of any row or column is unity, e.g.

$$r_{21}^2 + r_{22}^2 + r_{23}^2 = 1 = r_{13}^2 + r_{23}^2 + r_{33}^2 \text{ etc.}$$

The sum of the products of adjacent elements of any two rows or columns is equal to zero, e.g.

$$(r_{11} \cdot r_{12}) + (r_{21} \cdot r_{22}) + (r_{31} \cdot r_{32}) = 0$$

The square of each element is equal to the square of its co-factor,

e.g. $$(r_{32})^2 = (r_{11} \cdot r_{23} - r_{21} \cdot r_{13})^2$$

All the above relationships stem from six conditions that relate the nine elements. Hence the matrix contains only three independent parameters. The properties of this matrix were most thoroughly investigated by Euler in the eighteenth century.

A.3 The matrix of direction cosines (R_d)

In Fig A.1, **OP** is any vector with co-ordinates (x, y, z) in the lower case system of axes and (X, Y, Z) in the upper case system, such that the direction of the X-axis makes the angles $\theta_{11}, \theta_{12}, \theta_{13}$ with the x, y and z directions respectively. Similarly, angles $\theta_{21}, \theta_{22}, \theta_{23}$ and $\theta_{31}, \theta_{32}, \theta_{33}$ are generated by the Y and Z axes.

By the simple addition of vector quantities, we can write

$$X = x \cos \theta_{11} + y \cos \theta_{12} + z \cos \theta_{13}$$

$$Y = x \cos \theta_{21} + y \cos \theta_{22} + z \cos \theta_{23}$$

$$Z = x \cos \theta_{31} + y \cos \theta_{32} + z \cos \theta_{33}$$

Writing $C_{11} = \cos \theta_{11}$ etc., we have

$$\begin{pmatrix} X \\ Y \\ Z \end{pmatrix} = \begin{pmatrix} C_{11} & C_{12} & C_{13} \\ C_{21} & C_{22} & C_{23} \\ C_{31} & C_{32} & C_{33} \end{pmatrix} \begin{pmatrix} x \\ y \\ z \end{pmatrix}$$

the elements being the direction cosines as defined above.

$$\text{The matrix } R_d = \begin{pmatrix} C_{11} & C_{12} & C_{13} \\ C_{21} & C_{22} & C_{23} \\ C_{31} & C_{32} & C_{33} \end{pmatrix} \tag{A.2}$$

is an orthogonal matrix and so the nine direction cosines are related by six condition equations.

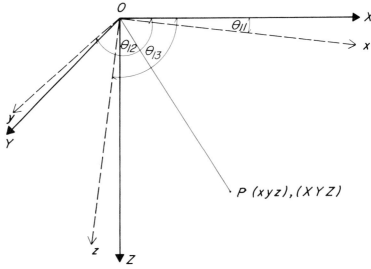

Figure A.1. Direction Cosines of Axes

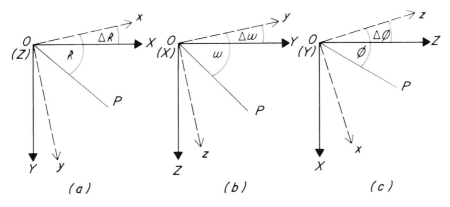

Positive Direction INTO THE PAPER in all cases.

Figure A.2. Separate Rotations About Three Axes

A.4 The Euler matrices (R_E)

The elements of this set of matrices are functions of the three angles of rotation about the fixed set of axes. These may be evaluated from elementary considerations in the following way.

If we consider a two-dimensional situation as illustrated in Fig A.2(a), the co-ordinates (x, y) of any point P are to be expressed in the (X, Y) system. This requires a positive (clockwise) rotation $\Delta\kappa$ of the axes (xy) in the XY plane, as shown.

We have therefore the following relationships:

$$x = OP \cos \kappa: \quad y = OP \sin \kappa: \quad z = Z$$

$$X = OP \cos (\kappa - \Delta\kappa): \quad Y = OP \sin (\kappa - \Delta\kappa): \quad Z = z$$

∴ $$X = OP(\cos \kappa \cos \Delta\kappa + \sin \kappa \sin \Delta\kappa)$$

∴ $$Y = OP(\sin \kappa \cos \Delta\kappa - \sin \Delta\kappa \cos \kappa)$$

i.e. $$X = x \cos \Delta\kappa + y \sin \Delta\kappa$$

i.e. $$Y = y \cos \Delta\kappa - x \sin \Delta\kappa$$

hence

$$\begin{pmatrix} X \\ Y \\ Z \end{pmatrix} = \begin{pmatrix} \cos \Delta\kappa & \sin \Delta\kappa & 0 \\ -\sin \Delta\kappa & \cos \Delta\kappa & 0 \\ 0 & 0 & 1 \end{pmatrix} \begin{pmatrix} x \\ y \\ z \end{pmatrix} \qquad (A.3)$$

$$\mathbf{X} = R_{\Delta\kappa} \cdot \mathbf{x}$$

By successively rewriting equation (A.3) and changing the terms in cyclic order we have

$$\begin{pmatrix} Y \\ Z \\ X \end{pmatrix} = \begin{pmatrix} \cos \Delta\omega & \sin \Delta\omega & 0 \\ -\sin \Delta\omega & \cos \Delta\omega & 0 \\ 0 & 0 & 1 \end{pmatrix} \begin{pmatrix} y \\ z \\ x \end{pmatrix}$$

i.e. $$\begin{pmatrix} X \\ Y \\ Z \end{pmatrix} = \begin{pmatrix} 1 & 0 & 0 \\ 0 & \cos \Delta\omega & \sin \Delta\omega \\ 0 & -\sin \Delta\omega & \cos \Delta\omega \end{pmatrix} \begin{pmatrix} x \\ y \\ z \end{pmatrix} \qquad (A.4)$$

i.e. $$\mathbf{X} = R_{\Delta\omega} \cdot \mathbf{x}$$

273

$$\begin{pmatrix} Z \\ X \\ Y \end{pmatrix} = \begin{pmatrix} \cos \Delta\varphi & \sin \Delta\varphi & 0 \\ -\sin \Delta\varphi & \cos \Delta\varphi & 0 \\ 0 & 0 & 1 \end{pmatrix} \begin{pmatrix} z \\ x \\ y \end{pmatrix}$$

i.e.

$$\begin{pmatrix} X \\ Y \\ Z \end{pmatrix} = \begin{pmatrix} \cos \Delta\varphi & 0 & -\sin \Delta\varphi \\ 0 & 1 & 0 \\ \sin \Delta\varphi & 0 & \cos \Delta\varphi \end{pmatrix} \begin{pmatrix} x \\ y \\ z \end{pmatrix} \qquad (A.5)$$

i.e.
$$\mathbf{X} = R_{\Delta\phi} \cdot \mathbf{x}$$

In the general three-dimensional case the axes (xyz) will require three rotations such as $(\Delta\kappa, \Delta\varphi, \Delta\omega)$ in order to align the lower case axes with the upper case system. The magnitude of each rotation will depend to some extent on the order in which the rotations are applied. There are six possible combinations of the three rotation matrices, such as

$$R_E = R_\omega \cdot R_\theta \cdot R_K \qquad (A.6)$$

i.e.
$$R_E = (\text{primary } \omega) \cdot (\text{secondary } \varphi) \cdot (\text{tertiary } \kappa)$$

In the above example we would say that the kappa rotation is the tertiary matrix (it only operates on the original (xy) co-ordinates), the phi rotation is a secondary matrix (it operates on the results of the previous transformation) and the omega rotation is a primary matrix. This concept corresponds to the mechanical universal joint which needs to be constructed in such a way that it has a primary, secondary and a tertiary axis of rotation – see Fig 8.4.

The order of rotations quoted above is one commonly found in photogrammetric instruments and will therefore be used here as an example. The kappa is very rarely found to be other than a tertiary axis (but see section 9.3).

If we substitute in equation (A.6) we obtain

$$R_E = \begin{pmatrix} 1 & 0 & 0 \\ 0 & \cos \Delta\omega & \sin \Delta\omega \\ 0 & -\sin \Delta\omega & \cos \Delta\omega \end{pmatrix} \begin{pmatrix} \cos \Delta\varphi & 0 & -\sin \Delta\varphi \\ 0 & 1 & 0 \\ \sin \Delta\varphi & 0 & \cos \Delta\omega \end{pmatrix} \begin{pmatrix} \cos \Delta\kappa & \sin \Delta\kappa & 0 \\ -\sin \Delta\kappa & \cos \Delta\kappa & 0 \\ 0 & 0 & 1 \end{pmatrix}$$

$$R_E = \begin{pmatrix} \cos \Delta\varphi \cos \Delta\kappa & \cos \Delta\varphi \sin \Delta\kappa & -\sin \Delta\varphi \\ -\cos \Delta\omega \sin \Delta\kappa + \sin \Delta\omega \sin \Delta\varphi \cos \Delta\kappa & \cos \Delta\omega \cos \Delta\kappa + \sin \Delta\omega \sin \Delta\varphi \sin \Delta\kappa & \sin \Delta\omega \cos \Delta\varphi \\ \sin \Delta\kappa \sin \Delta\omega + \cos \Delta\omega \sin \Delta\varphi \cos \Delta\kappa & -\sin \Delta\omega \cos \Delta\kappa + \cos \Delta\omega \sin \Delta\varphi \sin \Delta\kappa & \cos \Delta\omega \cos \Delta\varphi \end{pmatrix}$$

$$(A.7)$$

If we now compare the elements of matrix R_E of equation (A.7) with those of matrix R_d of equation (A.2) we have a set of identities that provides just one example of a set of three independent parameters satisfying the conditions for an orthogonal transformation. We have

$$\left.\begin{aligned}
C_{11} &= \cos \Delta\varphi \cos \Delta\kappa \\
C_{12} &= \cos \Delta\varphi \sin \Delta\kappa \\
C_{13} &= -\sin \Delta\varphi \\
C_{21} &= -\cos \Delta\omega \sin \Delta\kappa + \sin \Delta\omega \sin \Delta\varphi \cos \Delta\kappa \\
C_{22} &= \cos \Delta\omega \cos \Delta\kappa + \sin \Delta\omega \sin \Delta\varphi \sin \Delta\kappa \\
C_{23} &= \sin \Delta\omega \cos \Delta\varphi \\
C_{31} &= \sin \Delta\omega \sin \Delta\kappa + \cos \Delta\omega \sin \Delta\varphi \cos \Delta\kappa \\
C_{32} &= -\sin \Delta\omega \cos \Delta\kappa + \cos \Delta\omega \sin \Delta\varphi \sin \Delta\kappa \\
C_{33} &= \cos \Delta\omega \cos \Delta\varphi
\end{aligned}\right\} \qquad (A.8)$$

In the above treatment we have considered the rotation of axes about their origin. If we wish to consider the rotation of the vector about its origin then we need only note the following points.

A positive rotation of the vector is equivalent to a negative rotation of the axes, hence

$$\theta' = -\theta$$

$$-\sin\theta' = \sin\theta$$

$$\cos\theta' = \cos\theta$$

We need therefore only change the terms in accordance with the above in order to obtain the required transformation matrix.

We note further that the use of any other sign convention for the system of axes selected requires similar changes of sign.

A.5 The approximate orthogonal matrix (R_A)

When the angles of rotation are small then it is sometimes sufficiently accurate to simplify an expression such as equation (A.7) by making the following substitutions:

$$\sin\theta = \theta$$

$$\cos\theta = 1$$

$$\sin^2\theta = 0 \quad \text{and etc.}$$

From equation (A.7) we therefore have

$$R_A \approx \begin{pmatrix} 1 & \Delta\kappa & -\Delta\varphi \\ -\Delta\kappa & 1 & \Delta\omega \\ \Delta\varphi & -\Delta\omega & 1 \end{pmatrix} \tag{A.9}$$

It will be seen that this equation does not fit exactly the condition for orthogonality. However, it is sufficiently correct for some photogrammetric purposes.

We note that the matrix for the rotation of the vector is the transpose of R_A^T, i.e.

$$R_A' = R_A^T = \begin{pmatrix} 1 & -\Delta\kappa & \Delta\varphi \\ \Delta\kappa & 1 & -\Delta\omega \\ -\Delta\varphi & \Delta\omega & 1 \end{pmatrix} \tag{A.10}$$

A.6 The Cayley and Rodrigues orthogonal matrices (R_C and R_R)

It was first shown by Cayley that a matrix of the form given below must always be orthogonal.

$$R_C = (I - S)(I + S)^{-1} \tag{A.11}$$

where I is a unit matrix and S is any real skew symmetric matrix of the form

$$S = \tfrac{1}{2}\begin{pmatrix} 0 & v & -\mu \\ -v & 0 & \lambda \\ \mu & -\lambda & 0 \end{pmatrix} \tag{A.12}$$

We note that

$$(I - S)^T = (I + S)$$

$$(I + S)^T = (I - S)$$

275

Also we have

$$R_C^T = (I - S)^{-1}(I + S)$$

Finally, we note that the pairs of commuting matrices below are also orthogonal

$$(I \mp S)(I \pm S)^{-1} = (I \pm S)^{-1}(I \mp S)$$

If we expand the expression for R_C in the form given in equation (A.11) we obtain a form of the orthogonal matrix first introduced by Rodrigues:

$$R_C = \begin{pmatrix} 1 & -\frac{1}{2}v & \frac{1}{2}\mu \\ \frac{1}{2}v & 1 & -\frac{1}{2}\lambda \\ -\frac{1}{2}\mu & \frac{1}{2}\lambda & 1 \end{pmatrix} \begin{pmatrix} 1 & \frac{1}{2}v & -\frac{1}{2}\mu \\ -\frac{1}{2}v & 1 & \frac{1}{2}\lambda \\ \frac{1}{2}\mu & -\frac{1}{2}\lambda & 1 \end{pmatrix}^{-1}$$

hence

$$R_C = \begin{pmatrix} 1 & -\frac{1}{2}v & \frac{1}{2}\mu \\ \frac{1}{2}v & 1 & -\frac{1}{2}\lambda \\ -\frac{1}{2}\mu & \frac{1}{2}\lambda & 1 \end{pmatrix} \frac{1}{\Delta}\mathrm{Adj} \begin{pmatrix} 1 & \frac{1}{2}v & -\frac{1}{2}\mu \\ -\frac{1}{2}v & 1 & \frac{1}{2}\lambda \\ \frac{1}{2}\mu & -\frac{1}{2}\lambda & 1 \end{pmatrix}$$

where

$$\Delta = |I + S| = 1 + \tfrac{1}{4}(v^2 + \mu^2 + \lambda^2).$$

$$R_C \equiv R_R = \frac{1}{\Delta} \begin{pmatrix} 1 + \frac{1}{4}(-v^2 - \mu^2 + \lambda^2) & -v + \frac{1}{2}\mu\lambda & \mu + \frac{1}{2}v\lambda \\ v + \frac{1}{2}\mu\lambda & 1 + \frac{1}{4}(-v^2 + \mu^2 - \lambda^2) & -\lambda + \frac{1}{2}v\mu \\ -\mu + \frac{1}{2}v\lambda & \lambda + \frac{1}{2}v\mu & 1 + \frac{1}{4}(v^2 - \mu^2 - \lambda^2) \end{pmatrix}$$

$$(A.13)$$

which is the form of the equation first obtained by Rodrigues. It is of interest and importance because it is devoid of trigonometrical terms.

The elements of the above may be rearranged slightly to produce a form that is more amenable to some methods of calculation:

$$R_R = \frac{1}{\Delta} \begin{pmatrix} 2 - \Delta & -v & \mu \\ v & 2 - \Delta & -\lambda \\ -\mu & \lambda & 2 - \Delta \end{pmatrix} + \frac{1}{2\Delta} \begin{pmatrix} \lambda^2 & \mu\lambda & v\lambda \\ \mu\lambda & \mu^2 & v\mu \\ v\lambda & v\mu & v^2 \end{pmatrix}$$

If we write $(2 - \Delta) = \Delta'$ then we have

$$R_R = \frac{1}{\Delta} \begin{pmatrix} \Delta' & -v & \mu \\ v & \Delta' & -\lambda \\ -\mu & \lambda & \Delta' \end{pmatrix} + \tfrac{1}{2}\Delta \begin{pmatrix} \lambda \\ \mu \\ v \end{pmatrix}(\lambda \; \mu \; v) \qquad (A.14)$$

The considerable advantage of using the forms R_C and R_R of an orthogonal matrix lies in the fact that given the co-ordinates of at least three non-collinear points in both systems of axes, the three unknown parameters of rotation can be evaluated from a set of linear equations derived from R_C. Knowing these, the nine elements of R_R can readily be calculated. Using this matrix, the transformation of all other points can then be carried out.

276

Now, if we consider a transformation given by

$$\begin{pmatrix} X' \\ Y' \\ Z' \end{pmatrix} = R_C^T \begin{pmatrix} X \\ Y \\ Z \end{pmatrix} \tag{A.15}$$

i.e.

$$\begin{pmatrix} X' \\ Y' \\ Z' \end{pmatrix} = (I - S)^{-1}(I + S) \begin{pmatrix} X \\ Y \\ Z \end{pmatrix}$$

i.e.

$$(I - S)\begin{pmatrix} X' \\ Y' \\ Z' \end{pmatrix} = (I + S) \begin{pmatrix} X \\ Y \\ Z \end{pmatrix}$$

If we let $\Delta X = (X' - X)$ etc. we obtain from the above:

$$\Delta X = v(Y' + Y) \quad - \mu(Z' + Z)$$
$$\Delta Y = -v(X' + X) \quad\quad\quad\quad + \lambda(Z' + Z)$$
$$\Delta Z = \quad\quad\quad +\mu(X' + X) - \lambda(Y' + Y)$$

We can write the above equations in the form

$$\begin{pmatrix} \Delta X \\ \Delta Y \\ \Delta Z \end{pmatrix} = \begin{pmatrix} 0 & -(Z' + Z) & (Y' + Y) \\ (Z' + Z) & 0 & -(X' + X) \\ -(Y' + Y) & (X' + X) & 0 \end{pmatrix} \begin{pmatrix} \lambda \\ \mu \\ v \end{pmatrix} \tag{A.16}$$

This is a singular skew symmetric matrix and we therefore need a third point in order to provide six equations in all from which the three parameters can be calculated. (The first point is of course the common point of origin about which the rotation must take place.)

The parameters (λ, μ, v) of this case are not the elements of rotation $(\Delta\kappa, \Delta\varphi, \Delta\omega)$ of the Euler forms of the transformation mentioned earlier. A rotation can also be described by the rotation θ about a single axis of rotation of known direction. The parameters (λ, μ, v) are functions of this angle and the direction cosines (l, m, n) of the direction of the axis. They are related by the following expression,

$$\begin{pmatrix} \lambda \\ \mu \\ v \end{pmatrix} = 2 \tan \theta/2 \begin{pmatrix} l \\ m \\ n \end{pmatrix}$$

It will be realised that when $\theta = \pi$ the above expressions cannot be used, for in such a case at least one of the parameters λ, μ, v must be infinitely large. This is, however, not a situation so often encountered in photogrammetric processes.

A.7 The Schut form of the orthogonal matrix (R_S)

It was first shown by E. H. Thompson[A.1] that with a suitable choice of three parameters it was possible to derive three exact linear equations from which values of the parameters could be determined. The form of these equations (A.16) was given in the previous section.

277

However, the equations are not suitable if the required rotation is 180°. A set of four homogeneous equations that do not suffer from this restriction was therefore developed by G. H. Schut.[A.2] This same author's paper [A.3] on the subject of the construction of orthogonal matrices is also a valuable source of reference on this topic. The linear equations of Schut were derived in a number of different ways and the treatment given below is one of the methods.

It can be shown that an equation of the form

$$R_S = (dI - S)^{-1}(dI + S) \tag{A.17}$$

is an orthogonal matrix of a family with properties identical to those of R_C in equation (A.11). In this case I is a unit matrix as before and the form of S is given by

$$S = \begin{pmatrix} 0 & -c & b \\ c & 0 & a \\ -b & a & 0 \end{pmatrix}$$

Substituting these values into the equation (A.17) and carrying out the matrix multiplication, we have

$$R_S = \begin{pmatrix} d & c & -b \\ -c & d & a \\ b & -a & d \end{pmatrix}^{-1} \begin{pmatrix} d & -c & b \\ c & d & -a \\ -b & a & d \end{pmatrix}$$

$$= \frac{1}{\Delta} \mathrm{Adj} \begin{pmatrix} d & c & -b \\ -c & d & a \\ b & -a & d \end{pmatrix} \begin{pmatrix} d & -c & b \\ c & d & -a \\ -b & a & d \end{pmatrix}$$

where $\dfrac{1}{\Delta} = \dfrac{1}{d^2 + a^2 + b^2 + c^2}$,

i.e. $\quad R_S = \dfrac{1}{d^2 + a^2 + b^2 + c^2} \begin{pmatrix} d^2 + a^2 - b^2 - c^2 & 2ab - 2cd & 2ac + 2bd \\ 2ab + 2cd & d^2 - a^2 + b^2 - c^2 & 2bc - 2ad \\ 2ac - 2bd & 2bc + 2ad & d^2 - a^2 - b^2 + c^2 \end{pmatrix}$

$$\tag{A.18}$$

If we put $d = 1$ then equation (A.18) is identical to equation (A.13) when

$$a = \lambda/2: \qquad b = \mu/2: \qquad c = \nu/2$$

This time the transformation has been written in the form

$$\begin{pmatrix} X' \\ Y' \\ Z' \end{pmatrix} = R_S \begin{pmatrix} X \\ Y \\ Z \end{pmatrix}$$

(cf. eqn. A.15), i.e.

$$\begin{pmatrix} d & c & -b \\ -c & d & a \\ b & -a & d \end{pmatrix} \begin{pmatrix} X' \\ Y' \\ Z' \end{pmatrix} = \begin{pmatrix} d & -c & b \\ c & d & -a \\ -b & a & d \end{pmatrix} \begin{pmatrix} X \\ Y \\ Z \end{pmatrix}$$

278

We then obtain from the above equality the following three homogeneous equations in the four parameters a, b, c, and d

$$\left.\begin{array}{l} -b(Z'+Z)+c(Y'+Y)+d(X'-X)=0 \\ a(Z'+Z) \qquad\qquad -c(X'+X)+d(Y'-Y)=0 \\ -a(Y'+Y)+b(X'+X) \qquad\qquad +d(Z'-Z)=0 \end{array}\right\} \qquad (A.19)$$

If we now multiply the first equation of (A.19) by the factor a/d, the second equation by b/d and the third equation by c/d and sum the result, we obtain a fourth equation

$$a(X'-X)+b(Y'-Y)+c(Z'-Z)=0 \qquad (A.20)$$

Combining the four equations we obtain

$$\begin{pmatrix} 0 & -(Z'+Z) & (Y'+Y) & (X'-X) \\ (Z'+Z) & 0 & -(X'+X) & (Y'-Y) \\ -(Y'+Y) & (X'+X) & 0 & (Z'-Z) \\ (X'-X) & (Y'-Y) & (Z'-Z) & 0 \end{pmatrix} \begin{pmatrix} a \\ b \\ c \\ d \end{pmatrix} = 0 \qquad (A.21)$$

However, for each point other than the common point of rotation we have in fact produced only two independent equations. As the equations are homogeneous we can only obtain the ratios of the four parameters. One of the parameters may therefore be made equal to unity. For example if d is made equal to 1 then R_S of equation (A.18) becomes identical to R_R^T (see equation (A.13)), when the appropriate substitutions are made for a, b, and c. By selecting which of the four parameters will be made equal to unity, the situation whereby one of the remainder is very large or of infinite value can be avoided. Hence the failure case for the three equations (A.16) can be avoided. By using all four equations for each point the normal equations produced take on a particularly simple form.

In the determination of the rotation elements each point other than the origin therefore provides four homogeneous observation equations. In an overdetermined situation a least squares evaluation of the ratios of the parameters is readily accomplished, for the resulting normal equations will also be homogeneous. The elements of this set of equations can be written out as follows:

$$S_{11} = \Sigma[(X'-X)^2+(Y'+Y)^2+(Z'+Z)^2]$$
$$S_{22} = \Sigma[(X'+X)^2+(Y'-Y)^2+(Z'+Z)^2]$$
$$S_{33} = \Sigma[(X'+X)^2+(Y'+Y)^2+(Z'-Z)^2]$$
$$S_{44} = \Sigma[(X'-X)^2+(Y'-Y)^2+(Z'-Z)^2]$$

$$S_{12} = \Sigma[-2(YX'+Y'X)]$$
$$\quad = S_{21}$$
$$S_{13} = \Sigma[-2(XZ'+X'Z)]$$
$$\quad = S_{31}$$
$$S_{23} = \Sigma[-2(ZY'+ZY)]$$
$$\quad = S_{32}$$
$$S_{14} = \Sigma[2(ZY'-Z'Y)]$$
$$\quad = S_{41}$$

$$S_{24} = \Sigma[2(XZ' - X'Z)]$$
$$= S_{42}$$
$$S_{34} = \Sigma[2(YX' - Y'X)]$$
$$= S_{43}$$

where the summation signs indicate the sum of values for all given points. After the evaluation of the parameters, their values are introduced into the equation (A.18). Full computational details of the procedure are given in Ref. A.5.

Appendix B: A specimen specification for aerial photography

1   Area ...

2   *Scale of Photography*
Vertical air photography will be undertaken at approximately ... scale from a flying height above mean ground level of feet or metres.

3   *The Camera*
The camera used shall be of a modern precision design fitted with a lens that is nominally free from distortion. The camera must have been recently calibrated by an approved authority and the certified calibration report provided shall show that the radial distortion of the image with reference to the principal point of autocollimation does not exceed 0.02 mm measured in the focal plane inclusive of any filters to be used. The interval between calibration and photography shall in no instance exceed one year. Unless otherwise stated the normal focal length of the lens shall be 6 in. and the format of the negative 9 in. square, i.e. 152.4 and 230 mm respectively.
   The camera calibration report shall include the following information:
(a)   Serial number of the camera optical unit.
(b)   Co-ordinates of the principal point of auto-collimation relative to the fiducial unit.
(c)   The radial distortion of the image reference to the point of autocollimation as zero and the principal distance at which these distortions apply.
(d)   The distances between all fiducial marks.
(e)   A certificate stating when and by whom the camera was calibrated.

4   *Photographic Flying Requirements*
(a)   The area(s) shall be covered by straight strips of photographs having fore and aft overlap of $60 \pm 5\%$ and lateral overlap of $25 \pm 10\%$. In the event of great variation in ground level on any one strip, a reasonable increase in the permitted overlap will be accepted in order to maintain a uniform height above sea level; the forward overlap must in no case fall below 55%. A corresponding increase in lateral overlap, which must in no case fall below 15%, may be accepted in similar circumstances.
(b)   Straight strips are defined as those in which the angle at the principal point of each photograph subtended between the homologues of the principal points of preceding photographs shall be between $175°$ and $180°$.
(c)   If the end of a strip of photography joins the end of another strip for any reason, the overlap of the two strips is to be at least three photographs.

5   *'Crabbing' of Photographs*
'Crab' shall not exceed 5% or be such that stereoscopic gaps in the photography result from it.

6   *Camera Tilt*
Camera tilt shall not exceed 2% (or 5% where the flight altitude is less than 2000 ft above mean ground level).

7	*Cloud and Shadow*
Cloud or cloud shadow will not lie over a principal point of any photograph or over its homologue on either adjacent photograph. No single mass of cloud or cloud shadow will exceed 5% of the area of the photograph, and any area masked by cloud will be stereoscopically covered in the adjacent strip.

8	*Photograph Quality*
(a)	The negatives shall be on a freshly coated film of a recognised type.
(b)	The film base shall be of a type known to have low distortion characteristics.
(c)	Processing of the film shall be carried out by a means which will take fullest advantage of the low distortion characteristics of the film base.
(d)	Negatives shall be free from stains, scratches, finger or static marks, dirt and blemishes of all kinds.
(e)	All relevant collimating marks shall be distinct on every photograph.
(f)	The definition and contrast of the negatives shall be such that prints or diapositives shall show the maximum possible detail throughout the full range of tones over the whole photograph.
In areas where the photographic light is affected by local weather conditions or where the relief of the terrain causes deep shadows, some tolerance in the amount of detail discernible on the photographs will be accepted.

9	*Negative Number*
Each negative shall contain a strip so that when printing the following information may be read:
(a)	Reference number of job/sortie number, etc.
(b)	Date of photography.
(c)	Flying height.
(d)	Lens, focal length and number.
(e)	Negative number.

10	*Index Plots*
For each area photographed a comprehensive diagram or diagrams will be supplied on a transparent medium at a scale sufficient to show with reasonable accuracy the areas covered by each strip of photography. The film and negative numbers will be shown at the beginning and end of each strip, and the principal points of all of the photographs will be included if practicable.

11	*Material to be delivered*
The Contractors shall deliver to the Client ... set(s) of contact prints on double-weight semi-matt/glossy paper, covering the area photographed, together with ... set(s) of index plots. Prints to be made on an electronic printer or in such a manner that full advantage is taken of all detail appearing on the negative.

Bibliography

0.1   Thompson, M. M. (Ed.), *Manual of Photogrammetry* (2 vols). American Society of Photogrammetry, 1966.

0.2   (a) Longhurst, R. S., *Geometrical and Physical Optics*, Longman, London, 1970. (b) Jenkins, F. A. and White, H. W., *Fundamentals of Optics*, McGraw-Hill, New York, 1950.

0.3   Trorey, L. G., *Handbook of Aerial Mapping and Photogrammetry*. Cambridge University Press, 1952.

0.4   Crone, D. R., *Elementary Photogrammetry*. Edward Arnold, London, 1963.

0.5   Thompson, E. H., *Algebra of Matrices*. Adam Hilger, London, 1969.

0.6   Thompson, E. H., *Photogrammetry and Surveying. A Selection of Papers by E. H. Thompson* 1910–76. The Photogrammetric Society, London, 1977.

0.7   Laurila, S., *Electronic Surveying and Navigation*. Wiley, New York, 1975.

0.8   Kilford, K., *Elementary Air Survey*. Pitman, London, 1963.

0.9   Hallert, B., *Photogrammetry*. McGraw-Hill, New York, 1960.

0.10 Gruber, O. Von, *Photogrammetry: Collimated Lectures and Essays*. Chapman and Hall, London, 1932.

0.11 Cimerman, V. J. and Tomasegovic, Z., *Atlas of Photogrammetric Instruments*. Elsevier, Amsterdam, 1970.

0.12 Mikhail, E. F., *Observations and Least Squares*. Dun-Donnelley, New York, 1976.

0.13 *Military Engineering*. Volume XIII (School of Military Survey), Part 10: *Air Survey*. HMSO, 1972.

0.14 Burnside, C. D., *Electromagnetic Distance Measurement*. Aspects of Modern Land Surveying, Volume 1, Crosby Lockwood Staples, London, 1971.

1.1   Hallert, B., Swedish test fields for aerial photographs. *Photogrammetric Record* 5(25), April 1965.

1.2   Salmenpera, H., Camera calibration using a test field, Paper presented to Commission 1 of ISP Congress, Ottawa, *Photogrammetric Journal of Finland* 6, no. 1, 1972.

1.3   Recommended procedures for calibrating cameras and for related tests. *International Archives of Photogrammetry* 13, part 4, 1961.

1.4   Anculete, G. and Diaconescue, T. A., A methodology used in testing and calibrating photogrammetric cameras under normal working conditions. Paper presented to Commission 1 of ISP Congress, Helsinki, 1976.

1.5   Heimes, F. J., In-flight calibration of a survey aircraft system. Paper presented to Commission 1 of ISP Congress, Ottawa, 1972.

1.6   Brock, G. C., The status of the optical transfer function. *Photogrammetric Record* 7(42), October 1977.

1.7   Welch, R., Modulation transfer functions. *Photogrammetric Engineering*, March 1971.

1.8   Welch, R., Progress in the specification and analysis of image quality. Invited paper for Commission 1 of ISP Congress, Helsinki, 1976.

2.1 Woodrow, H. C., The use of colour photographs for large-scale mapping. *Photogrammetric Record* **5**(30), October 1967.

2.2 Smith, J. T. (Ed.), *Manual of Colour Photography*. American Society of Photogrammetry, 1968.

2.3 Brock, G. C., The possibilities for higher resolution in air survey photography. *Photogrammetric Record* **8**(47), April 1976.

2.4 Welch, R., and Halliday, J., Image quality controls for aerial photographs. *Photogrammetric Record* **8**(45), April 1975.

2.5 *Further Results of the Photogrammetric Tests of 'Oberriet'*. Commission C. OEEPE Publication no. 10, November 1975.

2.6 English, J. S. and Huggett, M. G., Doppler navigation systems for survey flying. *Photogrammetric Record* **4**(24), October 1964.

5.1 Anderson, R. O., Scale-point method of tilt determination. *Photogrammetric Engineering* **15**(2), 1949.

5.2 Church, E., *Revised Geometry of the Aerial Photograph*. Bulletin 15. Syracuse University Press, 1945.

5.3 Fagan, P. F., Photogrammetry in the National Survey. *Photogrammetric Record* **7**(40), October 1972.

5.4 Matthews, A. E. H., Revision of 1:2500 scale topographic maps. *Photogrammetric Record* **8**(48), October 1976.

5.5 Bean, R. K., The orthophotoscope and its development. *Canadian Surveyor* **22**, March 1968.

5.6 Meier, H. K., Theory and Practice of the Gigas–Zeiss orthoprojector. *Zeiss-Mitteilungen*, **4**(2), April 1966.

5.7 Bormann, G. E., Orthophoto attachment for the Wild A8 autograph. Paper presented to ASP–ACSM Convention, Washington, 1970.

5.8 Blachut, T. J., Methods and instruments for production and processing of orthophotos. Invited paper for Commission 11 of ISP Congress, Ottawa, 1972.

5.9 Loscher, W., Some aspects of orthophoto technology. *Photogrammetric Record* **5**(30), October 1967.

5.10 Meier, H. K., Orthoprojection systems and their practical potential. Presented Paper at Fourth National Survey Conference, Durban, 1970.

5.11 Visser, J. and others, Performance and applications of orthophotomaps. Paper presented to Commission 4 of ISP Congress, Ottawa, 1972.

6.1 Bedwell, C. H., The eye, vision and photogrammetry. *Photogrammetric Record* **7**(38), October 1971.

6.2 Palmer, D. A., Stereoscopy and photogrammetry. Invited paper for Commission 1, ISP Congress, Lisbon, 1964.

6.3 Zorn, H. C., Binocular vision. *ITC Textbook of Photogrammetry* **4**(2), 1967.

6.4 Thompson, E. H., Heights from parallax bar measurements. *Photogrammetric Record* **1**(4), 1954.

6.5 Thompson, E. H., Corrections to x-parallaxes. *Photogrammetric Record* **6**(32), October 1968.

6.6 Methley, B. D. F., Heights from parallax bar and computer. *Photogrammetric Record* **6**(35), April 1970.

8.1 Thompson, E. H., A new photogrammetric plotter: CPI. *Photogrammetric Record* **7**(38), October 1971.

8.2 Thompson, E. H., The CPI plotter: setting procedure and results. *Photogrammetric Record* **7**(39), April 1972. (Alternatively, see Ref 0.6).

8.3 Ligterink, G. H., Aerial triangulation by independent models: the co-ordinates of the perspective centre and their accuracy. *Photogrammetria* **26**(1), 1970.

8.4 Makarovic, A. (Ed.), *Testing of Instruments. ITC Textbook*, chap. IV 7, part A. ITC 1969.

8.5 Dowman, I. J., A working method for the calibration of plotting instruments using computers. *Photogrammetric Record* **7**(42), October 1972.

8.6 Borman, G. E., Features and design parameters of the Wild Aviomap system. Paper presented to Commission 11 of ISP Congress, Helsinki, 1976.

8.7 *Topographic Instructions. Multiplex Plotter Procedure.* Geological Survey, book 3. US Department of the Interior, 1960.

8.8 Thompson, E. H., The Thompson–Watts plotter model 2. *Photogrammetric Record* **4**(23), April 1964.

9.1 Schut, H. G., Construction of orthogonal matrices and their applications in analytical photogrammetry. *Photogrammetria* **15**(4), 1958–9.

9.2 Schut, G. H., On exact linear equations for the computation of the rotational elements of absolute orientation. *Photogrammetria* **17**(1), 1960.

9.3 Hallert, B., Two theorems from mathematical statistics concerning the application of the theory of errors to photogrammetry. *Photogrammetric Record* **3**(14), October 1959.

9.4 Cramer, H., *Mathematical Methods of Statistics.* Princeton and Uppsala, 1945.

9.5 Feller, W., *An Introduction to Probability Theory and its Applications*, 1957.

10.1 Miles, M. J., The theory of the analytical solution of the stereogram. *Photogrammetric Record* **4**(32), October 1968.

10.2 Helava, U. V., New principles for photogrammetric plotters. *Photogrammetria* **14**, 1958.

10.3 Helava, U. V., Analytical plotter in photogrammetric production line. *Photogrammetric Engineering* **24**, 1958.

10.4 Featuring analytical plotters. *Photogrammetric Engineering* **43**(11), November 1977.

10.5 Hobbie, D., The C100 Planicomp. Paper presented to Commission 11, ISP Congress, Helsinki, 1976.

10.6 Hobbie, D., The Zeiss G-2 Stereocord. *Photogrammetric Record* **8**(47), April 1976.

10.7 Schwidefsky, K., A New Precision Comparator. *Bildmessung und Luftbildwesen* **3**(28), 1960.

10.8 Schwebel, R., The New PK-1 Precision Comparator. *Bildmessung und Luftbildwesen* **4**(44), 1976.

11.1 Ackermann, F., Horizontal block adjustment with large numbers of points. *Bildmessung und Luftbildwesen* no. 4, 1970.

11.2 Schut, G. H., Development of programmes for strip and block adjustment at the National Research Council of Canada. *Photogrammetric Engineering* **30**(2),1964.

11.3 Schut, G. H., The use of polynomials in the three-dimensional adjustment of triangulated strips. *Canadian Surveyor* **16**(3), May 1962.

11.4 Schut, G. H., Polynomial transformation of strips versus linear transformation of models. A: Theory and experiments. *Photogrammetria* **22**, 1967.

11.5 Van den Hout, C. M. A., The Anblock method of planimetric block adjustment. *Photogrammetria* **12**, 1966.

11.6 Ackermann, F., Experience with applications of block adjustment for large scale surveys. *Photogrammetric Record* **7**(41), April 1973.

11.7 Ackermann, F., Ebner, H. and Klein, H., A programme package for block adjustment with independent models. *Information Relative to Cartography and Geodesy* series 11, no. 27. Frankfurt, 1972. (Nachrichten aus dem Karten und Vermessungwessen, Reihe (27).)

11.8 Blais, J. A. R., SPACE-M spatial photogrammetric adjustment for control extensions using independent models. Paper presented to Commision 111 of ISP Congress, Helsinki, 1976.

11.9 Amer, F., Digital block adjustment. *Photogrammetric Record* **4**(19), April 1962.

11.10 Ackermann, F., A method of analytical block adjustment for heights. *Photogrammetria* **19**(8), 1962.

11.11 Ebner, H., Theoretical accuracy models for block triangulation. *Bildmessung und Luftbildwesen* **40**(5), 1972.

11.12 Schut, G. H., Results of analytical triangulation and adjustment obtained in the International Test on Block Adjustment. Paper presented to Commission 111 of ISP Congress, London, 1960.

11.13 Boniface, P. R. J., Recent research into block adjustment methods of aircraft operating company. Paper presented to Commission 111 of ISP Symposium, London, 1971.

11.14 Ackermann, F. and others, Results of tests Oberschwaben. *Proceedings of the OEEPE Symposium on Experimental Research on Accuracy of Aerial Triangulation, Brussels, 1973.*

11.15 Gauthier, J. R. R., O'Donnell, J. H. and Low, B. A., The planimetric adjustment of very large blocks of models. *Canadian Surveyor* **27**(2), 1973.

11.16 Kubik, K. and Kure, J., ISP investigation into the accuracy of photogrammetric triangulation. Paper presented to Commission 111 of ISP Symposium, London, 1971.

11.17 Anderson, J. M., Erio, G. and Lee, C., Analytical bundle triangulation with large-scale photography: comparison with polynomial adjustment and experiments using added parameters. Paper presented to Commission 111 of ISP Symposium, Stuttgart, 1974.

11.18 Newby, P. R. T., Control for aerial triangulation. *Photogrammetric Record* **8**(46), October 1975.

11.19 Smith, A. D. N., Miles, M. J. and Verall, P., Analytical aerial triangulation block adjustment: the direct height solution incorporating tie strips. *Photogrammetric Record* **5**(29), April 1967.

11.20 Bauer & Muller, Height accuracy of blocks and bundle adjustment with additional parameters. Paper presented to Commission 111 of ISP Congress, Ottawa, 1972.

11.21 Brown, D. C., Evolution, application and potential of the bundle method of photogrammetric translation. Paper presented to Commission 111 of ISP Symposium, Stuttgart, 1974.

11.22 Brown, D. C., The bundle adjustment – progress and invited prospects. Invited paper for Commission 111 of ISP Congress, Helsinki, 1976.

11.23 Thompson, E. H., The theory of the method of least squares. *Photogrammetric Record* **4**(19), April 1962.

11.24 Thompson, E. H., Some remarks on the calculation of aerial triangulation. *Photogrammetric Record* **8**(48), October 1976.

12.1 Zarzycki, J. M., The use of horizon camera, Doppler navigator and statoscope in aerial triangulation. Paper presented to Commission 111 of ISP Symposium, Lisbon, 1964.

12.2 Loftstrom, K. K., Die letzten Fortschritte in der Horizont und Statoskopvermessung. *Schweiz. Zeitschrift fur Vermessungswesen und Kulturtechnik* Heft Nr 7, 1964.

12.3 Jerie, H. G. and Kure, J., *Data Analysis and Report on an Investigation into the*

Application of APR to Photogrammetric Mapping". ITC Publication A25/26, 1964.

12.4   Ackermann, F., Accuracy of statoscope data. Results from the OEEPE test Ober-schwaben. Paper presented to Commission 111 of ISP Symposium, Stuttgart, 1974.

12.5   Lines, J. D., Application of Airborne Profile Recording to 1:100000 Scale Mapping. Commonwealth Survey ·Officers' Conference Paper D3 1971. Directorate of Overseas Surveys, London.

12.6   Jerie, H. G., Theoretical height accuracy of strip and block triangulation with and without the use of auxiliary data. *Photogrammetria* **23**(1), 1968.

12.7   Zarzycki, J. M., The use of auxiliary data in aerial triangulation. *International Archives of Photogrammetry*, 1972.

12.8   Miles, M. J. and Smith, A. D. N., The use of airborne auxiliary data in the rigorous least squares adjustment of a block of aerial triangulation. *Photogrammetric Record* **6**(34), 1969.

12.9   Faig, W., Independent model triangulation with auxiliary vertical control. Paper presented to Commission 111 of ISP Congress, Helsinki, 1976.

12.10 Allam, M. M., Chaly, C. K. and Wong, C. K. Geometrical distribution of vertical control and the simultaneous adjustment of auxiliary data in independent model triangulation. Paper presented to Commission 111 of ISP Congress, Helsinki, 1976.

13.1   See Ref 2.5.

13.2   Eden, J. A., A new fast-working approach to analytical photogrammetry. *Photogrammetric Record* **5**(30), October 1967.

13.3   Van der Weele, A. J., Photogrammetric Observations. *Photogrammetric Record* **6**(33), April 1969.

A.1   Thompson, E. H., An exact linear solution of the problem of absolute orientation. *Photogrammetria* **15**, no. 4, 1958/59 (See also Ref A.4).

A.2   Schut, G. H., On exact linear equations for the computation of the rotational elements of absolute orientation. *Photogrammetria* **17**, no. 1, 1960/61.

A.3   Schut, G. H., Construction of orthogonal matrices and their application in analytical photogrammetry. *Photogrammetria* **15**, no. 4, 1958/59.

A.4   Thompson, E. H., The construction of orthogonal matrices. *Photogrammetric Record* **3**(13), April 1952.

A.5   Schut, G. H., *Formation of Strips from Independent Models*. Paper NRC 9695, National Research Council of Canada, 1967.

Plate 1. *Wild RC10 Universal Film Camera*

Plate 2. *Zeiss (Oberkochen) Universal Film Camera RMK 8.5/23*

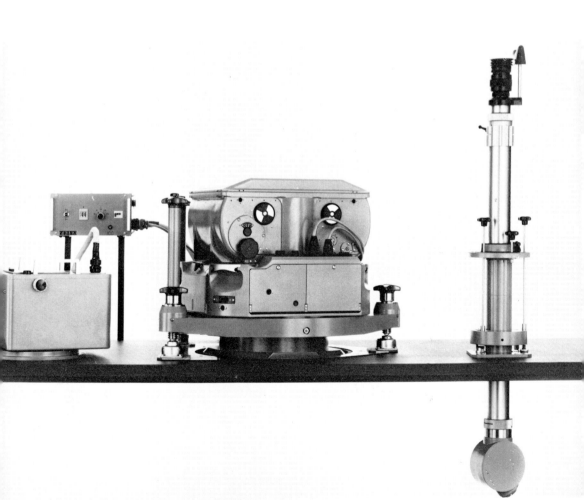

Plate 3. Zeiss (Oberkochen) Sketchmaster

Plate 4. Wild Stereoscope and Zeiss Snap Markers

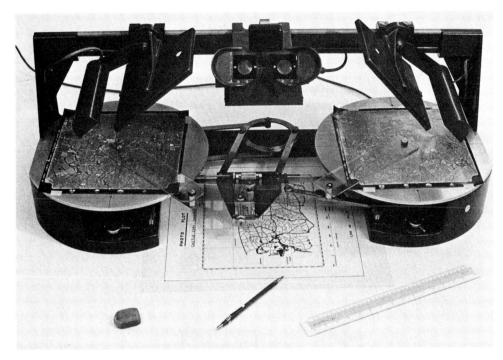

Plate 5. *Hilger-Watts Radial Line Plotter*

Plate 6. *Wild PUG 4 Point Transfer Device*

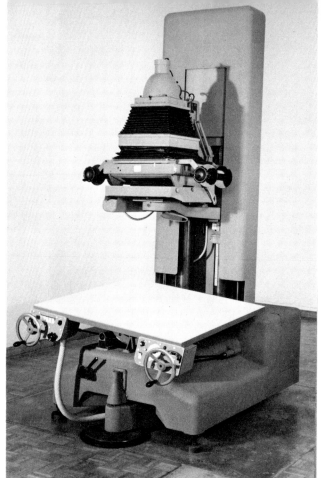

Plate 7. *Wild E4 Optical Rectifier–Enlarger*

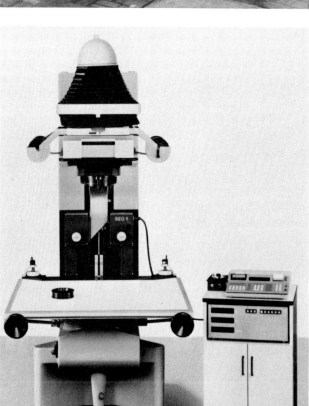

Plate 8. *Zeiss (Oberkochen) Rectifier SEG5*

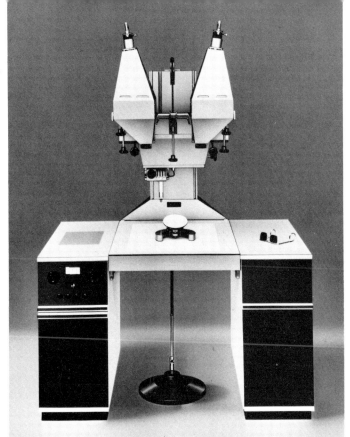

Plate 9. Zeiss (Oberkochen)
DP1 Optical
Projection Plotter

Plate 10. Kelsh Optical
Projection Plotter

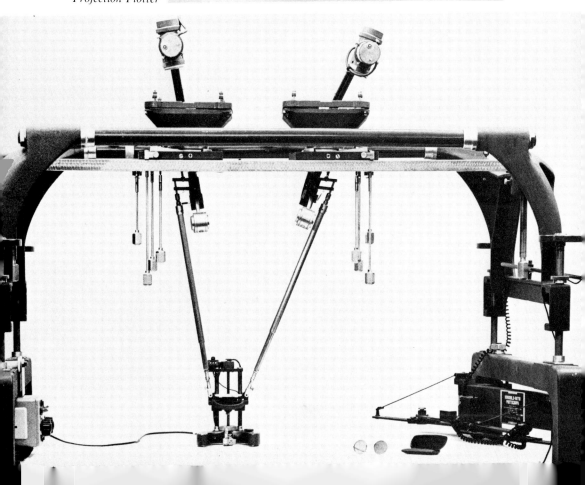

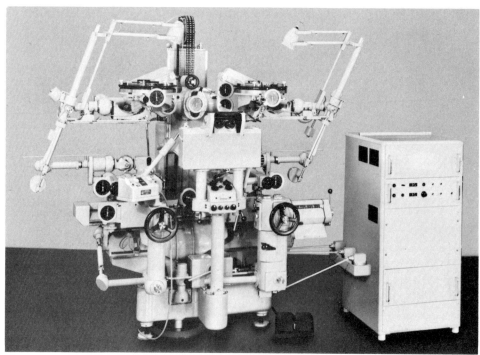

Plate 11. *Zeiss Stereoplanigraph Universal Instrument*
Plate 12. *Gigas–Zeiss GZ-1 Orthoprojector*

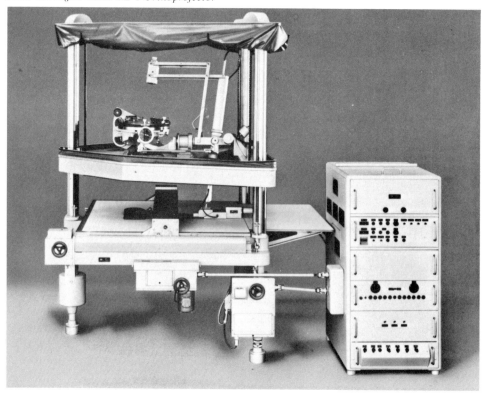

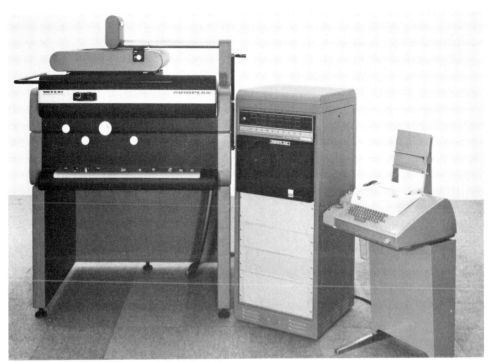

Plate 13. *Wild Avioplan OR1 Orthophoto Instrument*

Plate 14. *Zeiss (Jena) Topocart with Orthophoto Attachment*

Plate 15. *Wild Aviomap Plotting Instrument AM-H and Aviotab TA Plotting Table*

Pl.ate 16. *Wild A*10 *Autograph*

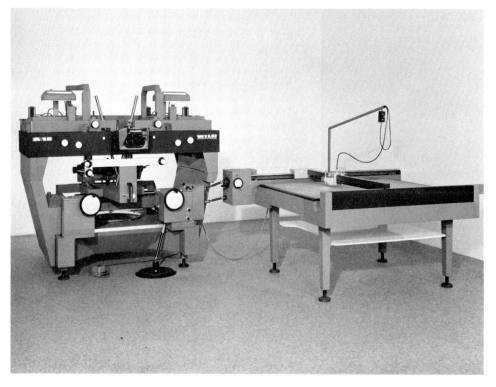

Plate 17. Thompson-Watts Model 2 Plotting Instrument

Plate 18. Kern PG2 Plotting Instrument

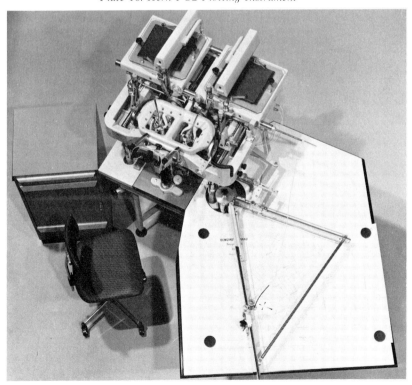

Plate 19. *Santoni Stereomicrometer Mark* 5

Plate 20. *Thompson CP*1 *Plotter*

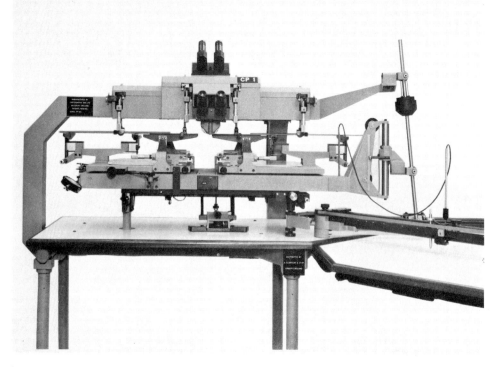

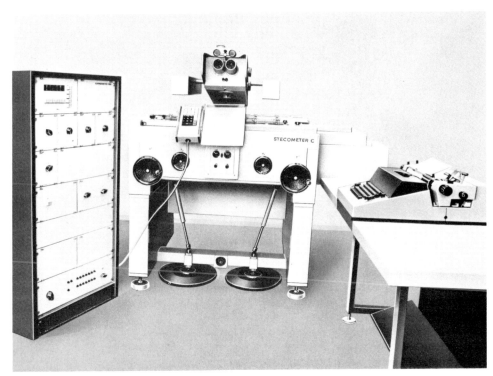

Plate 21. Zeiss (Jena) Stecometer C Stereocomparator

Plate 22. Zeiss (Oberkochen) PSK Stereocomparator

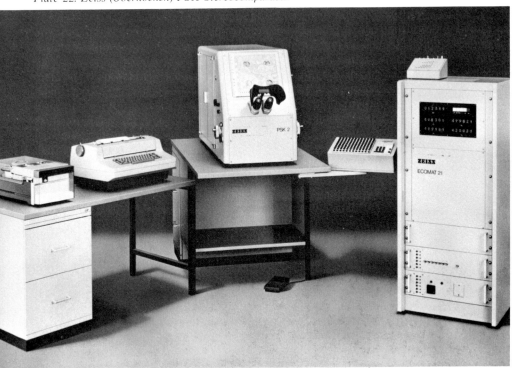

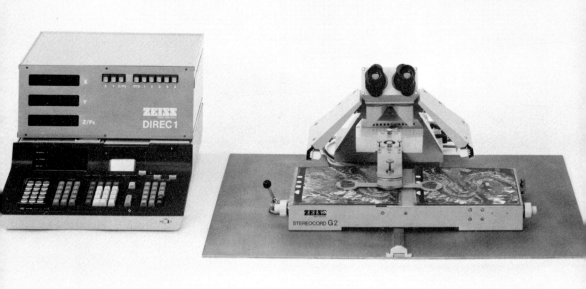

Plate 23. Zeiss (Oberkochen) Direc 1 Readout Unit and Stereocord G2

Plate 24. Kern Mk. 2 Monocomparator and Readout Unit

Index